Grundlehren der mathematischen Wissenschaften 230

A Series of Comprehensive Studies in Mathematics

Wilhelm Klingenberg

Lectures on Closed Geodesics

Springer-Verlag
Berlin Heidelberg New York 1978

Wilhelm Klingenberg
Mathematisches Institut der Universität Bonn, D-5300 Bonn

AMS Subject Classifications (1970):
Primary: 49 C 05, 49 F 15, 53 C 20, 55 D 35, 58 B 20, 58 E 05, 58 E 10, 58 F 05, 58 F 20
Secondary: 34 C 25, 49 B 05, 53 B 20, 55 C 30, 55 E 05, 57 A 20, 58 D 15, 58 F 15

ISBN 3-540-08393-6 Springer-Verlag Berlin Heidelberg New York
ISBN 0-387-08393-6 Springer-Verlag New York Heidelberg Berlin

Library of Congress Cataloging in Publication Data. Klingenberg, Wilhelm, 1924–. Lectures on closed geodesics (Grundlehren der mathematischen Wissenschaften; 230). Bibliography: p. Includes index. 1. Riemannian manifolds. 2. Curves on surfaces. I. Title. II. Series: Die Grundlehren der mathematischen Wissenschaften in Einzeldarstellungen; 230. QA649.K54. 516'.362. 77-13147.

Printed in Germany
Typesetting and printing: Oscar Brandstetter Druckerei KG, Wiesbaden. Binding: Konrad Triltsch, Würzburg.
2141/3140-543210

To the memory of my father

Preface

The question of existence of closed geodesics on a Riemannian manifold and the properties of the corresponding periodic orbits in the geodesic flow has been the object of intensive investigations since the beginning of global differential geometry during the last century.

The simplest case occurs for closed surfaces of negative curvature. Here, the fundamental group is very large and, as shown by Hadamard [Had] in 1898, every non-null homotopic closed curve can be deformed into a closed curve having minimal length in its free homotopy class. This minimal curve is, up to the parameterization, uniquely determined and represents a closed geodesic.

The question of existence of a closed geodesic on a simply connected closed surface is much more difficult. As pointed out by Poincaré [Po 1] in 1905, this problem has much in common with the problem of the existence of periodic orbits in the restricted three body problem. Poincaré [l.c.] outlined a proof that on an analytic convex surface which does not differ too much from the standard sphere there always exists at least one closed geodesic of elliptic type, i.e., the corresponding periodic orbit in the geodesic flow is infinitesimally stable.

During the following three decades, Birkhoff and Morse, on the one hand, and Lusternik and Schnirelmann, on the other, developed powerful new methods to prove the existence of one or possibly several closed geodesics on arbitrary Riemannian manifolds. The principal objects of these studies have always been Riemannian manifolds, for which the underlying differentiable manifold is a sphere. Take as the earliest example Jacobi's description [Ja] of the geodesics on an ellipsoid, in his Lectures on Dynamics in Königsberg, 1842/43. A later example is Morse's monograph [Mor 2] from 1934 where, among other things, he put much effort into an attempt to compute the $\mathbb{Z}_2$-Betti numbers of the space of unparameterized closed curves on the n-sphere.

There were, however, few results on the existence of more than one closed geodesic on such manifolds. Probably the most remarkable one is the theorem due to Lusternik and Schnirelmann [LS 1] in 1929. They showed that on a simply connected compact surface there exist at least three closed geodesics without self-intersections. Ellipsoids which are not too dissimilar from a round sphere provide examples of surfaces with exactly three such closed geodesics. Note that there are always infinitely many other prime closed geodesics on an ellipsoid; but they will generally have self-intersections.

In 1951, Lyusternik and Fet [LF] proved that at least one closed geodesic exists on every compact Riemannian manifold. Since the 1960's the theory of closed geodesics has been advanced considerably by the work of Alber, Švarc,

Fet, Gromoll and Meyer, Eliasson and the author. In hindsight it turned out to be of particular importance when the author indicated in 1965 [Kl 3] how to reformulate Morse's theory of the space of closed curves on a Riemannian manifold within the framework of Morse theory on Hilbert manifolds, which had been established in 1964 by Palais and Smale [PS].

A striking application of this new approach was made in 1969 by Gromoll and Meyer [GM 2]. They proved that, for every Riemannian metric on a compact simply connected differentiable manifold M, there exist infinitely many closed geodesics, provided that the sequence of the rational Betti numbers of the space ΛM of parameterized closed curves is unbounded.

As was shown recently by Vigué and Sullivan [VS], this hypothesis on the rational Betti numbers is satisfied if and only if the rational cohomology ring of the manifold M is not a truncated polynomial ring. Thus, the Gromoll-Meyer theorem does not apply to the case of Riemannian metrics on the sphere, i.e. it does not apply to the case which has been the principal object of study in the theory of closed geodesics during most of its earlier history.

In order to obtain results on the existence of infinitely many closed geodesics without any hypothesis concerning the topology of the underlying differentiable manifold, the author has employed the theory of Hamiltonian systems, and here, in particular, Poincaré's and Birkhoff's investigations of the Hamiltonian flow near a periodic orbit. (Note that periodic orbits of the geodesics flow are in 1 : 1 correspondence with the closed geodesics.) The outcome was the proof that, generically, there exist infinitely many prime closed geodesics on every compact Riemannian manifold with finite fundamental group, cf. [Kl 9], [Kl 16].

In the course of our work on the present manuscript we succeeded in solving one of the outstanding problems of the whole theory by proving that there always – not only generically – exist infinitely many prime closed geodesics on a compact Riemannian manifold with finite fundamental group. This result is new, even for convex surfaces — the case which was investigated by Poincaré at the very beginning of the theory of closed geodesics more than seven decades ago.

The proof of this result uses in an essential way the full structure of the Hilbert manifold of closed curves and the associated Morse complex. Only a minor part is played by the geodesic flow. It is to be expected, however, that future work on generic density properties of period orbits in particular will employ the Poincaré-Birkhoff theory of Hamiltonian systems to a greater extent.

The aim of these lectures is to present a reasonably complete, exhaustive and self-contained exposition of the methods and results relating to closed geodesics.

In Chapter 1 we give a concise description of the Hilbert manifold ΛM of closed curves on a Riemannian manifold M. ΛM and its structure are derived from M functorially and no information is lost by going from M to ΛM.

Chapter 2 develops the Morse-Lusternik-Schnirelmann theory on M. The canonical operations of S and of $\mathbb{Z}_2$ are explained. We conclude with a description of the Morse complex of ΛM which will play a crucial role later on.

In Chapter 3 we study the geodesic flow on the tangent bundle TM of a Riemannian manifold M. Besides offering a rather complete presentation of the index theorem for closed geodesics we are mainly interested in generic properties of the geodesic flow.

Chapter 4 is devoted to the principal results of the whole theory. After a proof of Fet's theorem, which involves the structure of the neighborhood of an unparameterized closed geodesic, we give a modified proof of a slightly extended version of the Gromoll-Meyer theorem. This theorem and Sullivan's theory of the minimal model of ΛM are then combined to show that there always exist infinitely many prime closed geodesics on M, if $\pi_1 M$ is finite. The crucial step in the proof is a lemma relating multiplicities of certain pairs of closed geodesics. We conclude the chapter with some generic existence theorems.

Chapter 5 begins with a detailed proof of the theorem of three closed geodesics – a long sought-after generalization of the classical result of Lusternik and Schnirelmann on closed surfaces of the type of S^2 to arbitrary simply connected manifolds. This is followed by miscellaneous results on manifolds of elliptic type, i.e. manifolds with finite fundamental group. The last paragraph is devoted to manifolds of hyperbolic type, such as those with negative sectional curvature, these manifolds having a very large fundamental group.

In the Appendix we prove the Lusternik-Schnirelmann theorem on the existence of three closed geodesics on a surface of genus 0. The proof uses the Lusternik-Schnirelmann approach to the closed geodesics problems, which is much more direct and elementary than Morse's approach in its refined version presented in these lectures. But then, the Morse theory of the Riemannian Hilbert manifold ΛM with the functional E is more powerful – witness the results of Chapter 4.

We have made the Appendix completely independent of the rest of these lectures. By filling in all the details and avoiding anything but the most basic facts of 2-dimensional Riemannian geometry and algebraic topology we present a proof of two classical theorems (besides the Lusternik-Schnirelmann Theorem we also prove the theorem of Lyusternik and Fet) in such a form as to fit into any course on Riemannian geometry in general.

These lectures grew out of courses which I have given in various places during the last ten years. I want to mention in particular Bonn, 1966/67 and 1972/73, Berkeley, 1971 and 1974, Bombay, 1972, and Minneapolis, 1975.

Here I wish to thank all those who enabled me, through generous invitations, to lecture and do research at various places and who, at the same time, through their comments, criticism and encouragement, helped to bring these lectures into their present form.

I am indebted to Wolfgang Ziller for his careful and critical reading of a major part of the manuscript and to Gudlaugur Thorbergsson for helping me in proof reading. For their tireless efforts in typing the various versions of the manuscript I want to thank Ruthie in Berkeley, Kathy in Minneapolis and – most of all – Cornelia in Bonn.

After long years of labor, I can only hope that my efforts will encourage a younger generation of mathematicians to take up where I left off and continue the attack on the numerous still unsolved problems in the theory of closed geodesics.

Bonn, December 1977 Wilhelm Klingenberg

Contents

Chapter 1. The Hilbert Manifold of Closed Curves

In this chapter we shall define the Hilbert manifold ΛM of H^1-maps of the parameterized circle S into a compact Riemannian manifold M.

We will show that ΛM carries an intrinsically defined Riemannian metric and a differentiable function E — the energy integral — such that grad E satisfies the condition (C) of Palais and Smale.

This condition allows an extension of the classical Morse-Lyusternik-Schnirelmann theory of a differentiable function on a Euclidean (i.e. finite dimensional) manifold to such a function on a Hilbert manifold.

1.1 Hilbert Manifolds

A *Hilbert manifold* M is a topological space with a countable base endowed with a differentiable atlas where the charts have their image in a fixed separable Hilbert space $\mathbb{M}$. We also say briefly that M is modeled after $\mathbb{M}$.

Just as for a Euclidean manifold M there is associated to M the tangent bundle of M:

$$\tau = \tau_M : TM \to M;$$

an atlas $(\phi_\alpha, U_\alpha)_{\alpha \in A}$ of M gives rise to the atlas $(T\phi_\alpha, TU_\alpha)_{\alpha \in A}$ of TM just as in the Euclidean case.

We also consider, for a given manifold M, vector bundles

$$\pi : E \to M$$

over M with the fibre $\mathbb{E}$ being a Banach space.

That is to say, we have an atlas for M such that, for each chart (ϕ, U) of that atlas, a local representation (Φ, ϕ, U) is given, i.e. a commutative diagram

$$\begin{array}{ccc} \pi^{-1}(U) & \xrightarrow{\Phi} & \phi(U) \times \mathbb{E} \\ \downarrow \pi & & \downarrow \mathrm{pr}_1 \\ U & \xrightarrow{\phi} & \phi(U) \end{array}$$

of morphisms with the following properties:

(i) the restriction Φ_p of Φ to the fibre $E_p = \pi^{-1}(p)$ over $p \in U$ is a topological linear isomorphism; and

(ii) if (Φ, ϕ, U) and (Φ', ϕ', U') are local representations, then the map

$$\Phi' \circ \Phi^{-1} : U \cap U' \to L(\mathbb{E};\mathbb{E}) \qquad p \mapsto \Phi'_p \circ \Phi_p^{-1}$$

is differentiable.

For details see [La], [El 3] and [FK].

If $(x, \xi) \in \phi(U) \times \mathbb{E}$ is a local representation of an element of E then we call ξ the *principal part* of this representation.

A local representation (Φ, ϕ, U) of $\pi: E \to M$ determines a local representation $(T\Phi, \Phi, \pi^{-1}(U))$ of the tangent bundle $\tau_E : TE \to E$

$$\begin{array}{ccc} T\pi^{-1}(U) & \xrightarrow{T\Phi} & \phi(U) \times \mathbb{E} \times \mathbb{M} \times \mathbb{E} \\ \downarrow{\scriptstyle \tau_E} & & \downarrow{\scriptstyle \mathrm{pr}_{1,2}} \\ \pi^{-1}(U) & \xrightarrow{\Phi} & \phi(U) \times \mathbb{E} \end{array}$$

If $(x, \xi) \in \phi(U) \times \mathbb{E}$ then we denote the elements in $\mathrm{pr}_{1,2}^{-1}(x, \xi)$ by

$$(x, \xi, y, \eta) \in \{(x, \xi)\} \times \mathbb{M} \times \mathbb{E}.$$

Of fundamental importance is the concept of a *connection* in a bundle $\pi : E \to M$. This is a map

$$K : TE \to E$$

such that for a local representation (Φ, ϕ, U) of E there exists a differentiable map

$$\Gamma_\phi : \phi(U) \to L(\mathbb{M}, \mathbb{E}; \mathbb{E})$$

such that the local representative $K_\phi := \Phi \circ K \circ T\Phi^{-1}$ of K

$$K_\phi : \phi(U) \times \mathbb{E} \times \mathbb{M} \times \mathbb{E} \to \phi(U) \times \mathbb{E}$$

is given by

$$(x, \xi, y, \eta) \mapsto (x, \eta + \Gamma_\phi(x)(y, \xi)).$$

The Γ_ϕ are called *Christoffel symbols*.

Note that K is a vector bundle morphism from τ_E into π over π, i.e. the following diagram is commutative:

$$\begin{array}{ccc} TE & \xrightarrow{K} & E \\ {\scriptstyle \tau_E}\downarrow & & \downarrow{\scriptstyle \pi} \\ E & \xrightarrow{\pi} & M \end{array}$$

and $K|T_\xi E \in L(T_\xi E;\, E_{\pi(\xi)})$. This follows immediately from the local representation of this diagram:

$$\begin{array}{ccc} (x,\xi,y,\eta) & \longrightarrow & (x,\eta+\Gamma_\phi(x)(y,\xi)) \\ \downarrow & & \downarrow \\ (x,\xi) & \longrightarrow & (x) \end{array}$$

Let K be a connection on $\tau: E \to M$. For $\xi \in E_p = \pi^{-1}(p)$ we define

$$T_{\xi v}E = \ker\,(T\pi: T_\xi E \to T_p M), \qquad \text{and}$$

$$T_{\xi h}E = \ker\,(K_\xi: T_\xi E \to E_p)$$

to be the *vertical* and the *horizontal* subspace of $T_\xi E$ respectively.

1.1.1 Proposition. *A connection K on $\pi: E \to M$ defines a splitting*

$$TE = T_h E \oplus T_v E$$

of the tangent bundle with

$$T_\xi E = T_{\xi h} E \oplus T_{\xi v} E.$$

More precisely, under the canonical identification of $T_{\xi v}E$ with E_p, $p = \pi(\xi)$, we can write this decomposition as

$$T_\xi E = (\mathrm{id} - K)\, T_\xi E + K T_\xi E.$$

Proof. By looking at the local representation K_ϕ of K we see that, if we identify $T_{\xi v}E$ with E_p, i.e. if we identify

$$\{(x,\xi,0,\eta)\} \in \{(x,\xi)\} \times \mathbb{M} \times \mathbb{E}$$

with

$$\{(x,\eta)\} \in \{x\} \times \mathbb{E},$$

then $K_\phi^2 = K_\phi$, $K_\phi T_\xi E = T_{\xi v} E$; that is to say, K_ϕ is a projection. □

The local representation of $T_{\xi_h}E$ is

$$\{(x, \xi, y, -\Gamma_\phi(x)(y, \xi)\} \in \{(x, \xi)\} \times \mathbb{M} \times \mathbb{E}.$$

Given a connection K on $\pi : E \to M$ we define the *covariant derivative* of a differentiable section $\xi : M \to E$ by

$$\nabla\xi := K \circ T\xi.$$

Note that $\nabla\xi$ is a section in the bundle

$$L(\tau; \pi) : L(TM; E) \to M.$$

Using the local representation (Φ, ϕ, U) of E we see that the principal part of $\nabla\xi$ is represented by

$$\nabla\xi_\phi(x) = D\xi_\phi(x) + \Gamma_\phi(x)(\ \ , \xi_\phi(x))$$

where $\xi_\phi : \phi(U) \to \mathbb{E}$ is the principal part of the local representation of ξ.

Note. Let $\Xi(M)$ and $\Xi_E(M)$ denote the space of sections of the bundles τ_M and π. Then a covariant derivative defines a map

$$\Xi(M) \times \Xi_E(M) \to \Xi_E(M)$$

$$(v, \xi) \mapsto \nabla\xi \,.\, v$$

which has a local representation given as above.

Whereas for Euclidean vector bundles over Euclidean manifolds such a map ∇ always defines a connection K, in our more general situation this need not always be true; see [FK] for further details. See also [El 3] for a more general setting.

For subsequent applications we consider the following special case of an induced bundle with an induced connection.

Let $S = [0,1]/\{0,1\}$ be the parametrized circle of length 1. Let $c : S \to M$ be a differentiable map into the base space M of a bundle $\pi : E \to M$. Then we have the induced bundle $c^*\pi$

$$\begin{array}{ccc} c^*E & \xrightarrow{\pi^*c} & E \\ \Big\downarrow{\scriptstyle c^*\pi} & & \Big\downarrow{\scriptstyle \pi} \\ S & \xrightarrow{c} & M \end{array}$$

with fibre $c^*\pi^{-1}(t) = E_{c(t)}$.

Let $\pi : E \to M$ have a connection $K_\pi : TE \to E$. Then $c^*\pi$ has a connection $K_{c^*\pi}$ defined by the commutative diagram:

$$\begin{array}{ccc} Tc^*E & \xrightarrow{T\pi^*c} & TE \\ \downarrow{\scriptstyle K_{c^*\pi}} & & \downarrow{\scriptstyle K_\pi} \\ c^*E & \xrightarrow{\pi^*c} & E \end{array}$$

Let ξ be a section in $c^*\pi$. Since S has the canonical coordinates $t \in [0,1]/\{0,1\}$, we denote by $\xi(t)$ the principal part of ξ. The canonical tangent vector to S will be denoted by ∂t. We then define

$$\nabla_c \xi := \nabla \xi \,.\, \partial t = K_{c^*\pi} \circ T\xi \,.\, \partial t.$$

Instead of $\nabla_c \xi$ we also write $\nabla \xi$.

Let (Φ, ϕ, U) be a local representation of π and assume that $c(t) \in U$ for some $t \in S$. We then set $\Phi \circ \pi^*c \circ \xi(t) = \xi_\phi(t)$.

The principal part $\nabla_c \xi_\phi(t)$ of the local representation of $\nabla_c \xi$ is then

$$\nabla_c \xi_\phi(t) = \dot{\xi}_\phi(t) + \Gamma_\phi(\phi \circ c(t))((\phi \circ c)^{\cdot}(t), \xi_\phi(t)).$$

To see this, we observe that $T\pi^*c \circ T\xi \,.\, \partial t$ is represented by

$$(\phi \circ c(t), \xi_\phi(t), (\phi \circ c) \,.\, (t), \dot{\xi}_\phi(t))$$

which belongs to the representation of $T_{c(t)}E$. Now apply K_π.

For a bundle $\pi : E \to M$ we have the associated bundle

$$L_s^2(\pi) : L_s^2(E) \to M$$

where the fibre $L_s^2(E)_p$ over p consists of the continuous symmetric bilinear maps, cf. [La]. Let $L_s^2(\mathbb{E})$ be the model of a fibre. It contains as an open subset $Ri(\mathbb{E})$ the *positive definite forms*, i.e. those forms which are $\geq \varepsilon$ (Hilbert metric on $\mathbb{E}$), for some $\varepsilon > 0$.

A *Riemannian metric* on $\pi : E \to M$ is a differentiable section

$$g : M \to L_s^2(E)$$

such that $g(p)$ is positive definite.

If we have a Riemannian metric g on $\tau : TM \to M$ then we call M a *Riemannian manifold* and we also call g a *Riemannian metric on M*.

Let g be a Riemannian metric on $\tau : E \to M$. We call a *connection* K on τ *Riemannian* if the following condition is satisfied:

For every open subset $U \subset M$ we have the relation

$$(*) \qquad Dg(\xi, \eta) \,.\, v = g(\nabla\xi \,.\, v, \eta) + g(\xi, \nabla\eta \,.\, v)$$

where v is an arbitrary section in $\tau_M|U$, ξ and η are arbitrary sections in $\pi|U$, and $\nabla\xi = K \circ T\xi$ is the covariant derivative determined by

$$K : T(E|U) \to E|U.$$

Note. If M has a differentiable partition of unity then in the previous definition it suffices to take only the open subset $U = M$ of M,cf. [FK].

Let K be a connection on $\tau : TM \to M$. Denote by

$$L_a^2(\tau; \tau) : L_a^2(TM; TM) \to M$$

the bundle associated to τ with fibre $L_a^2(\mathbb{M}; \mathbb{M})$: alternating bilinear continuous maps. The *torsion* of K is defined as a section T of $L_a^2(\tau; \tau)$ as follows. Consider a representation (T_ϕ, ϕ, U) of τ; the principal part of T

$$T_\phi : \phi(U) \to L_a^2(\mathbb{M}; \mathbb{M})$$

is given by

$$(**) \qquad T_\phi(x) = \Gamma_\phi(x)(u, v) - \Gamma_\phi(x)(v, u).$$

If $T \equiv 0$ then K is called *torsion free*.

1.1.2 Theorem. *On a Riemannian manifold there exists exactly one torsion free Riemannian connection, the so-called Levi-Cività connection.*

Proof. Let $(T\phi, \phi, U)$ be a local representation of $\tau : TM \to M$. By

$$g_\phi : \phi(U) \to L_s^2(\mathbb{M})$$

we denote the principal part of the corresponding local representation of the Riemannian metric g.

Consider for $(u, v, w) \in \mathbb{M} \times \mathbb{M} \times \mathbb{M}$ the relation

$$(***) \qquad g_\phi(x)(\Gamma_\phi(x)(u, v), w) = \tfrac{1}{2}(Dg_\phi(x) \,.\, u(v, w) + Dg_\phi(x) \,.\, v(u, w) - Dg_\phi(x) \,.\, w(u, v))$$

with $x \in \phi(U)$. This defines a continuous map

$$\Gamma_\phi : \phi(U) \to L^2(\mathbb{M}; \mathbb{M}).$$

It remains to check that the $\Gamma_\phi, \Gamma_{\phi'}$, satisfy the so-called *transformation formula* for two charts (ϕ, U) and (ϕ', U').

$$D(\phi'\circ\phi^{-1})\Gamma_\phi=\Gamma_{\phi'}\circ\big(D(\phi'\circ\phi^{-1})\times D(\phi'\circ\phi^{-1})\big)+ D^2(\phi'\circ\phi^{-1}).$$

This formula guarantees that the connection maps

$$K_\phi:\phi(U\cap U')\times\mathbb{M}\times\mathbb{M}\times\mathbb{M}\to\phi(U\cap U')\times\mathbb{M},\text{ and}$$

$$K_{\phi'}:\phi'(U'\cap U)\times\mathbb{M}\times\mathbb{M}\times\mathbb{M}\to\phi'(U'\cap U)\times\mathbb{M}$$

are related by

$$(\dagger)\qquad T\phi^{-1}\circ K_\phi\circ TT\phi=T\phi'^{-1}\circ K_{\phi'}\circ TT\phi'$$

and therefore give a globally defined connection $K: TTM\to TM$. Indeed, (†) is satisfied since, for (x,ξ,y,η) and (x',ξ',y',η') local representations of TTM, we have

$$x'=\phi'\circ\phi^{-1}(x),\ \xi'=D(\phi'\circ\phi^{-1})\xi,\ y'=D(\phi'\circ\phi^{-1})y,$$
$$\eta'=D(\phi'\circ\phi^{-1})\eta+D^2(\phi'\circ\phi^{-1})(y,\xi).$$

The definition of Γ_ϕ shows at once that the local representations T_ϕ of the torsion tensor is $\equiv 0$. It also shows that the connection is Riemannian since

$$(****)\qquad Dg_\phi(x)\,.\,u(v,w)=g_\phi(x)\big(\Gamma_\phi(x)(u,v),w\big) +g_\phi(x)\big(v,\Gamma_\phi(x)(u,w)\big)$$

which is the local version of the formula (*).

The uniqueness of the Levi-Cività connection follows from the observation that (**) and (****) imply (***).

Finally, we observe that if $\pi:E\to M$ has a Riemannian metric g and $c^*\pi:c^*E\to S$ is the bundle induced by a map $c:S\to M$, then we have an induced Riemannian metric c^*g on $c^*\pi$ given by $c^*g=g\circ(\pi^*c\times\pi^*c)$. □

1.2 The Manifold of Closed Curves

We denote by M a compact (Euclidean) manifold endowed with a Riemannian metric g, which we also denote by $\langle\,,\rangle$.

Let ∇ be the covariant derivative on TM, derived from the Levi-Cività connection.

Note. A large part of the construction which immediately follows would also work for a non-compact Riemannian manifold. For the finer aspects of the theory, however, compactness is necessary. We therefore restrict ourselves to compact M.

As in (1.1), we denote by S the *parameterized circle* $[0,1]/\{0,1\}$. Then we set

$$C^0(S, M) := \textit{set of } C^0\textit{-maps of } S \textit{ into } M;$$

$$C^\infty(S, M) := \textit{set of } C^\infty\textit{-maps of } S \textit{ into } M;$$

$$H^1(S, M) := \textit{set of } H^1\textit{-maps of } S \textit{ into } M.$$

Here a map $c: S \to M$ is called H^1 if it is absolutely continuous and the derivative $\dot{c}(t)$ (which is defined almost everywhere) is square integrable with respect to the Riemannian metric on M:

$$\int_S \langle \dot{c}(t), \dot{c}(t) \rangle_{c(t)} dt < \infty.$$

Note that

$$C^\infty(S, M) \subset H^1(S, M) \subset C^0(S, M).$$

We endow $C^0(S, M)$ with the compact open topology. Preparatory to constructing a topological structure (even the structure of a Hilbert manifold) on $H^1(S, M)$ we consider, for $c \in C^\infty(S, M)$, the bundle

$$c^*\tau : c^*TM \to S$$

and denote by

$$H^r(c^*TM)$$

the set of H^r-sections in $c^*\tau$, $r = 0,1$. Here, an H^0-section ξ is a square integrable section, i.e.

$$\langle \xi, \xi \rangle_0 := \int_S \langle \xi(t), \xi(t) \rangle_t dt < \infty.$$

Thus, $H^0(c^*TM)$ becomes a separable Hilbert space.

We define the scalar product on $H^1(c^*TM)$ by

$$\langle \xi, \xi \rangle_1 := \langle \xi, \xi \rangle_0 + \langle \nabla\xi, \nabla\xi \rangle_0$$

where $\nabla\xi(t) = \nabla_c \xi(t)$ is the almost everywhere defined covariant derivative, cf. (1.1).

We also consider

$$C^0(c^*TM) := \text{set of continuous sections}$$

and endow this vector space with the norm

$$\|\xi\|_\infty := \sup_t |\xi(t)|.$$

The norms on $H^r(c^*TM)$ derived from the scalar product will be denoted by $\| \ \|_r$, $r=0,1$.

1.2.1 Proposition. *The inclusions*

$$H^1(c^*TM) \hookrightarrow C^0(c^*TM) \hookrightarrow H^0(c^*TM)$$

are continuous. More precisely

(i) *if* $\xi \in C^0$, *then* $\|\xi\|_0 \leqslant \|\xi\|_\infty$; *and*

(ii) *if* $\xi \in H^1$, *then* $\|\xi\|_\infty^2 \leqslant 2\|\xi\|_1^2$.

Proof.

(1) $\|\xi\|_0^2 = \int_S \langle \xi(t), \xi(t) \rangle \, dt \leqslant \int_S \max |\xi(t)|^2 \, dt = \|\xi\|_\infty^2$.

(2) Choose t_1 such that $|\xi(t)| \leqslant |\xi(t_1)|$, for all t. Then

$$\|\xi\|_\infty^2 = |\xi(t)|^2 + \int_t^{t_1} \frac{d}{dt} |\xi(t)|^2 \, dt \leqslant$$

$$|\xi(t)|^2 + 2\int_S |\xi(t)| \, |\nabla\xi(t)| \, dt \leqslant$$

$$\langle \xi, \xi \rangle_0 + \langle \xi, \xi \rangle_0 + \langle \nabla\xi, \nabla\xi \rangle_0 \leqslant 2\|\xi\|_1^2. \quad \square$$

Consider finite dimensional vector bundles over S

$$\pi_j : E_j \to S, \ 1 \leqslant j \leqslant k,$$
$$\phi : F \to S$$

with fibre $\mathbb{E}_j$ and $\mathbb{F}$ respectively. These bundles will be endowed with a Riemannian metric. We then have the associated bundle

$$L(\pi_1, \ldots, \pi_k; \phi) : L(E_1, \ldots, E_k; F) \to S$$

of multilinear maps, i.e. the fibre consists of the vector space of multilinear maps

$$L(\mathbb{E}_1, \ldots, \mathbb{E}_k; \mathbb{F}).$$

The Riemannian metrics on the π_j and ϕ determine a norm on this bundle such that

$$|L(\xi_1, \ldots, \xi_k)| \leqslant |L| \, |\xi_1| \ldots |\xi_k|.$$

Consider the canonical inclusions

$$(*)\qquad H^1(L(E_1, E_2, \ldots, E_k; F)) \hookrightarrow L(H^0(E_1), H^1(E_2), \ldots, H^1(E_k); H^0(F)),$$

and

$$(**)\qquad H^1(L(E_1, E_2, \ldots, E_k; F)) \hookrightarrow L(H^1(E_1), H^1(E_2), \ldots, H^1(E_k); H^1(F))$$

given by

$$A = (A(t)) \mapsto \{\tilde{A} : (\xi_1, \ldots, \xi_k) \mapsto (A(t) \, . \, (\xi_1(t), \ldots, \xi_k(t))\}.$$

1.2.2 Proposition. *The inclusions* (*) *and* (**) *are continuous and linear. More precisely,*

$$\|\tilde{A}(\xi_1, \xi_2, \ldots, \xi_k)\|_0^2 \leqslant 2^k \|A\|_1^2 \|\xi_1\|_0^2 \|\xi_2\|_1^2 \ldots \|\xi_k\|_1^2, \textit{ and}$$

$$\|\tilde{A}(\xi_1, \xi_2, \ldots, \xi_k)\|_1^2 \leqslant \text{const} \, \|A\|_1^2 \|\xi_1\|_1^2 \|\xi_2\|_1^2 \ldots \|\xi_k\|_1^2.$$

Proof. (1) We have

$$\begin{aligned}\|\tilde{A}(\xi_1, \xi_2, \ldots, \xi_k)\|_0^2 &\leqslant \|A\|_\infty^2 \|\xi_1\|_0^2 \|\xi_2\|_\infty^2 \ldots \|\xi_k\|_\infty^2 \\ &\leqslant 2^k \|A\|_1^2 \|\xi_1\|_0^2 \|\xi_2\|_1^2 \ldots \|\xi_k\|_1^2\end{aligned}$$

since we can take out from within the integral the square of the maximum norm $\| \ \|_\infty^2$ for all but one element and then apply (1.2.1).

(2) Note that

$$\begin{aligned}&\nabla(A(\xi_1, \xi_2, \ldots, \xi_k)) = (\nabla A)(\xi_1, \xi_2, \ldots, \xi_k) + \\ &A(\nabla\xi_1, \xi_2, \ldots, \xi_k) + A(\xi_1, \nabla\xi_2, \ldots, \xi_k) + \cdots \\ &+ A(\xi_1, \xi_2, \ldots, \nabla\xi_k).\end{aligned}$$

Using the relation $(\sum_1^l a_j)^2 \leqslant l \sum_1^l a_j^2$ and the technique employed in (1) we find that

$$\begin{aligned}\|\nabla\tilde{A}(\xi_1, \xi_2, \ldots, \xi_k)\|_0^2 &\leqslant (k+1) 2^k (\|\nabla A\|_0^2 \|\xi_1\|_1^2 \|\xi_2\|_1^2 \ldots \|\xi_k\|_1^2 \\ &+ \|A\|_1^2 \|\nabla\xi_1\|_0^2 \|\xi_2\|_1^2 \ldots \|\xi_k\|_1^2 + \cdots + \|A\|_1^2 \|\xi_1\|_1^2 \|\xi_2\|_1^2 \ldots \|\nabla\xi_k\|_0^2) \\ &\leqslant \text{const} \, \|A\|_1^2 \|\xi_1\|_1^2 \|\xi_2\|_1^2 \ldots \|\xi_k\|_1^2. \quad \square\end{aligned}$$

1.2.3 Proposition. *Let $\mathcal{O}$ be an open set in the total space of*

$$\pi : E \to S$$

(dim $\mathbb{E} < \infty$), *such that $\mathcal{O}_t := \mathcal{O} \cap \pi^{-1}(t) \subset E_t$ is $\neq \emptyset$, for all $t \in S$. π shall have a Riemannian metric and Riemannian connection.*

Claim. *$H^1(\mathcal{O})$: = set of $\xi \in H^1(E)$ with $\xi(t) \in \mathcal{O}_t$ for all $t \in S$, is open in $H^1(E)$.*

Proof. Let $\xi \in H^1(\mathcal{O})$. There exists $\varepsilon > 0$ such that $\eta \in H^1(E)$ and $|\eta(t) - \xi(t)|^2 < 2\varepsilon^2$ for all $t \in S$ implies that $\eta(t) \in \mathcal{O}_t$, for all $t \in S$. From (1.2.1), it therefore follows that $\|\eta - \xi\|_1 < \varepsilon$ implies that $\eta \in H^1(\mathcal{O})$.

1.2.4 Proposition. *Let $\pi: E \to S$ and $\mathcal{O} \subset E$ as in* (1.2.3). *Let $\phi: F \to S$ be a bundle* (dim $\mathbb{F} < \infty$) *with Riemannian metric and Riemannian connection. Assume that*

$$f: \mathcal{O} \to F$$

is a differentiable fibre map, i.e. $\phi \circ f = \pi$.

Claim. *The induced map*

$$\tilde{f}: H^1(\mathcal{O}) \to H^1(F) \quad ; \quad (\xi(t)) \mapsto (f \circ \xi(t))$$

is continuous.

Proof. If $\|\eta - \xi\|_1$ tends to zero, then so does $\|\eta - \xi\|_\infty$ and $\|\nabla\eta - \nabla\xi\|_0$ and $\|\tilde{f}(\eta) - \tilde{f}(\xi)\|_0 \leqslant \|\tilde{f}(\eta) - \tilde{f}(\xi)\|_\infty$.

We observe that if $\dot{\eta}(t) = \dot{\eta}(t)_h + \dot{\eta}(t)_v$ is the decomposition into the horizontal and the vertical part, $\dot{\eta}(t)_h$ is locally of the form $(t, \eta(t), \partial t, -\Gamma_t(\partial t, \eta(t))$, i.e. it depends only on $\eta(t)$, whereas $\dot{\eta}(t)_v$ can be identified with $\nabla\eta(t)$. Therefore, if we denote by

$$Df(\eta(t)) = D_1 f(\eta(t)) + D_2 f(\eta(t))$$

the decomposition into the restriction of Df to the horizontal and the vertical space, we have

$$\begin{aligned}\nabla(f \circ \eta)(t) - \nabla(f \circ \xi)(t) &= D_2 f(\eta(t)) \,.\, \nabla\eta(t) - D_2 f(\xi(t)) \,.\, \nabla\xi(t) \\ &= D_2 f(\eta(t)) \,.\, (\nabla\eta(t) - \nabla\xi(t)) \\ &\quad + (D_2 f(\eta(t)) - D_2 f(\xi(t)) \,.\, \nabla\xi(t))\end{aligned}$$

modulo terms which tend to zero as $\|\xi - \eta\|_\infty$ tends to zero. Hence, also $\|\nabla\tilde{f}(\eta) - \nabla\tilde{f}(\xi)\|_0$ tends to zero. □

We can now prove the following slightly extended version of a fundamental *Lemma of Palais* [Pa 2]:

1.2.5 Lemma. *Let $f: \mathcal{O} \subset E \to F$ (with the previous notation) be a differentiable fibre map.*

Claim. *The map* $\tilde{f}: H^1(\mathcal{O}) \to H^1(F)$ *is differentiable with* $D\tilde{f} = (D_2 f)^{\sim}$.

Proof. From (1.2.4) we know that $f^{\sim}$ is continuous. The Taylor formula gives

$$f(\eta(t)) - f(\xi(t)) - D_2 f(\xi(t)) \,.\, (\eta(t) - \xi(t)) \\ = r(\xi(t), \eta(t)) \,.\, (\eta(t) - \xi(t))$$

where

$$r(\xi(t), \eta(t)) = \int_0^1 D_2 f(\xi(t) + s(\eta(t) - \xi(t)) ds - D_2 f(\xi(t))$$

is a fibre map of $\mathcal{O}' \times \mathcal{O}' \subset \mathcal{O} \times \mathcal{O} \subset E \times E$, $\mathcal{O}'$ convex, into the bundle $L(\pi, \phi)$: $L(E, F) \to S$.

From (1.2.4) we have that the associated map

$$r^{\sim}: H^1(\mathcal{O}' \times \mathcal{O}') \to H^1(L(E, F))$$

is continuous and

$$\|\tilde{f}(\eta) - \tilde{f}(\xi) - (D_2 f^{\sim}(\xi)) \,.\, (\eta - \xi)\|_1 = \|r^{\sim}(\xi, \eta) \,.\, (\eta - \xi)\|_1$$

$$\leqslant \text{const } \|r(\tilde{\xi}, \tilde{\eta})\|_1 \ \|\eta - \xi\|_1, \text{ and } \|r(\xi, \eta)\|_1 \to 0 \text{ as } \|\xi - \eta\|_1 \to 0,$$

$$\text{since } r^{\sim}(\xi, \xi) = 0.$$

Hence, $\tilde{f}$ is differentiable with $D\tilde{f} = (D_2 f)^{\sim}$.

In the same manner one shows that $D^r \tilde{f} = (D_2^r f)^{\sim}$. □

We recall the following standard result of Riemannian geometry, see e.g. [GKM] or [Mi 2].

1.2.6 Lemma. *Let* M *be a compact Riemannian manifold. For* $\varepsilon > 0$ *denote by* $\mathcal{O}_\varepsilon$ *the open* ε*-neighborhood of the* 0*-section of* $\tau: TM \to M$, *i.e.* $\mathcal{O}_\varepsilon = \{\xi \in TM, |\xi| < \varepsilon\}$. *We also write briefly* $\mathcal{O}$ *instead of* $\mathcal{O}_\varepsilon$.

Claim. *There exists* $\varepsilon > 0$ *such that the map*

$$(\tau, \exp): \mathcal{O}_\varepsilon \to M \times M$$

$$\xi \mapsto (\tau(\xi), \exp(\xi))$$

is a diffeomorphism onto an open neighborhood of the diagonal of $M \times M$.

In particular, $\exp |(\mathcal{O}_p := \mathcal{O} \cap T_p M)$ *is injective.* □

With such a $\mathcal{O} \subset TM$ we put, for $c: S \to M$ of class C^∞,

$$\mathcal{O}_c := c^* \mathcal{O} \subset c^* TM.$$

$\mathcal{O}_c$ is an open neighborhood of the 0-section of $c^*\tau$. We define

$$(\dagger) \qquad \exp_c : H^1(\mathcal{O}_c) \to H^1(S, M)$$

$$\text{by } \xi = (\xi(t)) \mapsto (\exp(\tau^* c\xi(t))).$$

1.2.7 Proposition. *The map* $(\dagger)$ *is injective with*

$$im \exp_c = \{e \in H^1(S, M);\ e(t) \in \exp(\mathcal{O} \cap T_{c(t)}M)\}.$$

Proof. Immediate from the definition. □

We put

$$\mathscr{U}(c) = \exp_c H^1(\mathcal{O}_c).$$

1.2.8 Lemma. *Let* $c, d \in C^\infty(S, M)$. *Then*

$$\exp_d^{-1} \circ \exp_c : \exp_c^{-1}(\mathscr{U}(c) \cap \mathscr{U}(d)) \to \exp_d^{-1}(\mathscr{U}(d) \cap \mathscr{U}(c))$$

is a diffeomorphism.

Proof. For each $t \in S$ define

$$\mathcal{O}_{c,d,t} := \mathcal{O}_{c,t} \cap (\exp \circ \tau^* c)^{-1} \circ (\exp \circ \tau^* d)\, \mathcal{O}_{d,t}$$

and $\mathcal{O}_{c,d} = \bigcup_t \mathcal{O}_{c,d,t}$, if $\mathcal{O}_{c,d,t} \neq \emptyset$ for all $t \in S$. Otherwise put $\mathcal{O}_{c,d} = \emptyset$.

$\mathcal{O}_{c,d}$ is an open subset of $\mathcal{O}_c$ and

$$H^1(\mathcal{O}_{c,d}) = \exp_c^{-1}(\mathscr{U}(c) \cap \mathscr{U}(d)).$$

The map

$$f_{d,c} := (\exp \circ \tau^* d)^{-1} \circ (\exp \circ \tau^* c) : \mathcal{O}_{c,d} \to d^* TM$$

is a fibre map and

$$\exp_d^{-1} \circ \exp_c = \tilde{f}_{d,c}.$$

Therefore the lemma follows from (1.2.5). □

1.2.9 Theorem. $H^1(S, M)$ *is a Hilbert manifold; the differentiable structure is given by the natural atlas*

$$(\exp_c^{-1}, \mathscr{U}(c)),\ c \in C^\infty(S, M).$$

Note. The model of $H^1(S, M)$ is any one of the equivalent separable infinite dimensional Hilbert spaces $H^1(c^* TM)$, $c \in C^\infty(S, M)$.

Proof. (1) The charts of the natural atlas are modeled on the separable Hilbert space represented by $H^1(c^*TM)$.

(2) The sets $\mathcal{U}(c)$, $c\in C^\infty(S, M)$, form an open covering of $H^1(S, M)$ since every $d\in H^1(S, M)$ can be approximated by a $c\in C^\infty(S, M)$ in the d_∞-metric such that $d\in\mathcal{U}(c)$, where

$$d_\infty(c, c') := \sup_{t\in S} d_M\big(c(t), c'(t)\big)$$

and $d_M(\ ,\ )$ is the Riemannian distance on M.

(3) From (1.2.8) we know that the natural atlas is of class C^∞.

To see that $H^1(S, M)$ has a countable base it suffices to show that the natural atlas has a countable subatlas. For this purpose we show that, for each integer $l>0$, the set

$$H^1(S, M)^l := \{c\in H^1(S, M);\ \textstyle\int_S \langle\dot c(t), \dot c(t)\rangle\, dt < 2l\}$$

can be covered by a finite subset of the natural atlas.

To see this we choose $\varepsilon>0$ as in (1.2.6) and let $m=m(\varepsilon, l)$ be an integer for which $18l<m\varepsilon^2$. Then we have, for $e\in H^1(S, M)^l$:

$$d_M(e_{j-1}, e_j)^2 \leqslant \left(\int_{(j-1)/m}^{j/m} |\dot e(t)|\, dt\right)^2$$

$$\leqslant 1/m \int_S \langle\dot e(t), \dot e(t)\rangle\, dt \leqslant 2l/m \leqslant \varepsilon^2/9$$

with $e_j := e(j/m)$.

It follows that $e|[(j-1)/m, j/m]$ lies entirely in a $(\varepsilon/3)$-ball.

There exists a finite set P of points on M such that the $(\varepsilon/3)$-balls around these points will cover M. Given $e\in H^1(S, M)^l$, we can find a sequence $\{p_1, \ldots, p_m\}$ in P such that $e_j\in B_{\varepsilon/3}(p_j)$.

For each of the finitely many sequences $\{p_1, \ldots, p_m\}$ of m elements in P we choose a curve $c\in C^\infty(S, M)\subset H^1(S, M)^l$ with $c(j/m)=p_j$. Then $e\in\mathcal{U}(c)$ for one of these c's. □

Remarks. A differentiable map

$$f\colon M\to N$$

induces a differentiable map

$$H^1(S, f)\colon H^1(S, M)\to H^1(S, N)$$
$$c(t)\mapsto f\circ c(t).$$

That is to say, for $H^1(S,\ )$, we have constructed a functor from the category of compact differentiable manifolds and maps into the category of Hilbert manifolds and differentiable maps.

In particular, the differentiable structure of $H^1(S, M)$ depends only on the differentiable structure on M.

We conclude this section with

1.2.10 Theorem. *The inclusion*

$$H^1(S, M) \hookrightarrow C^0(S, M)$$

is a homotopy equivalence.

Proof. We consider M to be a closed submanifold of some Euclidean space $\mathbb{E}$. Let N be an open tubular neighborhood of M in $\mathbb{E}$. M is a strong deformation retract of N under a differentiable homotopy

$$h_s : N \to N, \; 0 \leqslant s \leqslant 1,$$

with $h_0 = \mathrm{id}$, $h_1(N) = M$.

h_s induces a homotopy equivalence $h_s^{\sim}$, $0 \leqslant s \leqslant 1$, of $H^1(S, N)$ with $H^1(S, M)$ as well as of $C^0(S, N)$ with $C^0(S, M)$ by putting $(h_s^{\sim} \circ c)(t) = h_s \circ c(t)$.

Thus, $H^1(S, M)$ is shown to be homotopically equivalent to the open set $U^1 := H^1(S, N)$ of the Hilbert space $H^1(S, \mathbb{E})$ and $C^0(S, M)$ is shown to be homotopically equivalent to the open set $U^0 := C^0(S, N)$ of the locally convex Banach space $C^0(S, \mathbb{E})$ with the maximum norm.

From (1.2.2) we know that the inclusion $H^1(S, \mathbb{E}) \hookrightarrow C^0(S, \mathbb{E})$ is continuous and has a dense image. A theorem of Palais [Pa 3] implies now that U^1 and U^0 are homotopically equivalent. □

1.3 Riemannian Metric and Energy Integral of the Manifold of Closed Curves

From now on we shall also write ΛM instead of $H^1(S, M)$, and sometimes even simply Λ.

Our first goal is to define two bundles

$$\alpha^r : H^r(H^1(S, M)^* TM) \to H^1(S, M), \; r = 0, 1,$$

where the fibre over $c \in H^1(S, M)$ consists of the H^r-vector fields along c.

To give a precise definition we consider the natural chart $(\exp_c^{-1}, \mathscr{U}(c))$ of ΛM, based at $c \in C^\infty(S, M)$. For $\xi \in \mathcal{O}$, with $\mathcal{O}$ as in (1.2), we consider the map

$$\nabla_2 \exp(\xi) : T_{\tau\xi} M \to T_{\exp \xi} M$$
$$\eta \mapsto T \exp(\xi) \circ (K | T_{\xi v} TM)^{-1} \,.\, \eta.$$

Note that $(K | T_{\xi v} TM) : T_{\xi v} TM \to T_{\tau\xi} M$ is the canonical identification of the tangent space of the fibre $T_{\tau\xi} M$ at ξ with the fibre $T_{\tau\xi} M$.

Clearly, $\nabla_2 \exp(\xi)$ is a linear isomorphism.
Now define, for $r=0,1$,

$$\Phi_{r,c}^{-1} : H^1(\mathcal{O}_c)\times H^r(c^*TM)\to(\alpha^r)^{-1}\mathcal{U}(c) \text{ by}$$
$$(\xi(t), \eta_c(t))\mapsto(\nabla_2 \exp(\tau^* c\xi(t)) \,.\, \eta_c(t)).$$

1.3.1 Theorem. *The maps*

$$(\Phi_{r,c}, \exp_c^{-1}, \mathcal{U}(c)),\ c\in C^\infty(S, M),$$

are local representations of a bundle

$$\alpha^r : H^r(H^1(S, M)^* TM)\to \Lambda M$$

with typical fibre being the separable Hilbert space $H^r(S, c^*TM)$.
In particular, α^1 *is canonically isomorphic to the tangent bundle of* ΛM

$$\tau_{\Lambda M} : T\Lambda M\to \Lambda M.$$

Proof. Consider first the case $r=1$. We see that

$$(*)\qquad \Phi_{1,d}\circ\Phi_{1,c}^{-1} : H^1(\mathcal{O}_{c,d})\times H^1(c^*TM)\to H^1(\mathcal{O}_{d,c})\times H^1(d^*TM)$$

is of the form $(\tilde{f}_{d,c}, (D_2 f_{d,c})^\sim)$ with $f_{d,c}$ as in (1.2.7).

Hence (*) behaves like a transition map between local representations of a bundle. Moreover, since $(D_2 f_{d,c})^\sim = D\tilde{f}_{d,c}$, this transition map is precisely the one which defines the tangent bundle of ΛM.

In the case $r=0$ we observe that the differentiable map

$$(D_2 f_{d,c})^\sim : H^1(\mathcal{O}_{c,d})\to H^1(L(c^*TM; d^*TM)),$$

followed by the continuous linear inclusion

$$H^1(L(c^*TM; d^*TM))\hookrightarrow L(H^0(c^*TM); H^0(d^*TM))$$

(see (1.2.2)) gives a differentiable map. □

We define, for $\xi\in\mathcal{O}\subset TM$,

$$\nabla_1 \exp(\xi) : T_{\tau\xi}M\to T_{\exp\xi}M \text{ by}$$
$$\eta\mapsto T\exp(\xi)\circ(T\tau|T_{\xi h}TM)^{-1} \,.\, \eta.$$

This clearly is a linear isomorphism.
We define

$$\theta : \mathcal{O}\to L(TM; TM) \text{ by}$$
$$\theta(\xi) := \nabla_2 \exp(\xi)^{-1}\circ\nabla_1 \exp(\xi).$$

For $c \in C^\infty(S, M)$ we consider the fibre map

$$\theta_c := (\tau^* c)^{-1} \circ \theta \circ \tau^* c \, . \, \partial c : \mathcal{O}_c \to c^* TM.$$

1.3.2 Proposition. *Taking the tangent field $\partial e(t) = \dot{e}(t)$ along a curve $e \in \Lambda M$ is a differentiable section in the bundle α^0.*

In the local representation $H^1(\mathcal{O}_c) \times H^0(c^* TM)$ of α^0 over $\mathcal{U}(c)$, the principal part of ∂

$$\partial_c : H^1(\mathcal{O}_c) \to H^0(c^* TM)$$

is given by

$$\partial_c \xi(t) := \nabla_c \xi(t) + \theta_c(\xi(t)).$$

Proof. Let $e(t) = \exp(\tau^* c \xi(t))$. Then

$$\partial e(t) = T \exp(\tau^* c \xi(t)) \, . \, (\tau^* c \xi(t)_h^{\dot{}} + \tau^* c \xi(t)_v^{\dot{}})$$

where, on the right-hand side, we have indicated the decomposition into the horizontal and the vertical component.

Since

$$T\tau \, . \, \tau^* c \xi(t)_h^{\dot{}} = \partial c(t); \; K \, . \, \tau^* c \xi(t)_v^{\dot{}} = \nabla(\tau^* c \xi(t))$$

we obtain

$$\begin{aligned}\partial e(t) &= \nabla_1 \exp(\tau^* c \xi(t)) \, . \, \partial c(t) + \nabla_2 \exp(\tau^* c \xi(t)) \, . \, \nabla(\tau^* c \xi(t)) \\ &= \nabla_2 \exp(\tau^* c \xi(t)) \circ \tau^* c \, . \, (\nabla_c \xi(t) + \theta_c(\xi(t))). \quad \square\end{aligned}$$

1.3.3 Theorem. *The bundle $\alpha^0 : H^0(H^1(S, M)^* TM) \to \Lambda M$ has a Riemannian metric which is characterized by the property that on $(\alpha^0)^{-1}(c) = H^0(c^* TM)$, $c \in C^\infty(S, M)$, the metric is given by the product $\langle \, , \rangle_0$. We therefore denote this metric by $\langle \, , \rangle_0$.*

Proof. Define

$$G : \mathcal{O} \subset TM \to L(TM; TM)$$

by

$$\langle G(\xi) \; , \quad \rangle_{\tau\xi} = \langle \nabla_2 \exp(\xi) \; , \nabla_2 \exp(\xi) \; \rangle_{\exp \xi} \, .$$

From the definition of $\mathcal{O}$ it follows that $G(\xi)$ is a positive definite self-adjoint operator of class C^∞.

$$G_c := (\tau^* c)^{-1} \circ G \circ (\tau^* c) : \mathcal{O}_c \to L(c^* TM; c^* TM)$$

is a fibre map and the composite map

$$G_c^{\sim} : H^1(\mathcal{O}_c) \to H^1(L(c^* TM \, ; c^* TM)) \hookrightarrow L(H^0(c^* TM); H^0(c^* TM))$$

is a positive definite self-adjoint operator of class C^∞. Thus, $\xi \in H^1(\mathcal{O}_c) \mapsto \langle G_c^{\sim}(\xi)\ ,\ \rangle_0$ is a Riemannian metric on $H^1(\mathcal{O}_c)$.

One also checks that this metric is independent of the choice of the local representation.

Finally, on $\xi = 0$, $G_c^{\sim} = \mathrm{id}$, i.e. on $H^0(c^*TM)$ the metric coincides with $\langle\ ,\ \rangle_0$. Since the $c \in C^\infty(S, M)$ are dense in $\Lambda M = H^1(S, M)$, this property characterizes the metric. □

1.3.4 Lemma. *The Levi-Cività connection K on the Riemannian tangent bundle $\tau_M : TM \to M$ of M induces a Riemannian connection K_{α^0} on $\alpha^0 : H^0(H^1(S, M)^*TM) \to \Lambda M$.*

Proof. Define

$$\Gamma : \mathcal{O} \to L^2(TM; TM) \text{ by}$$

$$2\langle G(\xi)\Gamma(\xi) . (\eta, \zeta), \theta\rangle_{\tau\xi} \equiv \langle D_2 G(\xi) . (\eta, \zeta), \theta\rangle_{\tau\xi}$$
$$+ \langle D_2 G(\xi) . (\zeta, \theta), \eta\rangle_{\tau\xi} - \langle D_2 G(\xi) . (\theta, \eta), \zeta\rangle_{\tau\xi}$$

identically in all $\eta, \zeta, \theta \in T_{\tau\xi} M$. Then

$$\Gamma_c = (\tau^* c)^{-1} \circ \Gamma \circ \tau^* c . (\tau^* c \times \tau^* c) : \mathcal{O}_c \to L^2(c^*TM; c^*TM)$$

is a fibre map. Hence, the composition

$$\Gamma_c^{\sim} : H^1(\mathcal{O}_c) \to H^1(L_s^2(c^*TM; c^*TM))$$
$$\hookrightarrow L(H^1(c^*M), H^0(c^*TM); H^0(c^*TM))$$

is differentiable (cf. (1.2.2)) and gives the Christoffel symbols of a Riemannian connection K_{α^0} of α^0; that the $\Gamma_c^{\sim}$ actually satisfy the transformation formula for Christoffel symbols follows from their definition; that K_{α^0} is Riemannian is contained in the identity

$$\langle (DG_c^{\sim}(\xi) . \eta)\zeta, \theta\rangle_0 = \langle G_c^{\sim}(\xi)\Gamma_c^{\sim}(\xi) . (\eta, \zeta), \theta\rangle_0$$
$$+ \langle G_c^{\sim}(\xi)\Gamma_c^{\sim}(\xi) . (\eta, \theta), \zeta\rangle_0$$

with $\eta \in H^1(c^*TM)$ and $\zeta, \theta \in H^0(c^*TM)$, which also follows immediately from the definitions; cf. also (****) at the end of (1.1). □

1.3.5 Proposition. *The covariant derivative $\nabla_{\alpha^0}\partial$ of the section ∂ in α^0 is a differentiable section in the bundle*

$$L(\alpha^1; \alpha^0) : L(H^1(H^1(S, M)^*TM); H^0(H^1(S, M)^*TM)) \to \Lambda M.$$

If $\xi \in H^1(\mathcal{O}_c)$ *is the local representative of* $e \in \mathcal{U}(c)$ *and* $\eta \in T_e\Lambda M$, *then the principal part of the local representative of* $\nabla_{\alpha^0}\partial e\,.\,\eta$ *is*

$$\nabla\eta(t) + D_2(\theta_c(\xi(t))\,.\,\eta(t) + \Gamma_c(\xi(t))\,.\,(\eta(t), \partial_c\xi(t))$$

where

$$\partial_c\xi(t) = \nabla\xi(t) + \theta_c(\xi(t)) \in H^0(c^*TM)$$

is the principal part of the local representative of ∂e, *cf.* (1.3.2), *and* $\eta(t) \in H^1(c^*TM)$ *is the principal part of the local representative of* η.

We shall write $\nabla\eta$ instead of $\nabla_{\alpha^0}\partial e\,.\,\eta$ if there will be no confusion. Note that for $e = c$, i.e. $\xi = 0$, $\nabla_{\alpha^0}\partial e\,.\,\eta$ indeed has $\nabla\eta$ as a local representative.

Proof. Locally, the principal part of the covariant derivative is given by

$$D(\partial_c(\xi(t))\,.\,\eta(t) + \Gamma_c(\xi(t))\,.\,(\eta(t), \partial_c\xi(t)).$$

Since ∇ is linear and $\theta_c : \mathcal{O}_c \to c^*TM$ is a fibre map and consequently $D\theta_c^{\sim} = (D_2\theta_c)^{\sim}$, it follows that the first summand is equal to

$$\nabla\eta(t) + D_2(\theta_c(\xi(t))\,.\,\eta(t). \quad \square$$

1.3.6 Theorem. *The tangent bundle* $\alpha^1 \equiv \tau_{\Lambda M}$ *of* ΛM *has a Riemannian metric which is characterized by the property that it coincides on* $T_c\Lambda M \cong H^1(c^*TM)$, $c \in C^\infty(S, M)$, *with the product* $\langle\,,\,\rangle_1$. *We therefore denote this metric by* $\langle\,,\,\rangle_1$.

Proof. We define the Riemannian metric on $T_e\Lambda M$ by

$$\langle\,,\,\rangle_0 + \langle \nabla_{\alpha^0}\partial e\,,\, \nabla_{\alpha^0}\partial e\rangle_0\,.$$

That this is a differentiable section in $L_s^2(\tau_{\Lambda M}) : L_s^2(T\Lambda M) \to \Lambda M$ follows from (1.3.3) and (1.3.5).

On $H^1(c^*TM)$, $c \in C^\infty(S, M)$, this metric clearly coincides with $\langle\,,\,\rangle_1$. Since the $c \in C^\infty(S, M)$ are dense in $\Lambda M = H^1(S, M)$, this characterizes the Riemannian metric. $\square$

We define an open neighborhood of the zero section of $\tau_{\Lambda M}$ by

$$\tilde{\mathcal{O}} := \{\eta \in TH^1(S, M);\ \eta(t) \in \mathcal{O} \subset TM\}$$

with $\mathcal{O}$ as in (1.2).

Clearly, for $c \in C^\infty(S, M)$, we have $\tilde{\mathcal{O}} \cap T_c\Lambda M \cong H^1(\mathcal{O}_c)$.

We will show that the maps

$$\exp_c : H^1(\mathcal{O}_c) \to \mathcal{U}(c)$$

can be pieced together to form a differentiable map

$$\exp^{\sim} : \tilde{\mathcal{O}} \to \Lambda M$$

which has the properties of the exponential map of a connection on $\tau_{\Lambda M}$. Note, however, that $\exp^{\sim}$ will not be the exponential map of the Levi-Cività connection derived from the Riemannian metric $\langle\ ,\ \rangle_1$ on ΛM. For further details see [FK].

1.3.7 Theorem. *The map*

$$\tau_{\Lambda M} \times \exp^{\sim} : \tilde{\mathcal{O}} \subset T\Lambda M \to \Lambda M \times \Lambda M$$
$$(\eta(t)) \mapsto (\tau_M \eta(t), \exp(\eta(t)))$$

is differentiable. It maps a sufficiently small open neighborhood of the zero section diffeomorphically onto an open neighborhood of the diagonal of $\Lambda M \times \Lambda M$.

Proof. Recall the local representation

$$\Phi_{1,c}^{-1} : H^1(\mathcal{O}_c) \times H^1(c^*TM) \to (\tau_{\Lambda M})^{-1}\mathcal{U}(c)$$
$$(\xi(t), \eta_c(t)) \mapsto (\nabla_2 \exp(\tau^* c\xi(t)) \cdot \eta_c(t))$$

for $\tau_{\Lambda M}$. Hence, the local representation of $\exp^{\sim}$ is

$$(\exp \circ \tau^* c)^{-1} \circ \exp \circ \nabla_2 \exp(\tau^* c\xi(t)) \cdot \eta_c(t)$$

and we see that this is a differentiable fibre map

$$\mathcal{O}_c \times c^*TM \to c^*TM$$

so that $\exp^{\sim}$ is differentiable.

$\tau_{\Lambda M} \times \exp^{\sim}$ maps the zero section of $\tau_{\Lambda M}$ 1 : 1 onto the diagonal Δ of $\Lambda M \times \Lambda M$. It only remains to be shown that its differential at the element $0 \in T_c\Lambda M$ is a linear isomorphism onto $T_{(c,c)}\Delta$.

To see this we observe that in the local representation

$$(\tau_{\Lambda M} \times \exp^{\sim})(0, \eta_c(t)) = (0, \eta_c(t))$$
$$(\tau_{\Lambda M} \times \exp^{\sim})(\xi(t), 0) = (\xi(t), \xi(t))$$

hence our claim holds. □

As an important corollary we have

1.3.8 Theorem. *For every $c \in \Lambda M$ there exists a natural chart based at c*

$$(\exp_c^{-1}, \mathcal{U}(c)), \ \textit{where}$$
$$\exp_c := \exp^{\sim} | \mathcal{O} \cap T_c\Lambda M. \quad \square$$

Note. Here we should possibly have restricted $\tilde{\mathcal{O}} \cap T_c\Lambda M$ to a smaller open neighborhood of $0 \in T_c\Lambda M$. In any case, an $e \in \mathcal{U}(c)$ is represented by the element $\eta(t) \in H^1(TM)$ satisfying

$$\eta(t) \in \mathcal{O} \cap T_{c(t)}M, \qquad \exp \eta(t) = e(t).$$

We now define a function $E: \Lambda M \to \mathbb{R}$ on ΛM, the so-called *energy integral*, by

$$E(c) := \tfrac{1}{2}\langle \partial c, \partial c \rangle_0 .$$

1.3.9 Lemma. *The function E is differentiable with*

$$DE(c) \,.\, \eta = \langle \partial c, \nabla \eta \rangle_0 .$$

Proof. The differentiability of E follows from (1.3.2) and (1.3.3).
To compute the differential of E we use the Riemannian connection ∇_{α^0} on α^0:

$$D\langle \partial c, \partial c \rangle_0 \,.\, \eta = 2\langle \partial c, \nabla_{\alpha^0} \partial c \,.\, \eta \rangle_0$$

and recall that we have put $\nabla_{\alpha^0} \partial c \,.\, \eta = \nabla \eta$. □

1.3.10 Complement. *In the local representation of α^1 over $\mathscr{U}(c)$, E is given by*

$$E_c := E \circ \exp_c : H^1(\mathcal{O}_c) \to \mathbb{R}$$
$$\xi \mapsto \tfrac{1}{2}\langle G_c^{\sim}(\xi) \,.\, \partial_c \xi, \partial_c \xi \rangle_0$$

and

$$DE_c(\xi) \,.\, \eta = \langle G_c^{\sim}(\xi) \,.\, \partial_c \xi, \nabla \eta + (D_2 \theta_c)^{\sim}(\xi) \,.\, \eta \rangle_0$$
$$+ \tfrac{1}{2}\langle (D_2 G_c)^{\sim}(\xi) \,.\, (\eta, \partial_c \xi), \partial_c \xi \rangle_0 .$$

Proof. We have

$$DE_c(\xi) \,.\, \eta = \langle G_c^{\sim}(\xi) \,.\, \partial_c \xi, \nabla_{\alpha^0}(\partial_c \xi) \,.\, \eta \rangle_0 .$$

Substitute here the expression (1.3.5) for the second factor and use the identity at the end of the proof of (1.3.4), setting $\zeta = \theta = \partial_c \xi$. □

The importance of the function E lies in the fact that it allows the characterization of the *closed geodesics* in ΛM, i.e. the differentiable maps $c: S \to M$ satisfying $\nabla \dot{c} = 0$, $\dot{c} \neq 0$.

1.3.11 Theorem. *$c \in \Lambda M$ is a closed geodesic or a constant map if and only if it is a critical point of E, i.e. $DE(c) = 0$.*

Proof. If c is differentiable, then we can write

$$DE(c) \,.\, \eta = \langle \partial c, \nabla \eta \rangle_0 = -\langle \nabla \partial c, \eta \rangle_0 .$$

Hence, $\nabla \partial c = 0$ implies $DE(c) = 0$.

Assume now that $DE(c) \,.\, \eta = 0$ for all $\eta \in T_c \Lambda M$. Then c is a weak solution of $\nabla \partial c = 0$. From the theory of elliptic operators it follows that c is differentiable, i.e. a closed geodesic or constant.

If we wish to avoid this theory, we can give a direct proof based on the so-called *Lemma of Du Bois-Raymond:*

Let $\xi(t)$ be a parallel H^1-vector field (not necessarily periodic) along $c(t)$, i.e. $\nabla\xi(t)=0$, satisfying $\xi(1)=\zeta(1)$, where $\zeta(t)$ is the H^1-vector field defined by

$$\nabla\zeta(t)=\partial c(t),\ \zeta(0)=0.$$

Put $\zeta(t)-t\xi(t)=\eta(t)$. Then

$$\eta(0)=\eta(1)=0;\ \nabla\eta(t)=\partial c(t)-\xi(t).$$

Moreover,

$$\langle\xi,\partial c-\xi\rangle_0=\langle\xi,\nabla\eta\rangle_0=\int_S\frac{d}{dt}\langle\xi(t),\eta(t)\rangle\,dt=0$$

and

$$0=DE(c)\,.\,\eta=\langle\partial c,\nabla\eta\rangle_0.$$

Hence

$$\langle\partial c-\xi,\partial c-\xi\rangle_0=0$$

i.e. $\partial c(t)=\xi(t)$ is of class H^1.

To see that $\partial c(1)=\partial c(0)$ one applies the same argument, using $c(t+1/2)$ instead of $c(t)$. □

1.4 The Condition (C) of Palais and Smale and its Consequences

We begin by collecting a few properties of ΛM, E, and their structures, which can be viewed as generalizations for the non-linear case (i.e. for manifolds) of well-known results of the Hilbert space $H^1(S,\mathbb{E})$, $\mathbb{E}$ a Euclidean space.

1.4.1 Proposition. *Let* $c\in\Lambda M$. *Then*

$$d_M^2(c(t_0),c(t_1))\leqslant|t_1-t_0|\,2E(c).$$

Proof. $d_M^2(c(t_0),c(t_1))\leqslant\left(\int_{t_0}^{t_1}|\dot c(t)|\,dt\right)^2\leqslant\left|\int_{t_0}^{t_1}1^2dt\right|\cdot\int_{t_0}^{t_1}|\dot c(t)|^2dt.$ □

1.4.2 Proposition. *Let* $c,c'\in\Lambda M$. *Then*

$$d_\infty^2(c,c')\leqslant 2d_\Lambda^2(c,c')$$

where d_Λ *is the distance on* ΛM *derived from the Riemannian metric* $\langle\ ,\ \rangle_1$.

Note. This generalizes (1.2.1).

Proof. It suffices to show that for every differentiable curve $\kappa(s)$, $0 \leqslant s \leqslant 1$, in Λ, with $\kappa(0)=c$, $\kappa(1)=c'$, we have

$$d_\infty^2(c, c') \leqslant 2L^2(\kappa).$$

Here we exclude the possibility that c and c' belong to different connected components of Λ, in which case we define the distance to be ∞.

Consider

$$\begin{aligned} \tilde{\kappa} : S \times I &\to M \\ (t, s) &\mapsto \kappa(s)(t). \end{aligned}$$

For fixed $t = t_0$ is $\kappa(s)(t_0)$ a differentiable curve, since the map

$$\begin{aligned} t : H^1(S, M) &\to M \\ c &\mapsto c(t) \end{aligned}$$

is differentiable, as can be seen from its local representation. Choose t_0 such that

$$d_\infty(c, c') = d_M(c(t_0), c'(t_0)).$$

Then

$$\begin{aligned} d_\infty^2(c, c') = d_M^2(c(t_0), c'(t_0)) &\leqslant \left(\int_0^1 \left| \frac{\partial \tilde{\kappa}}{\partial s}(t_0, s) \right| ds \right)^2 \\ &\leqslant \left(\int_0^1 \max_t \left| \frac{\partial \tilde{\kappa}}{\partial s}(t, s) \right| ds \right)^2 \\ &\leqslant 2 \left(\int_0^1 \left\| \frac{\partial \tilde{\kappa}}{\partial s}(s) \right\|_1 ds \right)^2 = 2L(\kappa)^2. \quad \square \end{aligned}$$

1.4.3 Proposition. *Let $c, c' \in \Lambda M$. Then*

$$\left| \sqrt{2E(c)} - \sqrt{2E(c')} \right| \leqslant d_\Lambda(c, c').$$

Note. This is the non-linear analog of

$$\left| \, \|\nabla \xi\|_0 - \|\nabla \xi'\|_0 \right| \leqslant \|\xi - \xi'\|_1 .$$

Proof. We join c and c' by a differentiable curve $\kappa(s)$, $0 \leqslant s \leqslant 1$, in ΛM, as in the proof of (1.4.2). We assume: $\kappa(s) \notin \Lambda^0 M$ for all s. The remaining cases can be taken care of using a limit argument, cf. also [FK]. Then

$$\frac{d}{ds}\|\partial\kappa(s)\|_0 = \frac{1}{\|\partial\kappa(s)\|_0}\int_S\left\langle\frac{\partial\tilde\kappa}{\partial t}(t,s), \nabla\frac{\partial\tilde\kappa}{\partial s}(t,s)\right\rangle dt$$

$$\leqslant \frac{1}{\|\partial\kappa(s)\|_0}\left(\int_S\left|\frac{\partial\tilde\kappa}{\partial t}(t,s)\right|^2 dt\right)^{1/2}\left(\int_S\left|\nabla\frac{\partial\tilde\kappa}{\partial s}(t,s)\right|^2 dt\right)^{1/2}$$

$$\leqslant \left\|\frac{d\kappa}{ds}\right\|_1.$$

Integration yields

$$\sqrt{2E(\kappa(1))} - \sqrt{2E(\kappa(0))} \leqslant L(\kappa).$$

Since this holds for every κ and also under permutation of c and c' we have proved the proposition. □

1.4.4 Lemma. *The inclusion*

$$\Lambda M = H^1(S, M) \hookrightarrow C^0(S, M)$$

is continuous and compact, i.e. the image of every bounded subset of ΛM has compact closure in $C^0(S, M)$.

Proof. The continuity follows from (1.4.2). The compactness follows from the theorem of Arzela-Ascoli, once we have established the following two properties with $K \subset \Lambda M$ bounded and connected:

(i) the inclusion $K \hookrightarrow C^0(S, M)$ is equicontinuous; and

(ii) for $t_0 \in S$, the set $K(t_0) := \{c(t_0), c \in K\}$ is relatively compact in M.

Now, (ii) is obvious since M is compact. To see (i) we observe that, for a fixed $c_0 \in K$, there exists $k_0 > 0$ such that $d_\Lambda(c, c_0) < k_0$ for all $c \in K$. Then, from (1.4.3), the existence of $\kappa > 0$ such that $E|K \leqslant \kappa$ follows. (1.4.1) now implies equicontinuity. □

As a first application of the lemma we obtain:

1.4.5 Theorem. *$\Lambda M = H^1(S, M)$ is a complete metric space with respect to the metric d_Λ.*

Proof. Let $\{c_m\}$ be a Cauchy sequence in Λ. The inclusion of this sequence in the complete space $C^0(S, M)$ shows that there exists a limit $c_0 \in C^0(S, M)$ for this sequence.

Since c_0 can be approximated by a differentiable curve c, we can assume that, for all sufficiently large m, $c_m \in \mathscr{U}(c)$. Put $\exp_c^{-1}(c_m) = \xi_m \in H^1(\mathscr{O}_c)$.

The theorem now follows from the observation that the local representations

$$\langle G(\xi)\,,\ \rangle_0 + \langle G(\xi)\nabla, \nabla\ \rangle_0$$

of the Hilbert metric of $T_{\exp_c\xi}\Lambda M$ in $T_c\Lambda M$, $\xi\in H^1(\mathcal{O}_c)$, are all equivalent to the Hilbert metric $\langle\, ,\rangle_1$ in $T_c\Lambda M\cong H^1(c^*TM)$. Therefore, the Cauchy sequence $\{\xi_m\}$ in the metric representing the distance d_Λ is also a Cauchy sequence in the standard metric $\langle\, ,\rangle_1$ of $H^1(c^*TM)$. Since the latter space is complete, our theorem follows. □

For every real κ we denote by Λ^κ or $\Lambda^\kappa M$ and $\Lambda^{\kappa-}$ or $\Lambda^{\kappa-}M$ the set of $c\in\Lambda$ satisfying $E(c)\leqslant\kappa$ and $E(c)<\kappa$ respectively.

Note that $\Lambda^0 M$ consists of the constant maps $c_p: S\to p\in M$. The following proposition provides some information concerning the canonical inclusions $M\hookrightarrow\Lambda^0 M\hookrightarrow\Lambda M$.

1.4.6 Proposition. *The inclusion*

$$i: M\to\Lambda M: p\mapsto c_p$$

is an isometric, totally geodesic embedding.

Proof. Clearly, $c_p\in C^\infty(S, M)$ and the natural chart $(\exp_p^{-1}, \mathcal{U}(c_p))$, based at c_p, represents curves $c\in\mathcal{U}(p)$ by vector fields $\xi(t)\in H^1(c_p^*TM)$ with $\tau^*c_p\xi(t)\in T_pM$. Put

$$T_p\Lambda^0 := \{\xi_0\in T_p\Lambda;\ \xi_0=\text{const}\}$$

$$T_p^\perp\Lambda^0 := \{\xi\in T_p\Lambda;\ \langle\xi, \xi_0\rangle_1=\langle\xi, \xi_0\rangle_0=0, \text{ all } \quad \xi_0\in T_p\Lambda^0\}.$$

This is an orthogonal decomposition of $T_{c_p}\Lambda$, and the elements $\xi(t)\in H^1(\mathcal{O}_{c_p})$ with $\tau^*c_p\xi(t)\in T_p\Lambda^0$ are precisely the coordinates of the elements of $\Lambda^0 M$; thus $\Lambda^0 M$ is a submanifold.

The inclusion $M\to\Lambda^0 M;\ p\to c_p$ is isometric, since on $T_{c_p}\Lambda^0 M$, $\langle\, ,\rangle_1$ coincides with $\langle\, ,\rangle_p$. Finally, $\Lambda^0 M$ is totally geodesic because, for $c, c'\in\Lambda^0 M$, (1.4.2) can be refined to $d_\infty(c, c')=d_M(c, c')=d_\Lambda(c, c')$. □

With the help of the Riemannian metric on ΛM we define the *gradient field of* E by

$$\langle\operatorname{grad} E(c), \eta\rangle_1 = DE(c)\,.\,\eta=\langle\partial c, \nabla\eta\rangle_0,$$

identical for all $\eta\in T_c\Lambda M$.

If c is differentiable, we can apply partial integration and characterize grad $E(c)(t)$ as the periodic solution of

$$\nabla^2\eta(t)-\eta(t)=\nabla\partial c(t).$$

We now come to the crucial result which allows an extension of the classical theory of a function and its critical points on a Euclidean manifold to a Hilbert manifold. This is the so-called *Condition* (C) of Palais and Smale, cf. [PS]:

"Let $\{c_m\}$ be a sequence on ΛM such that

(i) the sequence $\{E(c_m)\}$ is bounded,

(ii) the sequence $\{\|\mathrm{grad}\, E(c_m)\|_1\}$ tends to zero.

Then $\{c_m\}$ has limit points and any limit point is a critical point of E."

Note. Condition (C) should be viewed as a substitute for the fact that ΛM is not locally compact.

1.4.7 Theorem. *ΛM, with its Riemannian metric and the function E, satisfies Condition (C).*

Proof. (cf. [El 2]). From (1.4.1) it follows that the sequence $\{c_m\}$ is equicontinuous. Since $\{c_m(t_0)\}$ is relatively compact then, according to Arzela-Ascoli's theorem, $\{c_m\}$ has a limit element c_0 in $C^0(S, M)$.

As in the proof of (1.4.5), we can assume that all elements of a subsequence (which we denote again by $\{c_m\}$) belong to the domain $\mathscr{U}(c)$ of a natural chart $(\exp_c, \mathscr{U}(c))$, $c \in C^\infty(S, M)$, and form a Cauchy sequence in the metric d_∞ of $C^0(S, M)$. We will show that the elements $\{\xi_m := \exp_c^{-1} c_m\}$ of $H^1(\mathscr{O}_c)$ form a Cauchy sequence in the metric d_Λ of ΛM.

To see this we first observe that $\|\xi_l - \xi_m\|_0$ tends to zero, since

$$\|\xi_l - \xi_m\|_0 \leqslant \|\xi_l - \xi_m\|_\infty .$$

From the local representation (1.3.2) of $\partial_c \xi$ we have

$$\|\nabla \xi_m\|_0 \leqslant \|\partial_c \xi_m\|_0 + \|\theta_c^{\sim}(\xi_m)\|_0$$

and

$$\kappa \|\partial_c \xi_m\|_0^2 \leqslant \langle G_c^{\sim}(\xi_m) \,.\, \partial_c \xi_m, \partial_c \xi_m \rangle_0 = 2E(c_m)$$

for some $\kappa > 0$. Hence $\|\xi_m\|_1^2$ is bounded.

From (1.3.2) we have

$$\|\nabla \xi - \nabla \eta\|_0 \leqslant \|\partial_c \xi - \partial_c \eta\|_0 + \|\theta_c^{\sim}(\xi) - \theta_c^{\sim}(\eta)\|_0 .$$

Therefore, it suffices to show that

$$\|\partial_c \xi_l - \partial_c \xi_m\|_0 \to 0 \quad \text{for} \quad l, m \to \infty .$$

To see this we consider that, using (1.3.2) and (1.3.10),

$$\begin{aligned} 2\kappa \|\partial_c \xi - \partial_c \eta\|_0^2 \leqslant & \langle G_c^{\sim}(\xi) \,.\, (\partial_c \xi - \partial_c \eta), (\partial_c \xi - \partial_c \eta) \rangle_0 \\ & + \langle G_c^{\sim}(\eta) \,.\, (\partial_c \xi - \partial_c \eta), (\partial_c \xi - \partial_c \eta) \rangle_0 \\ & = 2DE(\xi) \,.\, (\xi - \eta) - 2DE_c(\eta) \,.\, (\xi - \eta) \end{aligned}$$

$$-\langle (G_c^{\sim}(\xi) - G_c^{\sim}(\eta)) \,.\, (\partial_c \xi + \partial_c \eta), (\partial_c \xi - \partial_c \eta) \rangle_0$$

$$+2\langle G_c^{\sim}(\xi) \,.\, \partial_c \xi, \theta_c^{\sim}(\xi) - \theta_c^{\sim}(\eta) - (D_2 \theta_c)^{\sim}(\xi) \,.\, (\xi - \eta) \rangle_0$$

$$-2\langle G_c^{\sim}(\eta) \,.\, \partial_c \eta, \theta_c^{\sim}(\xi) - \theta_c^{\sim}(\eta) - (D_2 \theta_c)^{\sim}(\eta) \,.\, (\xi - \eta) \rangle_0$$

$$-\langle (D_2 G_c)^{\sim}(\xi) \,.\, (\xi - \eta, \partial_c \xi), \partial_c \xi \rangle_0$$

$$+\langle (D_2 G_c)^{\sim}(\xi) \,.\, (\xi - \eta, \partial_c \eta), \partial_c \eta \rangle_0 .$$

Now, as $\|\xi_m\|_1$, and therefore $\|\partial_c \xi_m\|_0$, is bounded and $\|\xi_l - \xi_m\|_0$ tends to zero as well as $\|DE_c(\xi_l)\|_1$, each of the six lines above tends to zero respectively, if we replace ξ, η with ξ_l, ξ_m.

This completes the proof of (1.4.7). □

By $Cr\Lambda$ we denote the set of critical elements of E in $\Lambda = \Lambda M$. As a first consequence of (1.4.7) we have:

1.4.8 Lemma. *Let $\kappa > 0$ and put $Cr\Lambda \cap E^{-1}(\kappa) = K$. Let $\mathscr{U}$ be an open neighborhood of K in Λ.*

Claim. *There exists $\varepsilon = \varepsilon(\kappa, \mathscr{U}) > 0$ and $\eta = \eta(\varepsilon) > 0$ such that*

$$c \in (\Lambda^{\kappa+\varepsilon} - \Lambda^{(\kappa-\varepsilon)-}) \cap \complement\, \mathscr{U}$$

implies

$$\|\text{grad } E(c)\|_1 \geqslant \eta .$$

Proof. Assume that this were not the case. Then this would mean that there exists a sequence $\{c_m\}$ in $\complement\,\mathscr{U}$ with $\lim \|\text{grad } E(c_m)\|_1 = 0$, $\lim E(c_m) = \kappa$. From (1.4.7) we have the existence of a limit point of this sequence which belongs to K – a contradiction. □

Note. A particularly important case is $K = \emptyset$, where we can take $\mathscr{U} = \emptyset$.
Another immediate consequence of (1.4.7) is

1.4.9 Proposition. *$Cr\Lambda \cap E^{-1}(\kappa)$ is compact as is $Cr\Lambda \cap \Lambda^{\kappa}$, for every $\kappa \geqslant 0$.* □

There exists, just as for finite dimensions, for a sufficiently small interval containing $0 \in \mathbb{R}$, the integral curve $\phi_s c$ of the vector field – grad E on Λ starting at c, cf. [La].

1.4.10 Lemma.

(i) $$\frac{d}{ds} E(\phi_s c) = -\|\text{grad } E(\phi_s c)\|_1^2, \quad \textit{hence} \quad E(\phi_{s_1} c) - E(\phi_{s_0} c)$$

$$= -\int_{s_0}^{s_1} \|\text{grad } E\|_1^2 \, ds \leqslant 0 \quad \textit{for} \quad s_1 \geqslant s_0 ;$$

(ii) $$d_\Lambda^2(\phi_{s_1}c, \phi_{s_0}c) \leqslant \left|\int_{s_0}^{s_1} \|\operatorname{grad} E\|_1^2\, ds\right| \cdot |s_1 - s_0|$$
$$= |E(\phi_{s_1}c) - E(\phi_{s_0}c)|\; |s_1 - s_0| \leqslant E(c)\,|s_1 - s_0|$$

for $0 \leqslant s_0, s_1$.

Proof. (i) Using

$$\frac{d\phi_s(c)}{ds} = -\operatorname{grad} E(\phi_s c)$$

we have

$$\frac{d}{ds} E(\phi_s c) = DE(\phi_s c) \,.\, -\operatorname{grad} E(\phi_s c) = -\|\operatorname{grad} E(\phi_s c)\|_1^2;$$

(ii) $$d_\Lambda^2(\phi_{s_1}c, \phi_{s_0}c) \leqslant \left(\int_{s_0}^{s_1} \left\|\frac{d\phi_s c}{ds}\right\|_1 ds\right)^2 \leqslant \left|\int_{s_0}^{s_1} \|\operatorname{grad} E\|_1^2\, ds\right| \cdot |s_1 - s_0|. \quad \square$$

As a consequence we have:

1.4.11 Theorem. *The map*

$$\phi_s : \Lambda \to \Lambda;\; c \to \phi_s c$$

is defined for all $s \geqslant 0$.

Note. Here we essentially use the fact that $E(\phi_s c)$ is bounded from below by 0. If $E(\phi_s c)$ is bounded also from above then the same argument will show that $\phi_s c$ is defined also for all $s \leqslant 0$.

Proof. Assume that there is a $c \in \Lambda$ such that $\phi_s c$ is defined only for $s < s^+$, $s^+ < \infty$. Consider a sequence $\{s_m\}$ of real numbers in $[0, s^+[$ with $\lim s_m = s^+$. From (1.4.10) (ii), it follows that $\{\phi_{s_m} c\}$ is a Cauchy sequence, hence, according to (1.4.5), $\lim \phi_{s_m} c$ exists, which we define to be $\phi_{s^+} c$. Therefore, $\phi_s c$ can also be defined in $[0, s^+ + \varepsilon[$, some $\varepsilon > 0$. $\square$

The following lemma is essential for the proof of the existence of critical points $\neq$ const.

1.4.12 Lemma. *Assume that* $\kappa \geqslant 0$ *is not a critical value, i.e.* $\mathrm{Cr}\Lambda \cap E^{-1}(\kappa) = \emptyset$. *Then* $\kappa > 0$ *and there exist* $\varepsilon > 0$ *and* $s_0 \geqslant 0$ *such that*

$$\phi_s \Lambda^{\kappa+\varepsilon} \subset \Lambda^{(\kappa-\varepsilon)-}$$

for all $s \geqslant s_0$.

Proof. From (1.4.8) we have, with $\mathscr{U} = \emptyset$, the existence of $\varepsilon > 0$ and $\eta > 0$ such that $\|\xi(c)\|_1 \geqslant \eta$ for c satisfying $\kappa - \varepsilon \leqslant E(c) \leqslant \kappa + \varepsilon$.

If $E(c) < \kappa - \varepsilon$, then also $E(\phi_s c) < \kappa - \varepsilon$, for all $s \geq 0$. We derive a contradiction from the assumption that there is a c, $\kappa - \varepsilon \leq E(c) \leq \kappa + \varepsilon$, with $E(\phi_s c) \geq \kappa - \varepsilon$ for all s, $0 \leq s \leq s_0$ with $s_0 > 2\varepsilon/\eta^2$.

Indeed, from (1.4.10) (i) it would then follow that

$$E(\phi_{s_0} c) = E(c) - \int_0^{s_0} \|\text{grad}\,(\phi_s c)\|_1^2\, ds \leq \kappa + \varepsilon - \eta^2 s_0 < \kappa - \varepsilon. \quad \square$$

For subsequent applications we show:

1.4.13 Proposition. *Consider the differentiable section $\omega : \Lambda M \to T^*\Lambda M$ given by $\omega \,.\, \eta = -\langle \partial c, \eta \rangle_0$, for $\eta \in T_c\Lambda$.*

Claim. *The vector field u, representing ω with respect to the Riemannian metric $\langle\ ,\ \rangle_1$, i.e. $\langle u(c), \eta \rangle_1 = -\langle \partial c, \eta \rangle_0$, is given by*

$$u(c) = \nabla\, \text{grad}\, E(c) - \partial c.$$

It satisfies the relations:

$$\nabla u = \text{grad}\, E, \quad \langle u, \text{grad}\, E \rangle_1 = 0.$$

Proof. It suffices to prove our claim for $c \in C^\infty(S, M)$. In this case we find, for the expression above, $\nabla u = \text{grad}\, E$, thus, using our formula for grad E following (1.4.6), $\nabla^2 u(c) - u(c) = \partial c$, i.e. $\omega \,.\, \eta = \langle u, \eta \rangle_1$.

To obtain the last statement we note that, for an arbitrary $w \in T_c\Lambda$, $2\langle w, \nabla w \rangle_0 = \int_S d\langle w(t), w(t) \rangle = 0$. Setting $\nabla u = \text{grad}\, E$, and our claim holds. $\square$

1.4.14 Proposition. *There exist $\varepsilon > 0$ and $\alpha = \alpha(\varepsilon) > 0$ such that*

$$\|\text{grad}\, E(c)\|_1^2 \geq \alpha 2 E(c)$$

for all $c \in \Lambda^\varepsilon$.

Proof. From (1.4.13) we obtain

$$\|\text{grad}\, E(c)\|_1^2 = \langle \partial c, \nabla\, \text{grad}\, E(c) \rangle_0 = 2E(c) + \langle \partial c, u(c) \rangle_0.$$

We shall derive an estimate of the form

$$(*) \qquad |\langle \partial c, u(c) \rangle_0| \leq \beta \sqrt{2E(c)}\, \|\text{grad}\, E(c)\|_1$$

for some $\beta < 4$ and for $E(c)$ sufficiently small. Using

$$\sqrt{2E(c)}\, \|\text{grad}\, E(c)\|_1 \leq \|\text{grad}\, E(c)\|_1^2 + E(c)/2$$

we then have that

$$\|\text{grad}\, E(c)\|_1^2 \geq 2E(c) - \beta(\|\text{grad}\, E(c)\|_1^2 + E(c)/2),$$

which is the desired result.

To prove (*) we first assume that M has Euclidean coordinates in a neighborhood of $c(0)=c(1)$. From $\nabla u(c)=\operatorname{grad} E(c)$ we obtain

$$u(t)=u(0)+\int_0^t \operatorname{grad} E\circ c(s)\,ds.$$

Set $\operatorname{grad} E\circ c(s)=y(s)$, $\int_0^t y(s)\,ds=z(t)$. Since

$$\|z\|_\infty=\max_{0\leqslant t\leqslant 1}|z(t)|\leqslant\max_{0\leqslant t\leqslant 1}|y(t)|=\|y\|_\infty\leqslant 2\|y\|_1,$$

(cf. (1.2.1)), we obtain, using the fact that $c(1)=c(0)$,

$$|\langle\partial c, u\rangle_0|=|\int_S\langle\dot c(t), u(0)\rangle dt+\int_S\langle\dot c(t), z(t)\rangle dt|$$

$$=|\int\langle\dot c(t),z(t)\rangle dt|\leqslant\sqrt{2E(c)}\,\|z\|_\infty\leqslant 2\sqrt{2E(c)}\,\|\operatorname{grad} E(c)\|_1$$

i.e. (*) holds with $\beta=2$.

In the general case, where Euclidean coordinates may not exist locally, we observe that, for $E(c)$ sufficiently small, the curve c belongs to a small neighborhood of its origin $c(0)$ in M : $L(c)\leqslant\sqrt{2\mathrm{E}(c)}$.

Hence, the normal coordinates based at $c(0)$ will differ arbitrarily little from Euclidean coordinates when $E(c)$ is chosen sufficiently small. This is uniformly true for all c on M since M is compact.

Thus we can prove (*) and hence (1.4.14). □

1.4.15 Theorem. *There exists $\varepsilon>0$ such that, for every $c\in\Lambda^\varepsilon M$, the trajectory $\phi_s c$ of the* $-\operatorname{grad}$ *E-flow starting at c, has a unique limit element which belongs to $\Lambda^0 M$, the set of point curves. The trajectories are of finite length. In particular, $\Lambda^0 M$ is a strong deformation retract of $\Lambda^\varepsilon M$.*

Remark. This was proved by Karcher [Ka 2]. Eliasson [El 4] gave another proof which is closely related to the construction of the stable manifold of $\Lambda^0 M$, cf. (2.5).

Proof. Choose $\varepsilon>0$ as in (1.4.14) and such that there are no critical points in $\Lambda^\varepsilon-\Lambda^0$. The latter condition will be satisfied for all sufficiently small ε since $E(c)\leqslant\varepsilon$ implies that $L(c)\leqslant\sqrt{2E(c)}<\sqrt{2\varepsilon}$, and there are no closed geodesics on a compact Riemannian manifold of arbitrarily small length, cf. [GKM].

For $c\in\Lambda^\varepsilon$ we have, in accordance with (1.4.14),

$$L\,(\text{trajectory }\phi_s c)=\int_0^\infty\left\|\frac{d}{ds}\phi_s c\right\|_1 ds=\int_0^\infty\|\operatorname{grad} E(\phi_s c)\|_1\,ds$$

$$=\int_0^\infty\|\operatorname{grad} E\|_1^{-1}\cdot\left(-\frac{dE}{ds}(\phi_s c)\right)ds$$

$$\leqslant\alpha^{-1/2}\int_0^{E(c)}(2E)^{-1/2}dE=\alpha^{-1/2}\sqrt{2E(c)}$$

since $E(\phi_\infty c)=0$.

Therefore, since all limit points of the trajectory $\phi_s c$ must be in $\Lambda^0 M$, there can be at most one.

We now define a deformation $\phi_\tau : \Lambda^\varepsilon \to \Lambda$, $0 \leqslant \tau \leqslant 1$, by associating, for c, the point $\phi_{s(\tau)} c$ with $s(\tau) = \tan(\pi\tau/2)$ if $c \notin \Lambda^0$. In case $c \in \Lambda^0$ we define $\phi_\tau c$ by c.

It remains to be shown that $\phi_\tau c$ depends continuously on τ and c. To see this, it suffices to observe that, according to our estimates, the length of the trajectory $\phi_s c$ becomes arbitrarily small uniformly for all c, provided $E(c)$ is sufficiently small. Hence, if we have a limit point $\phi_1 c \in \Lambda^0$ and a neighborhood $B_{2\delta}(\phi_1 c)$ of radius 2δ given, then we choose $\eta > 0$ small enough so that the trajectory of any point of Λ^η has length $< \delta$. Thus we need only make sure that there exists a neighborhood of c with the property that, for all of its points, the trajectory enters $\Lambda^\eta \cap B_\delta(\phi_1 c)$. But this can clearly be done, using the continuous dependence of the endpoints of trajectories for finite time s. □

1.4.16 Corollary. *The $\Lambda^\varepsilon M$, $\varepsilon \to 0$, form a fundamental system of neighborhoods for $\Lambda^0 M$.*

Proof. From the proof of (1.4.15) we know that, for $c \in \Lambda^\varepsilon M$, $\varepsilon > 0$ sufficiently small, $d_\Lambda(c, \Lambda^0 M)$ is less than a given $\rho > 0$. Conversely, (1.4.3) shows that for $d_\Lambda(c, \Lambda^0 M)$ sufficiently small, c will belong to a given $\Lambda^{\varepsilon'} M$, $\varepsilon' > 0$. □

The geometry of the Riemannian Hilbert manifold $(\Lambda M, \langle\, ,\rangle_1)$ has not been studied in great detail yet. For instance, the geometric meaning of geodesics on ΛM, viewed as 1-parameter families of closed H^1-curves on M, is not clear.

We saw, in (1.4.5), that $(\Lambda M, \langle\, ,\rangle_1)$ is a complete metric space. The theorem of Hopf and Rinow (cf. [GKM]) that on a complete, finite dimensional connected Riemannian manifold any two points can be joined by a minimizing geodesic, in general does not hold for Riemannian Hilbert manifolds. Eliasson has shown [unpublished], however, that on $(\Lambda M, \langle\, ,\rangle_1)$ the Hopf-Rinow Theorem does hold.

We include here a result on the connection between closed geodesics on M and those on $(\Lambda M, \langle\, ,\rangle_1)$:

1.4.17 Proposition. *Let $c = (c(t))$ be a closed geodesic. Then the curve $\kappa_c(s)$, $s \in S$, given by $\kappa_c(s) = e^{2\pi i s} \cdot c = (c(t+s))$, $t \in S$, is a closed geodesic of ΛM of length $L(c)$.*

Proof. Since $\nabla\partial c = 0$, we have $\|\kappa_c'(s)\|_1^2 = \langle \partial c, \partial c\rangle_0 = L(c)^2$, i.e. $\kappa_c(s)$ is parameterized proportional to arc length and has length $L(c)$.

It remains to show that κ_c locally minimizes the distance. Choose $\varepsilon > 0$ sufficiently small. Let $\lambda(s)$, $0 \leqslant s \leqslant \varepsilon$, be an arbitrary curve in Λ from $\kappa_c(0) = \lambda(0)$ to $\varkappa_c(\varepsilon) = \lambda(\varepsilon)$. Set $\lambda(s)(t) = \lambda^\sim(t, s)$. Since $c(t+s)$, $0 \leqslant s \leqslant \varepsilon$, is the curve of minimal length from $c(t)$ to $c(t+\varepsilon)$ we have

$$\int_0^\varepsilon |\partial\lambda^\sim(t, s)/\partial s|\, ds \geqslant \int_0^\varepsilon |\dot c(t+s)|\, ds.$$

Thus

$$L(\lambda) \geqslant \int_0^\varepsilon \int_0^1 |\partial\lambda^\sim(t, s)/\partial s|\, dt\, ds \geqslant \text{length}\ \kappa_c|[0, \varepsilon]. \quad \square$$

We conclude this chapter by drawing attention to a certain generalization of the theory of closed geodesics, i.e. the theory of isometry-invariant geodesics. This theory was developed by Grove [Gro 1], [Gro 2].

Chapter 2. The Morse-Lusternik-Schnirelmann Theory on the Manifold of Closed Curves

As a first application of the results of Chapter 1 we are going to develop the Lusternik-Schnirelmann theory of $(\Lambda M, \langle\,,\,\rangle_1, E)$. In particular, we shall prove the existence of at least one closed geodesic on an arbitrary compact Riemannian manifold, following Lyusternik and Fet.

In the second section we study the canonical S-action and $\mathbb{Z}_2$-action on ΛM and are thus led to the space ΠS^n of unparameterized closed curves on M. We continue with a detailed description of the space ΓS^n of circles on the standard n-sphere S^n. Those homology classes of ΠS^n, which can be represented by cycles belonging to ΓS^n, have played a fundamental role since the very beginning of the theory of closed geodesics. As we shall see in Chapters 4 and 5, this has not changed until today.

Section 4 is devoted to a detailed exposition of the Morse theory proper on $(\Lambda M, \langle\,,\,\rangle_1, E)$. The hypothesis there is that all closed geodesics belong to non-degenerate critical submanifolds. In the last section we present a refinement of this theory by introducing the Morse complex. This complex contains the complete relevant structure of $(\Lambda M, \langle\,,\,\rangle_1, E)$ in a pure and condensed form.

2.1 The Lusternik-Schnirelmann Theory on ΛM

We continue with the concepts and notation introduced in Chapter 1. Recall, in particular, that we have the semigroup of E-decreasing deformations

$$\phi_s : \Lambda M \to \Lambda M;\ c \mapsto \phi_s c,\ s \geqslant 0.$$

given by integrating the vector field $-\operatorname{grad} E$.

We call a non-empty set $\mathscr{A}$ of non-empty subsets $A \subset \Lambda = \Lambda M$ with $E|A$ bounded, a *ϕ-family* if it is closed under ϕ_s, i.e.

$$A \in \mathscr{A} \quad \text{implies that} \quad \phi_s A \in \mathscr{A}, \quad \text{for all } s \geqslant 0.$$

Examples of ϕ-families

(i) The elements $\{\phi_s c;\ s \geqslant 0\}$ of a trajectory form a ϕ-family. More generally, if A is any non-empty subset of Λ with $E|A$ bounded, then the set $\{\phi_s A, s \geqslant 0\}$ is a ϕ-family.

(ii) The set of all elements in a connected component of Λ is a ϕ-family.

(iii) Consider a free homotopy class of continuous maps of the k-sphere into Λ. The set of images is a ϕ-family because $f: S^k \to \Lambda$ and $\phi_s \circ f: S^k \to \Lambda$ are in the same homotopy class.

(iv) Let w be a singular homology class $\neq 0$ of Λ (any coefficients). For each singular cycle $u \in w$ denote by $|u|$ the union of the images of the singular simplicies of u. Then $E \| u|$ is bounded, and $\phi_s u$ is homologous to u; hence, $\{|u|, u \in w$ (i.e. u a cycle representing w)$\}$ is a ϕ-family.

Let $\alpha \in \mathbb{R}$ and choose $\varepsilon > 0$ such that there are no critical values of E in $]\alpha, \alpha+\varepsilon]$. Then we call a non-empty set $\mathscr{A}$ of non-empty sets $A \subset \Lambda$ a *ϕ-family of Λ* mod Λ^α if $E|A$ is bounded, $A \in \mathscr{A}$ implies that $\phi_s A \in \mathscr{A}$ for all $s \geqslant 0$, and $A \in \mathscr{A}$ implies that $A \not\subset \Lambda^{\alpha+\varepsilon}$.

For $\alpha < 0$ we obtain the concept of a ϕ-family of Λ.

An example of a ϕ-family of Λ mod Λ^α is given by the carriers $|u|$ of the cycles u of a non-trivial homology class of Λ mod Λ^α, α as above. Here the case $\alpha = 0$ is of particular interest.

We define the *critical value of a ϕ-family $\mathscr{A}$ of Λ* mod Λ^α by

$$\kappa_{\mathscr{A}} := \inf_{A \in \mathscr{A}} \sup E|A.$$

That $\kappa_{\mathscr{A}}$ is also a critical value of E is given by the following theorem.

2.1.1 Theorem. *Let $\kappa = \kappa_{\mathscr{A}}$ be the critical value of a ϕ-family $\mathscr{A}$ of Λ* mod Λ^α. *Then $\kappa > \alpha$ and there exists $c \in Cr\Lambda$ with $E(c) = \kappa$.*

2.1.2 Complement. *Assume that the elements A of the ϕ-family $\mathscr{A}$ are compact. Let $\mathscr{U}$ be an open neighborhood of the critical set*

$$K = Cr\Lambda \cap E^{-1}(\kappa).$$

Then there exists an $A' \in \mathscr{A}$ such that

$$\phi_s A' \subset \mathscr{U} \cup \Lambda^{\kappa-}$$

for all $s \geqslant 0$.

Note. From the complement we see that if K is decomposed into finitely many disjoint closed subsets, then there must be one, say K', such that, for every open neighborhood $\mathscr{U}'$ of K', we have

$$\phi_s A' \cap (\mathscr{U}' \cup \Lambda^{\kappa-}) \neq \phi \quad \text{but} \quad \phi_s A' \not\subset \Lambda^{\kappa-}$$

for all $s \geqslant 0$. We then say that the representative A' of the family $\mathscr{A}$ remains hanging at K'.

Proof. Assume $\kappa \leqslant \alpha$. Since for every $\varepsilon > 0$, there exists an $A \in \mathscr{A}$ with $A \subset \Lambda^{\kappa+\varepsilon}$, we obtain a contradiction to the definition of a ϕ-family of Λ mod Λ^{α}.

Assume that $Cr\Lambda \cap E^{-1}(\kappa)$ is empty. From (1.4.12) we then have the existence of $\varepsilon > 0$ and $s_0 \geqslant 0$ such that $\phi_{s_0} A \subset \Lambda^{(\kappa-\varepsilon)-}$, for any $A \subset \Lambda^{\kappa+\varepsilon}$. Since there are $A \in \mathscr{A}$ with $A \subset \Lambda^{\kappa+\varepsilon}$ for every $\varepsilon > 0$, we get a contradiction to the definition of $\kappa = \kappa_{\mathscr{A}}$. Thus we have proved the theorem.

For the proof of the complement it suffices, in the light of (1.4.15), to consider the case $\kappa > 0$.

We choose, for the given neighborhood $\mathscr{U}$ of K, open neighborhoods $\mathscr{V}$ and $\mathscr{W}$ satisfying

$$\mathscr{U} \supset \mathscr{V} \supset \mathscr{W} \supset K$$

and strictly contained in each other. That is to say, there exists $\rho > 0$ such that

$$d_\Lambda(c', c'') \geqslant \rho \quad \text{for} \quad c' \in \mathscr{V},\ c'' \notin \mathscr{U},$$

$$d_\Lambda(c, c') \geqslant \rho \quad \text{for} \quad c \in \mathscr{W},\ c' \notin \mathscr{V}.$$

That this is possible follows from the fact that K is compact and d_Λ is a metric on Λ.

According to (1.4.8) there exist $\varepsilon > 0$ and $\eta > 0$ such that $\|\text{grad } E(c)\|_1 \geqslant \eta$ for

$$(*) \qquad c \in \complement \mathscr{W} \cap (\Lambda^{\kappa+\varepsilon} - \Lambda^{(\kappa-\varepsilon)-}).$$

Assume that $\phi_s c$, $0 \leqslant s \leqslant s_0$, is a trajectory from a point

$$c \in \mathscr{W} \cap (\Lambda^{\kappa+\varepsilon} - \Lambda^{(\kappa-\varepsilon)-})$$

to a point $c' = \phi_{s_0} c \in \complement \mathscr{V}$ which lies completely in $(\Lambda^{\kappa+\varepsilon} - \Lambda^{(\kappa-\varepsilon)-})$. Since this trajectory must contain an arc of length $> \rho$ belonging to

$$\complement \mathscr{W} \cap (\Lambda^{\kappa+\varepsilon} - \Lambda^{(\kappa-\varepsilon)-})$$

we can estimate the decrease of E along the trajectory as follows by using (1.4.10):

$$E(c) - E(c_0) = \int_0^{s_0} \|\text{grad } E(\phi_s c)\|_1^2 \, ds \geqslant \eta^2 s_0 \geqslant \eta^2 \rho^2 / (\kappa + \varepsilon).$$

Hence, by taking $\varepsilon > 0$ sufficiently small, we can assume that the trajectory $\phi_s c$ of an element $c \in \mathscr{W} \cap \Lambda^{(\kappa+\varepsilon)-}$ does not leave the set $(\mathscr{V} \cup \Lambda^{(\kappa-\varepsilon)-}) \cap \Lambda^{(\kappa+\varepsilon)-}$ because otherwise it would lose an amount $> 2\varepsilon$ of its E-value and therefore enter $\Lambda^{(\kappa-\varepsilon)-}$.

It is clear that at the same time we can assume that the trajectory $\phi_s c'$ of an element $c' \in \mathscr{V} \cap \Lambda^{(\kappa+\varepsilon)-}$ will remain in $(\mathscr{U} \cup \Lambda^{(\kappa-\varepsilon)-}) \cap \Lambda^{(\kappa+\varepsilon)-}$.

Now choose $A \in \mathscr{A}$ with $A \in \Lambda^{(\kappa+\varepsilon)-}$. Then, for every $c \in A$, there is an $s_0 = s(c) \geqslant 0$ such that the trajectory $\phi_s c$, $s \geqslant s_0$, lies in $(\mathscr{V} \cup \Lambda^{(\kappa-\varepsilon)-}) \cap \Lambda^{(\kappa+\varepsilon)-}$. Indeed, since $\|\text{grad } E(c)\|_1$ is bounded away from zero in the set (*), there will be an $s_0 \geqslant 0$ with $\phi_{s_0} c \in \mathscr{W} \cap \Lambda^{\kappa+\varepsilon}$, and we can now apply our previous result.

With $s_0 = s(c)$ as above, we can find an open neighborhood $\mathscr{U}(c)$ of c such that

$$\phi_{s_0} \mathscr{U}(c) \subset (\mathscr{V} \cup \Lambda^{\kappa-}) \cap \Lambda^{(\kappa+\varepsilon)-}.$$

As we saw above, we then have

$$\phi_s \mathscr{U}(c) \subset (\mathscr{U} \cap \Lambda^{(\kappa+\varepsilon)-}) \cap \Lambda^{\kappa-}$$

for all $s \geqslant s_0$. Since A is compact, a finite number of such $\mathscr{U}(c)$ cover A. Therefore we can assume that there is an $s_0 \geqslant 0$ with

$$\phi_s A \subset \mathscr{U} \cup \Lambda^{\;-}$$

for all $s \geqslant s_0$. Now put $A' := \phi_{s_0} A$. This completes the proof of the complement. □

As the first application of the previous theorem we obtain:

2.1.3 Theorem. *Assume that M is not simply connected. Corresponding to every conjugacy class $\neq 1$ of the fundamental group $\pi_1 M$ of M we have a connected component Λ' of $\Lambda = \Lambda M$ which does not contain the set Λ^0 of point curves.*

Claim. *The function E assumes its infimum κ' on Λ', and $E^{-1}(\kappa') \cap \Lambda'$ consists of closed geodesics.*

Proof. Recall that the set Λ' is a ϕ-family. Its critical value κ' is positive since a curve c with $E(c)$, and hence $L(c)$ (the length of c), sufficiently small will be null homotopic.

If $c \in E^{-1}(\kappa') \cap \Lambda'$ were not a critical point of E, then $\|\text{grad } E(c)\|_1 > 0$ and $E(\phi_s c) < \kappa'$, but $\phi_s c \in \Lambda'$ – a contradiction. □

Note. It is possible (and customary) to prove the previous theorem in a completely different manner, without any reference to the theory of the space ΛM. Instead, one considers the universal Riemannian covering manifold $\tilde{M}$ of M. Any non-trivial element τ of the fundamental group $\pi_1 M$ operates as a fixed-point free isometry on M. One shows that the function $f(\tilde{p}) :=$ distance from $\tilde{p}$ to $\tau\tilde{p}$ assumes its infimum on $\tilde{M}$. Let ω be this infimum. Then $\omega > 0$. Let $\tilde{p} \in M$ such that the distance $d(\tilde{p}, \tau\tilde{p}) = \omega$. Let $\tilde{c}$ be a geodesic from $\tilde{p}$ to $\tau\tilde{p}$ of length ω. Then the projection of $\tilde{c}$ into M gives a closed geodesic c on M having minimal E-value among the closed curves homotopic to c.

This proof, in special cases, is due to Hadamard [Ha] and Cartan [Ca 2] (whose proof contains an error). See also Busemann [Bu]. Berger [Be] gives a complete proof.

To prove the existence of a closed geodesic also in the case $\pi_1 M = 0$ we first observe:

2.1.4 Lemma. *The map*

$$\gamma : \Lambda M \to M : c \to c(0) = c(1)$$

is homotopy equivalent to a Serre-fibration with fibre of the type ΩM = loop space of M.

γ has a section by associating with $p \in M$ the point curve $c_p : S \to p \in M$.

Proof. Recall that Serre-fibration means that a homotopy in the base can be lifted to a homotopy in the total space. Consider the commutative diagram:

$$\begin{array}{ccc} H^1(S, M) = \Lambda M & \xrightarrow{i} & C^0(S, M) \\ \downarrow \gamma & & \downarrow \gamma \\ M & \xrightarrow{\text{id}} & M . \end{array}$$

For the fibration on the right-hand side the lemma is well known, see e.g., [Sp], [Šv]. According to (1.2.10), i is a homotopy equivalence. □

2.1.5 Corollary. *The exact homotopy sequence of the fibration γ splits*

$$\ldots \pi_k \Omega \to \pi_k \Lambda \underset{i_k}{\overset{\curvearrowleft}{\to}} \pi_k M \to \pi_{k-1} \Omega \ldots .$$

Hence, for $k \geqslant 1$

$$\pi_k \Lambda M \cong \pi_k M + \pi_{k+1} M .$$

Proof. This is a standard result of algebraic topology, see [Šv], [Sp]. □

We now can prove the celebrated theorem of Lyusternik and Fet [LF] and [Fe 1]. Our proof is similar to the one given by Olivier [Ol]. Cf. also the appendix for a proof using only the most basic facts of Lusternik-Schnirelmann theory.

2.1.6 Theorem. *Let $\pi_1 M = 0$. Then there exists a closed geodesic on M.*

Proof. A well-known result in algebraic topology asserts the existence of a smallest $k+1 \geqslant 2$ such that

$$\pi_{k+1}(M) = H_{k+1}(M; \mathbb{Z}) \neq 0$$

or, for a given prime field $\mathbb{Z}_p$, of a number $l+1 \geqslant 2$ such that

$$\pi_{l+1}(M) \otimes \mathbb{Z}_p = H_{l+1}(M, \mathbb{Z}_p) \neq 0 .$$

From (2.1.5), we have that $\pi_k \Lambda M = \pi_{k+1} M \neq 0$. So we now take as a ϕ-family $\mathscr{A}$ the set of images of maps $f: S^k \to \Lambda M$ in a non-trivial homotopy class.

We claim that $\kappa = \kappa_{\mathscr{A}} > 0$. Indeed, otherwise there would exist, for any given $\varepsilon > 0$, an element f in the homotopy class satisfying

$$f(S^k) \subset \Lambda^\varepsilon.$$

But then we would have from (1.4.15) that $f: S^k \to \Lambda M$ is homotopic to a map $f': S^k \to \Lambda^0 M \cong M$, i.e. the homotopy class would be trivial — a contradiction. □

We would like to give a slightly different proof of (2.1.6) which avoids the use of (1.4.15) and constitutes an adaptation of the original proof of Lyusternik and Fet [LF], [Fe 1].

We begin with the following construction. Fix a hemisphere H^{k+1} on S^{k+1} and let $D^k \subset \partial H^{k+1}$ be a closed half-equator of S^{k+1}. For each $p \in D^k$, define an element

$$a_p : S \to S^{k+1}$$

of ΛS^{k+1} as follows: If $p \in D^k - \partial D^k$, then a_p is the parameterized circle on S^{k+1} which starts at p and goes out into H^{k+1} and lies in the plane orthogonal to the $\mathbb{R}^k$ containing ∂D^k. If $p \in \partial D^k$ we let a_p be the constant map $S \to p$.

We have thus defined a map

$$\delta^k : (D^k, \partial D^k) \to (\Lambda S^{k+1}, \Lambda^0 S^{k+1});\ p \mapsto a_p.$$

We can assume that we have a differentiable map

$$f: S^{k+1} \to M$$

which is homotopically non-trivial.

f induces a differentiable map

$$f^* := \Lambda f \circ \delta^k : (D^k, \partial D^k) \to (\Lambda M, \Lambda^0 M).$$

2.1.7 Proposition. *The homotopy $\phi_s \circ f^*$, $s \geqslant 0$, of f^* induced by the map $\phi_s : \Lambda \to \Lambda$, (1.4.11), induces a homotopy f_s, $s \geqslant 0$ of $f = f_0$ as follows:*

for $q \in \partial D^k \subset S^{k+1}$, set *$f_s(q) = f(q)$;*

for $q \in S^{k+1} - \partial D^k$ there exists a well-determined $p \in D^k - \partial D^k$ and $t \in S$ such that $q = a_p(t)$, so set $f_s(q) = \phi_s f^* a_p(t)$.

The proof is clear. □

2.1.8 Theorem. *Let*

$$f: S^{k+1} \to M$$

be homotopically non-trivial. Then the critical value κ of the ϕ-family $\{\phi_s \circ \Lambda f \circ \delta^k(D^k)\}$ is greater than zero, i.e. there exists a closed geodesic c with $E(c) = \kappa$.

Proof. Assume that the critical value κ is zero. Then, for every $\varepsilon > 0$, there exists $s_0 \geq 0$ with $\phi_{s_0} \circ \Lambda f \circ \delta^k(D^k) \subset \Lambda^\varepsilon$. From (2.1.7) it follows that f is homotopic to a map f_{s_0}. But f_{s_0} carries all circles and point curves $a_p(t)$, $p \in D^k$, on S^{k+1} into curves $f_{s_0} a_p(t)$ with E-value $\leq \varepsilon$.

If ε is sufficiently small, these curves are in a disc-like neighborhood of their initial point $f_{s_0} a_p(0)$ which can be retracted to $f_{s_0} a_p(0)$. Hence, f is homotopic to a map $S^{k+1} \to \operatorname{im} f_{s_0} D^k$, i.e. homotopic to zero – a contradiction. □

Note. In the previous theorems we have constructed a closed geodesic for every non-trivial homotopy class of M. In general, there will be many different such homotopy classes. However, this does not necessarily mean that in this manner we shall obtain many different closed geodesics.

First of all, to get several closed geodesics one should construct them with homology classes rather than homotopy classes. The concept of subordinated homology classes, stemming from Lusternik and Schnirelmann [LS 1, 2], which we shall explain below, does give a method for constructing different closed geodesics.

We should be aware, however, of the possibility that different closed geodesics c and c' might just be different coverings of an underlying closed geodesic c_0 – that is to say, there might be a closed geodesic $c_0 = c_0(t)$ such that $c(t) = c_0(mt)$, $c'(t) = c_0(m't)$, m, m' integers ≥ 1, $m \neq m'$.

We continue to consider homology and cohomology with an arbitrary field of coefficients.

2.1.9 Lemma. *There exists a linear map*

$$H_*(\Lambda, \Lambda^0) \otimes H^*(\Lambda - \Lambda^0) \to H_*(\Lambda, \Lambda^0)$$

which is induced from the cap product

$$H_*(\Lambda - \Lambda^0, \Lambda^\varepsilon - \Lambda^0) \otimes H^*(\Lambda - \Lambda^0) \xrightarrow{\cap} H_*(\Lambda - \Lambda^0, \Lambda^\varepsilon - \Lambda^0)$$

on the triple $(\Lambda, \Lambda^\varepsilon, \Lambda^0)$; *here,* $\varepsilon > 0$ *is assumed to be so small that* Λ^0 *is a deformation retract of* Λ^ε.

Proof. We assume the existence of the cap product in the triple $(\Lambda, \Lambda^\varepsilon, \Lambda^0)$, cf. [Stee]. Since Λ^0 is a strong deformation retract of Λ^ε, the exact homology sequence of the triple $(\Lambda, \Lambda^\varepsilon, \Lambda^0)$ yields $H_*(\Lambda, \Lambda^0) = H_*(\Lambda, \Lambda^\varepsilon)$, and excision yields $H_*(\Lambda, \Lambda^\varepsilon) = H_*(\Lambda - \Lambda^0, \Lambda^\varepsilon - \Lambda^0)$. □

Let w', w be non-zero homology classes of Λ mod Λ^0, of dimension k and $k+l$, $l > 0$, respectively. We say that w' is *subordinate* to w if there exists an l-dimensional cohomology class ζ of $\Lambda - \Lambda^0$ such that $w' = w \cap \zeta$.

2.1.10 Theorem. *Let w' be subordinate to w as homology classes of Λ* mod Λ^0. *Then the critical values κ and κ' of w and w' satisfy $\kappa \geq \kappa' > 0$. If $\kappa = \kappa'$ then the set of critical elements in $E^{-1}(\kappa)$ has non-vanishing homology in dimensions $k =$* dim $w -$ dim $w' > 0$. *Since Λ is locally contractible, this implies that this set has covering dimension $\geq k$.*

Proof. κ and κ' must be >0 since the cycles in w resp. w' form a ϕ-family of Λ mod Λ^0.

For every cycle $u\in w$, there exists a cocycle $\eta\in\zeta$ such that $u' := u\cap\eta$ is a cycle representing w'. Hence $\kappa\geqslant\kappa'$.

Now assume that $\kappa=\kappa'$. Put $Cr\Lambda\cap E^{-1}(\kappa)=K$. For any open neighborhood $\mathscr{U}$ of K in $\Lambda-\Lambda^0$, there exists $u\in w$ with $|u|\subset\mathscr{U}\cup\Lambda^{\kappa-}$, cf. (2.1.2). We can choose a subdivision of the singular simplices σ of u, which is so fine that either $|\sigma|\subset\mathscr{U}$ or $|\sigma|\subset\Lambda^{\kappa-}$.

We claim that $\zeta|\mathscr{U}\neq 0$, i.e. since ΛM is locally contractible, $Cr\Lambda\cap E^{-1}(\kappa)$ carries a non-trivial cycle of dim ζ. Indeed, otherwise there would exist a cocycle $\eta\in\zeta$ with carrier $\eta\cap\mathscr{U}=\emptyset$. But then

$$|u\cap\eta|\cap\mathscr{U}=\emptyset, \quad \text{i.e.} \quad |u\cap\eta|\subset\Lambda^{\kappa-}$$

and therefore $\kappa'<\kappa$. □

2.1.11 Corallary. *Let w_i, $1\leqslant i\leqslant r$, be pairwise subordinated homology classes of Λ mod Λ^0, dim $w_i=k_i$, $0\leqslant k_1<\cdots<k_r$.*

Let κ_i be the critical value of w_{k_i}. Then

$$0<\kappa_1\leqslant\cdots\leqslant\kappa_r$$

and there exists a closed geodesic on the E-level κ_i. Moreover, if $\kappa_j=\kappa_{j+j'}$, $j'>0$, then the set of closed geodesic at E-level κ_j has covering dimension $\geqslant k_{j+j'}-k_j$.

Note. If $w_1,\ldots,w_r$ are pairwise subordinated classes as in (2.1.11), $w_{i-1}=w_i\cap\zeta_{i-1}$, then we have from the formula

$$(w\cap\zeta)\cap\zeta'=w\cap(\zeta\cup\zeta'):$$

$$w_1=w_r\cap(\zeta_1\cup\cdots\cup\zeta_{r-1}),$$

that is to say, there exist $(r-1)$ cohomology classes of $\Lambda-\Lambda^0$ of positive dimension having cup product $\neq 0$.

One calls the supremum of the number of cohomology classes of a space which have a cup product $\neq 0$ *cup-length* of that space. Here, all cohomology classes except one shall have positive dimension. This concept is closely related to the concept of the category of a topological space; actually, the cup-length is a lower bound for the category, cf. [Sch], [Pa 4].

We see from the previous results that if there are $(r-1)$ elements ζ_i, $1\leqslant i\leqslant r-1$, in $H^*(\Lambda M-\Lambda^0 M)$, dim $\zeta_i>0$, with a product $\zeta_1\cup\cdots\cup\zeta_{r-1}\neq 0$, and if there exists $w\in H_*(\Lambda M, \Lambda^0 M)$ such that $w\cap(\zeta_1\cup\cdots\cup\zeta_{r-1})\neq 0$, then there are at least r different closed geodesics on M. Here we assume that the critical values κ_i occurring in (2.1.11) are all distinct. Otherwise we should exclude the possibility that the set of closed geodesics at E-level, $\kappa_j=\kappa_{j+j'}$, $j'>0$, does not simply consist of the $O(2)$-orbit of a single closed geodesic c, i.e. of the geodesics of the form $c(\pm t+r)$, cf. (2.2) for more details. Indeed, we shall not count closed geodesics as being really different if they are distinguished from one another only in their parametrization. In (2.2.11) we shall give a formulation which avoids this possibility.

In addition, we recall our remarks preceding (2.1.9): Different closed geodesics might very well be just different coverings of the same underlying closed geodesic. See also [FK] for further comments.

2.2 The Space of Unparameterized Closed Curves

In this section we shall consider spaces of unparameterized closed curves on a manifold M; they will be obtained from the space ΛM by the quotient of a continuous S-action and a continuous $O(2)$-action so as to give oriented and unoriented, unparameterized closed curves, respectively.

The idea of considering unparameterized closed curves goes back to the beginning of the theory of closed geodesics, cf. [LS 1] and [Mor 2]. However, in the older approach one identified parameterized closed curves which differed by a much larger group of parameterizations, e.g. by the group of diffeomorphisms of the circle S. In our approach we shall identify curves c and c' only if there exists an $r\in S$ with $c'(t)=c(t+r)$ or $=c(-t+r)$. This means that we have at our disposal the results of the theory of a circle action on a Hilbert manifold. Note, however, that the circle action is not differentiable, but only of class H^1.

For this reason, our approach seems more natural, particularly since the different definitions of a space of closed curves all yield the same homotopy type because the group of orientation preserving diffeomorphisms of the circle can be retracted onto $SO(2)=S$.

As we shall see, the Riemannian structure of ΛM and the energy integral E are compatible with the S-action and the orientation reversing involution on ΛM. Generally speaking, this forces the function E to have more critical points than would otherwise be necessary, or what amounts to a closely related phenomenon: The space of unparameterized closed curves is bound to have "more" homology than the space ΛM of parameterized curves, at least for $\mathbb{Z}_2$-coefficients. This fact was already observed by Morse in his (incidentally incorrect) computation of the "circular connectivities" cf. [Mor 2] and also Bott [Bo 1].

Consider the map

$$\tilde{\chi}: S\times\Lambda M\to\Lambda M$$

$$(z,c)\mapsto z\,.\,c$$

with $z\,.\,c(t)=c(t+r)$ where $z=e^{2\pi ir}\in S$.

2.2.1 Lemma. *The map $\chi^{\sim}$ defines a continuous circle action on ΛM. For every fixed $z \in S$, the map*

$$\chi_z^{\sim}: \Lambda M \to \Lambda M : c \mapsto z \,.\, c$$

is an isometry which leaves the function E invariant.

Proof. Consider the natural charts based at an element $c \in C^\infty(S, M)$ and the element $z \,.\, c \in C^\infty(S, M)$. Then the map $\chi_z^{\sim}$ is represented by the linear isomorphism

$$(\xi(t)) \in H^1(c^*TM) \mapsto (\xi(t+r)) \in H^1(z \,.\, c^*TM).$$

From the definitions of E and the Riemannian metric $\langle \, , \rangle_1$ on ΛM we see at once that $E(z \,.\, c) = E(c)$ and $\langle D\chi_z^{\sim}, D\chi_z^{\sim} \rangle_1 = \langle \, , \rangle_1$.

It remains to show that $\chi^{\sim}$ is continuous. Since $\chi_z^{\sim}$ is an isometry, we have

$$d_\Lambda(c, z \,.\, e) \leqslant d_\Lambda(c, z \,.\, c) + d_\Lambda(c, e)$$

$$d_\Lambda(e, z \,.\, e) \leqslant d_\Lambda(c, z \,.\, c) + 2d_\Lambda(e, c).$$

Hence, it suffices to show that $d_\Lambda(c, z \,.\, c)$ tends to zero as $z \to 1$, $c \in C^\infty(S, M)$. But since

$$\exp_c^{-1}(c) = 0 \in H^1(c^*TM)$$

and the map $z \in S \mapsto z \,.\, c \in \Lambda M$ is represented by

$$e^{2\pi i r} \mapsto ((\pi^* c)^{-1} \exp_{c(t)}^{-1} c(t+r)) \in H^1(c^*TM)$$

one sees that $\|\exp_c^{-1}(z \,.\, c)\|_1 \to 0$ for $z \to 1$, cf. (1.3.5).

Finally, we observe that $\chi^{\sim}$ cannot be differentiable, since for every $c \in \Lambda$, the map $z \in S \mapsto z \,.\, c \in \Lambda$ would then have to be differentiable. In particular, the tangent vector $d\, z \,.\, c/dz$ at $z = 1$ would have to be an element of $T_c\Lambda$. But this vector is equal to ∂c, which need not be an element of $T_c\Lambda$. □

2.2.2 Corollary 1. *If c is a critical point of E then the whole orbit $S \,.\, c$ consists of critical points.*

This is an immediate consequence of (2.2.1). □

The *isotropy group of c* with respect to the S-action $\chi^{\sim}$, $I^{\sim}(c)$, is defined as the subgroup formed by the $z \in S$ with $z \,.\, c = c$.

2.2.3 Corollary 2. *$I^{\sim}(c)$ is a closed subgroup of S, i.e. either $I^{\sim}(c) = S$, which is equivalent to $c \in \Lambda^0 M$ = space of point curves, or $I^{\sim}(c)$ is a cyclic group of finite order.*

Proof. Observe that $\chi_c^{\sim}: S \to \Lambda : z \mapsto z \,.\, c$ is continuous. Hence $I^{\sim}(c) = (\chi_c^{\sim})^{-1}(c)$ is closed. $\square$

We call $c \in \Lambda M$ *m-fold covered* or *of multiplicity m*, if $I^{\sim}(c) = \mathbb{Z}_m$. If $m=1$ we also call c *prime*.

Remark. $I^{\sim}(c) = \mathbb{Z}_m$ implies that $c(t+1/m) = c(t)$, for all $t \in S$.

If $c \in \Lambda$, we define the *m-fold covering of c*, c^m, by $c^m(t) := c(mt)$. Clearly, $I^{\sim}(c^m)$ contains $\mathbb{Z}_m$.

If $I^{\sim}(c) = \mathbb{Z}_m$, then c_0, defined by $c_0(t) = c(t/m)$, is prime and $c = c_0^m$. c_0 is called the *underlying prime curve of c*.

We define the space of *unparameterized oriented closed curves* on M, $\tilde{\Pi} M$, as the quotient space of ΛM with respect to the S-action $\chi^{\sim}$. Let

$$\tilde{\pi} : \Lambda M \to \tilde{\Pi} M = \Lambda M /_{\chi^{\sim}} S$$

be the quotient map.

$\tilde{\Pi} = \tilde{\Pi} M$ will be endowed with the quotient topology, i.e. a set $B \subset \tilde{\Pi}$ will be open if and only if the counterimage of B under $\tilde{\Pi}$ is open.

Clearly, $E : \tilde{\Pi} M \to \mathbb{R}$ is a continuous function. We call $\tilde{c} \in \tilde{\Pi}$ *critical* if $\tilde{c}$ is the image under $\tilde{\pi}$ of a critical point of Λ.

Another way of describing the space $\tilde{\Pi}$ would be to say that it consists of the orbits of the S-action $\chi^{\sim}$.

In particular, the critical points of $\tilde{\Pi}$ are the S-orbits of critical points in Λ.

2.2.4 Theorem. *The E-decreasing deformation ϕ_s of the space ΛM induces an E-decreasing deformation $\tilde{\psi}_s$ of $\tilde{\Pi} M$ which is characterized by the commutativity of the following diagram:*

$$\begin{array}{ccc} \Lambda M & \xrightarrow{\phi_s} & \Lambda M \\ \downarrow \tilde{\pi} & & \downarrow \tilde{\pi} \\ \tilde{\Pi} M & \xrightarrow{\tilde{\psi}_s} & \tilde{\Pi} M. \end{array}$$

Proof. From the fact that the S-action $\chi^{\sim}$ is an isometry leaving E invariant we have

$$\operatorname{grad} E(z \,.\, c) = Dz \,.\, \operatorname{grad} E(c).$$

This implies that $\phi_s z \,.\, c = z \,.\, \phi_s c$; hence the theorem. $\square$

Consider the *orientation reversing* map

$$\theta : \Lambda M \to \Lambda M \,;\; c(\mathrm{t}) \mapsto c(1-t).$$

Clearly, $\theta^2 = id$, i.e. θ defines a $\mathbb{Z}_2$-action on ΛM.

2.2.5 Theorem. *θ is an isometry leaving E invariant. Moreover*

$$z \,.\, \theta(c) = \theta(z^{-1} \,.\, c),$$

i.e. θ carries orbits of the S-action $\chi^{\sim}$ into orbits and hence defines an involution on $\tilde{\Pi}M$ which we also denote by θ.

Proof. The local representation of θ in the natural coordinates based at c and θc, respectively, is given by

$$\xi(t) \mapsto \xi(1-t),$$

i.e. a linear isometry. Moreover, one sees at once that $E(\theta c) = E(c)$.

To prove the last statement we observe that

$$z \,.\, (\theta c)(t) = (\theta c)(t+r) = c(1-t-r)$$

$$= z^{-1} \,.\, c(1-t) = (\theta z^{-1} \,.\, c)(t)$$

with $z = e^{2\pi i r}$. □

We define the space of (*unoriented*) *unparameterized closed curves on M* by

$$\Pi M := \tilde{\Pi} M /_{\theta} \mathbb{Z}_2$$

with the induced topology.

Let

$$\theta : \tilde{\Pi} M \to \Pi M$$

be the quotient map. The composite map $\theta \circ \tilde{\pi}$ will be denoted by π, i.e. we also have the map

$$\pi : \Lambda M \to \Pi M.$$

Using the S-action $\chi^{\sim}$ and the involution θ, we define an $O(2)$-action χ on ΛM as follows:

$$\chi(\alpha, c) = \begin{cases} z \,.\, c, & \text{if} \quad \alpha = z \in SO(2) \cong S \\ z \,.\, \theta c, & \text{if} \quad \alpha = z\sigma \in O(2) - SO(2) \end{cases}$$

where $\sigma : \mathbb{R}^2 \to \mathbb{R}^2$ is the reflection on the x-axis:

$$(x, y) \mapsto (x, -y).$$

Clearly, ΠM is canonically isomorphic to the quotient space $\Lambda M /_{\chi} O(2)$.

We can interpret ΠM as the space of orbits in ΛM under the $O(2)$-action χ.

In particular, since E is constant on the orbits, E can be viewed as a continuous function on ΠM. The critical points of E in ΠM are the orbits of critical points in ΛM under the $O(2)$-action χ. So a non-constant critical point in ΠM determines two disjoint immersions of the circle:

$$z \in S \mapsto z \,.\, c \quad \text{and} \quad z \,.\, \theta(c).$$

We have the following counterpart of (2.2.4):

2.2.6 Theorem. *The deformation ϕ_s of ΛM induces a deformation ψ_s of ΠM which is characterized by the commutativity of the following diagram:*

$$\begin{array}{ccc} \Lambda M & \xrightarrow{\phi_s} & \Lambda M \\ \downarrow \pi & & \downarrow \pi \\ \Pi M & \xrightarrow{\psi_s} & \Pi M \end{array}$$

Proof. Since the $O(2)$-action on ΛM is an isometry which respects E we have, for $\alpha \in O(2)$,

$$D\alpha \,.\, \operatorname{grad} E(c) = \operatorname{grad} E(\alpha \,.\, c)$$

where we have defined

$$\alpha : \Lambda M \to \Lambda M \quad \text{by} \quad c \mapsto \chi(\alpha, c).$$

Hence $\phi_s \alpha \,.\, c = \alpha \,.\, \phi_s c$. □

We define, for any real number κ,

$$\tilde{\Pi}^\kappa := \{\tilde{c} \in \tilde{\Pi};\ E(\tilde{c}) \leq \kappa\}$$

$$\Pi^\kappa := \{\bar{c} \in \Pi;\ E(\bar{c}) \leq \kappa\}.$$

In other words, $\tilde{\Pi}^\kappa = \tilde{\pi} \Lambda^\kappa$, $\Pi^\kappa = \pi \Lambda^\kappa$.

Note that $\tilde{\Pi}^0$ and Π^0 are canonically isomorphic to $\Lambda^0 M \cong M$.

2.2.7 Theorem. *There exists $\varepsilon > 0$ such that $\tilde{\Pi}^0$ is a strong deformation retract of $\tilde{\Pi}^\varepsilon$ and Π^0 is a strong deformation retract of Π^ε.*

Proof. This follows from (1.4.15), and (2.2.4) and (2.2.6), respectively. □

The next lemma states the existence of a *slice* of the S-action $\chi^\sim$ on ΛM, cf. [Pa 1] for this concept.

2.2.8 Lemma. *Let $c \in \Lambda M - \Lambda^0 M$. Then there exists, for every $z \in S$, a slice $\Sigma(z \,.\, c)$, i.e. a local hypersurface of codimension 1 in ΛM having $z \,.\, c$ as center, such that*

(i) $z \,.\, \Sigma(c) = \Sigma(z \,.\, c)$;

(ii) *for every* $z \notin \tilde{I}(c) : \Sigma(z \,.\, c) \cap \Sigma(c) = \emptyset$;

(iii) *there exists an open neighborhood* U *of* 1 *on* S *such that the map*

$$U \times \Sigma(c) \to \Lambda M$$

$$(z, e) \mapsto z \,.\, e$$

is a homeomorphism onto an open neighborhood of c *in* ΛM;

(iv) $I^{\sim}(c)$ *operates as group of isometries on* $\Sigma(c)$, *endowed with the induced metric, leaving* c *invariant. In particular, the induced action of* $I^{\sim}(c)$ *on the tangent space of* $\Sigma(c)$ *at* c *is an orthogonal representation of* $I^{\sim}(c)$ *in a Hilbert space.*

Note. The importance of the slice lies mainly in the fact that $\tilde{\pi}|\Sigma(c)$ (or $\tilde{\pi}|\Sigma(z \,.\, c)$, any $z \in S$) maps $\Sigma(c)$ (or $\Sigma(z \,.\, c)$) onto an open neighborhood of $\tilde{\pi}(c)$. The map is given by the quotient of the isometric action of the isotropy group $I^{\sim}(c)$ of c on $\Sigma(c)$.

Proof. (cf. [Ka 2]). Consider the vector field $u(c)$ of (1.4.13), representing the form $\omega \,.\, \eta = -\langle \partial c, \eta \rangle_0$. Choose $\varepsilon > 0$ so small that the map $\exp_c := \exp \rceil T_c\Lambda$, restricted to the ε-ball in $T_c\Lambda$, is an injective diffeomorphism. Let H_ε be the intersection of the hypersurface $\omega = 0$ in $T_c\Lambda$ with this ε-ball. Put $\Sigma(c) := \exp_c H_\varepsilon$.

Since our definitions are all equivariant, we see that (i) holds if we define $\Sigma(z \,.\, c)$ in the same way. Since, moreover, S is operating as a group of isometries, we also obtain (iv).

Now consider, for a neighborhood U of 1 in S and a neighborhood $\mathscr{V}$ of c in ΛM, the function

$$f : (z, e) \in U \times \mathscr{V} \mapsto -\langle \partial c, T(z) \,.\, \exp_{z.c}^{-1} z \,.\, e \rangle_0 \in \mathbb{R}$$

with $T(z) = \exp_c^{-1} \circ \exp_{z.c}$ being a coordinate transformation. f is differentiable in z and f_z at $(1, c)$ is equal to $-\langle \partial c, \partial c \rangle_0 \neq 0$, $f(1, c) = 0$.

Hence, we have from the theory of implicit functions the existence of a locally defined continuous map $e \mapsto z(e)$ with $z(c) = 1$, such that $f(z(e), e) = 0$. This proves (ii) and (iii). □

2.2.9 Corollary. $\tilde{\Pi} M$ *and* ΠM *are locally contractible.*

Proof. If $c = c_p =$ a constant curve with image $p \in M$, then we can retract a neighborhood, first via the deformation ϕ_s onto a neighborhood of $p \in M$, and then onto p — again equivariantly with respect to the $SO(2)$ and $O(2)$ action, respectively.

Now let $E(c) > 0$. The canonical retraction along radial lines of $\Sigma(c)$ into c commutes with the isometric operation of $I^{\sim}(c)$ and hence gives a local retraction in $\tilde{\Pi}M$ at $\tilde{\pi}(c)$. Since this contraction commutes, in addition, with the isometry θ, it also gives a local contraction of ΠM at $\pi(c)$. □

We have on $\tilde{\Pi}M$ and ΠM the analogue of a ϕ-family. To simplify the exposition we restrict our attention to the space $\Pi = \Pi M$.

A *ψ-family* $\mathscr{B}$ of Π mod Π^α is defined as the image under the projection $\pi: \Lambda \to \Pi$ of a ϕ-family $\mathscr{A}$ of Λ mod Λ^α, cf. (2.1).

Equivalently a ψ-family can also be described as a ϕ-family which is closed under the χ-operation of $O(2)$.

We then have the following immediate consequence of (2.1.1):

2.2.10 Theorem. *Let $\kappa = \kappa_{\mathscr{B}}$ be the critical value of a ψ-family $\mathscr{B}$ of Π* mod Π^α, *i.e.*

$$\kappa_{\mathscr{B}} = \inf_{B \in \mathscr{B}} \sup E|B.$$

Then $\alpha < \kappa$ and there exist critical points at E-level κ. □

Recall that a critical point in Π can be viewed as the $O(2)$-orbit of a critical point in Λ.

The carriers of cycles in a non-trivial homology class of Π mod Π^α are examples of ψ-families.

As in (2.1), we have the concept of *subordinated homology classes of Π* mod Π^0. These are non-zero homology classes w and w' of Π mod Π^0 for which there exists a cohomology class ζ of $\Pi - \Pi^0$ of positive dimension such that $w' = w \cap \zeta$.

It is very important to note that the image under the canonical projection of a homology class of Λ mod Λ^0 is not always a homology class of Π mod Π^0.

Since we have shown that Π is locally contractible, we have the following analogue of (2.1.10):

2.2.11 Theorem. *Let w and w' be subordinated homology classes of Π* modΠ^0. *Then the critical values κ and κ' of w and w', respectively, satisfy $\kappa \geq \kappa' > 0$. If $\kappa = \kappa'$ then there exists on Π a set of critical elements in $E^{-1}(\kappa)$ of covering dimension* $\geq \dim \zeta > 0$. □

It is clear that we have as corollary the analogue to (2.1.11).

Finally we remark that the concept of subordinated homology classes is related to the concept of category, cf. [Sch], [Pa 4].

2.3 Closed Geodesics on Spheres

The simplest compact simply connected manifolds are certainly the spheres. Indeed, one may say that the theory of closed geodesics got started with Poincaré's attempt to prove the existence of a closed geodesic on a convex surface, cf. [Po 1]. Morse [Mor 2] considers this proof to be unsatisfactory.

A proof of the existence of a closed geodesic on an arbitrary surface of genus 0 was given by Birkhoff [Bi 1], cf. also [Bi 2]. Later, Lusternik and Schnirelmann [LS 1] showed that on such a surface there always exist three closed geodesics without self-intersection – a result which cannot be improved upon in general, since on an ellipsoid with three different axes, where the lengths of these axes do not differ from one another too much, the three plane ellipses on the ellipsoid are the only closed geodesics without self-intersection; see (5.1.2).

In [Bi 3], Birkhoff showed the existence of at least one closed geodesic on an n-dimensional analytic Riemannian manifold which, modulo its Riemannian structure, is a sphere.

Morse, on the one hand, and Lusternik and Schnirelmann, on the other, developed almost simultaneously more sophisticated topological tools to attack the problem of the existence of more than one closed geodesic on a n-dimensional Riemannian manifold, on a manifold diffeomorphic to the n-sphere in particular. Morse's main results are contained in his monograph [Mor 2] where, among other things, he made exhaustive studies of the homology of the space of unparameterized closed curves on the sphere. In particular, he was interested in the "circular connectivities" of S^n which are the same as the Betti numbers in $\mathbb{Z}_2$-homology of the space ΠS^n.

His results, however, were not correct, as was pointed out by Švarc [Šv]. He failed to take into consideration the fact that the group S does not act freely on ΛM. A similar mistake was repeated later by Bott [Bo 1] and Shizuma [Shi].

In this section we shall employ the previously developed methods to investigate the existence of several closed geodesics on a manifold of the homotopy type of the sphere S^n. These methods represent the combined efforts of Morse, Lusternik and Schnirelmann and, more recently, Alber and the author.

However, so far our theory does not yield the existence of more than one prime closed geodesic on such a manifold unless the Riemannian metric satisfies some restrictions on the curvature, cf. (2.3.8). Not until (4.3) shall we be able to prove the existence of an infinite number of prime closed geodesics by bringing into play some hitherto unobserved features for the Morse theory on ΛM. Cf. also (5.1) for the existence of $2n$-1 "short" closed geodesics.

We begin by exhibiting a distinguished subset in ΛS^n, i.e. the space AS^n *of parameterized circles on* S^n.

In order to define AS^n we assume S^n to be embedded in the standard way as the hypersurface of constant curvature equal to 1 in $\mathbb{R}^{n+1}$. A *circle on* S^n is either a point curve or the intersection of S^n with a 2-plane of $\mathbb{R}^{n+1}$ having distance <1 from the origin. When we speak of a parameterized circle we mean (provided the circle is not a point curve) that the parameter is chosen proportional to arc length. By A^0S^n we denote the subset of point circles isomorphic to S^n.

The space AS^n has as a subspace the *space* BS^n *of parameterized great circles*. Clearly, BS^n is isomorphic to the Stiefel manifold $V(2, n-1)$ of orthonormal 2-frames at the origin of $\mathbb{R}^{n+1}$. Hence, $BS^n = O(n+1)/O(n-1)$, dim $BS^n = 2n-1$; see e.g. [Mi 1].

Consider AS^n as a subspace of ΛS^n. The $O(2)$-action χ on ΛS^n leaves the subspaces AS^n and BS^n invariant. We call

$$\Gamma S^n := AS^n/_\chi O(2)$$

$$\Delta S^n := BS^n/_\chi O(2)$$

the space of (unparameterized) circles and of (unparameterized) great circles on S^n, respectively. We write $\Gamma^0 S^n$ for $A^0 S^n/O(2) \cong S^n$. ΔS^n is isomorphic to the Grassmann manifold $G(2, n-1)$ of 2-planes through the origin of $\mathbb{R}^{n+1}$. Hence, $\Delta S^n = O(n+1)/O(2) \times O(n-1)$, $\dim \Delta S^n = 2n-2$.

Define

$$\alpha : AS^n - A^0 S^n \to BS^n$$

by associating to a parameterized circle the parameterized great circle which is obtained by first carrying the plane of the circle by parallel translation into a plane through the origin of $\mathbb{R}^{n+1}$ and then blowing up the circle into a great circle.

The fibre $\alpha^{-1}(S^1)$ over a great circle $S^1 \in BS^n$ is an open $(n-1)$-disc, since the circles over S^1 are determined by the coordinate vector of their midpoint in $\mathbb{R}^{n+1}$. Thus, α defines a $(n-1)$-disc bundle over BS^n.

Since the map α commutes with the $O(2)$-operation χ on $AS^n - A^0 S^n$, we obtain as quotient a map

$$\gamma : \Gamma S^n - \Gamma^0 S^n \to \Delta S^n.$$

γ is the projection of a fibration over ΔS^n with fibre an $(n-1)$-disc bundle.

One sees at once that γ is the disc bundle of the canonical $(n-1)$-dimensional vector bundle γ^{n-1} over $\Delta S^n = G(2, n-1)$, cf. [Mi 1], [Sp]. From the theory of vector bundles, it then follows that the $\mathbb{Z}_2$-cohomology of Γ mod Γ^0 can be written as

$$H^*(\Gamma S^n, \Gamma^0 S^n) = y^{n-1} \cup H^*(G(2, n-1))$$

where y^{n-1} is the Thom class of the bundle γ^{n-1}.

We wish to describe in more detail the cohomology of $\Gamma S^n - \Gamma^0 S^n$, i.e. $H^*(G(2, n-1))$. We begin by defining the cycle $[a_1, a_2]$ of $G(2, n-1)$, where a_1, a_2 are integers satisfying $0 \leqslant a_1 \leqslant a_2 \leqslant n-1$.

$[a_1, a_2]$ consists of all the great circles on the subsphere

$$S^{a_2+1} := \left\{ \sum_0^{a_2+1} x_i^2 = 1 \right\}$$

of S^n which meet the subsphere

$$S^{a_1} := \left\{ \sum_0^{a_1} x_i^2 = 1 \right\}.$$

Clearly, $[a_1, a_2]$ is a cycle of dimension $a_1 + a_2$, since the subset of $[a_1, a_2]$ formed by the great circles not belonging to S^{a_1} is of the form $P^{a_1} \times D^{a_2}$, where P^{a_1} is the projective real space of dimension a_1 and the boundary (i.e. the set of great circles lying on S^{a_1}) has dimension $2a_1 - 2 \leqslant a_1 + a_2 - 2$.

2.3.1 Proposition. *The $n(n+1)/2$ cycles $[a_1, a_2]$ represent a basis of the $\mathbb{Z}_2$-homology of $G(2, n-1)$. Moreover, $G(2, n-1)$ contains disjoint submanifolds of the type $G(2, n-2)$ and P^{n-1} such that $G(2, n-1) - G(2, n-2)$ is a $(n-1)$-disc bundle π_{n-1} over P^{n-1}, and $G(2, n-1) - P^{n-1}$ is a 2-disc bundle π_1' over $G(2, n-2)$.*

2.3.2 Complement. *Denote by $(a_1, a_2) \in H^*(G(2, n-1))$, $0 \leqslant a_1 \leqslant a_2 \leqslant n-1$, the elements of the basis which is dual to the basis $\{[a_1, a_2]\}$. Then the cup-product satisfies:*

(i) $\quad (a_1, a_2) = (0, a_1) \cup (0, a_2) + (0, a_1 - 1) \cup (0, a_2 + 1);$ *and*

(ii) $\quad (0, a) \cup (a_1, a_2) = \Sigma (a_1 + i, a_2 + a - i),$

$$0 \leqslant i \leqslant \min (a, a_2 - a_1).$$

Here we put $(a, b) = 0$ if $0 \leqslant a \leqslant b \leqslant n-1$ is not satisfied.

Remark. The additive structure of $H_*(G(2, n-1))$ and $H^*(G(2, n-1))$ was determined by Ehresmann [Eh]. The multiplicative structure of $H^*(G(2, n-1))$ was determined by Chern [Ch].

An alternative way of describing the cohomology ring of $G(2, n-1)$ is as follows, cf. Borel [Bor 1]: Denote by $S(x_1, \ldots, x_p)$ the algebra of symmetric polynomials in $x_1, \ldots, x_p$, and by $S^+(x_1, \ldots, x_p)$ the subalgebra generated by the non-constant polynomials. Then $H^*(G(2, n-1))$ can be written as

$$S(u_1, u_2) \otimes S(u_3, \ldots, u_{n+1}) / S^+(u_1, u_2, u_3, \ldots, u_{n+1}).$$

The elementary symmetric polynomials $u_1 + u_2$ and $u_1 \cdot u_2$ are the Stiefel-Whitney classes w^1 and w^2 of the canonical $\mathbb{R}^2$-bundle over $G(2, n-1)$; they correspond to the elements (0,1) and (1,1). The elementary symmetric polynomials in the $u_3, \ldots u_{n+1}$ represent the Stiefel-Whitney classes $\bar{w}^1, \ldots, \bar{w}^{n-1}$ of the canonical $\mathbb{R}^{n-1}$-bundle over $G(2, n-1)$; they correspond to the elements $(0, 1), \ldots, (0, n-1)$.

Proof of (2.3.1). We consider $G(2, n-2)$ embedded into $G(2, n-1)$ by taking the great circles on

$$S^{n-1} = \left\{ \sum_0^{n-1} x_i^2 = 1 \right\}.$$

We embed P^{n-1} by taking the great circles on S^n passing through the x_n-axis.

Define

$$\pi_{n-1} : G(2, n-1) - G(2, n-2) \to P^{n-1}$$

by turning a great circle around its two intersection points with S^{n-1} into P^{n-1}. Define

$$\pi_1' : G(2, n-1) - P^{n-1} \to G(2, n-2)$$

by turning a great circle (if it does not already belong to S^{n-1}) around its intersection points with S^{n-1} into S^{n-1}.

Then the Thom isomorphism yields

$$H_*(G(2, n-1), G(2, n-2)) = H_{*-(n-1)}(P^{n-1})$$

$$H_*(G(2, n-1), P^{n-1}) = H_{*-2}(G(2, n-2)).$$

Under π_{n-1}, the cycles $[a_1, n-1]$ go into non-trivial cycles of P^{n-1} and actually represent a basis of the homology. Hence, by induction on n, we get the statement (2.3.1) from the exact homology sequence of the pair $(G(2, n-1), G(2, n-2))$. □

For the proof of (2.3.2) we refer to the literature [Ch], [Al 1]. We only remark that it is geometrically quite obvious that the intersection number between $[a_1, a_2]$ and $[n-1-a_2, n-1-a_1]$ is 1, whereas the intersection of $[a_1, a_2]$ with all other cycles of dimension $2n-2-a_1-a_2$ is zero mod 2.

2.3.3 Proposition. *The cup-length of* $G(2, n-1)$ *is equal to* $g(n) = 2n-s-1$, *with s determined by* $0 \leq s = n-2^k < 2^k$.

Remark. This was proved by Alber [Al 1]. Using the description of $H^*(G(2, n-1))$ given after (2.3.2), we shall show, with $w^1 = u_1 + u_2$ and $w^2 = u_1 \cdot u_2$, that

$$(w^1)^{2n-2s-2}(w^2)^s \neq 0.$$

Proof. We first derive the formula

$$(*) \qquad (0, 1)^{2^k-1} = (0, 2^k-1).$$

We proceed by induction. (*) is true for $k=1$; assume it holds for $k-1$ and write, using (2.3.2),

$$(0, 1)^{2^k-1} = (0, 1) \cup (0, 1)^{2^{k-1}-1} \cup (0, 1)^{2^{k-1}-1}$$

$$= (0, 1) \cup (0, 2^{k-1}-1) \cup (0, 2^{k-1}-1)$$

$$= (0, 1) \cup \sum_{0}^{2^{k-1}-1} (i, 2^k-2-i) = (0, 2^k-1).$$

Recall that we use $\mathbb{Z}_2$-coefficients. Hence,

$$(0,1)^{2n-2s-2}=(0,2^k-1)\cup(0,2^k-1)=\sum_{0}^{2^k-1}(i,2^{k+1}-2-i)\neq 0$$

and

$$(0,1)^{2n-2s-1}=(0,2^{k+1}-1)=0.$$

Multiplication with (1, 1) yields

$$(1,1)\cup(0,1)^{2n-2s-2}=\sum_{1}^{2^k}(i,2^{k+1}-i)$$

$$(1,1)^s\cup(0,1)^{2n-2s-2}=\sum_{s}^{2^k+s-1}(i,2^{k+1}+2s-2-i)=(n-1,n-1).$$

Hence, the cup-length of $G(2,n-1)$ is $\geq g(n)$. If it were $>g(n)$ we would have to multiply a number $>2n-2s-2$ of cocycles of dimension 1. Since (0, 1) is the only cocycle of dimension 1, this is impossible without obtaining zero. □

Recall that we wrote $\Gamma S^n-\Gamma^0 S^n$ for an $(n-1)$-disc bundle over $\Delta S^n=G(2,n-1)$. From (2.3.1) we therefore see that we obtain a basis for the homology of ΓS^n mod $\Gamma^0 S^n$ if we associate with every cycle $[a_1,a_2]$ of ΔS^n the cycle $\{a_1,a_2\}$ of ΓS^n mod $\Gamma^0 S^n$ which consists of the circles parallel to circles of $[a_1,a_2]$; i.e. $\{a_1,a_2\}=\gamma^{-1}[a_1,a_2]$, $\dim\{a_1,a_2\}=a_1+a_2+n-1$.

$$\{a_1,a_2\}\cap(a_1,a_2)=\{0,0\}$$

Combining this with (2.3.2) we obtain:

2.3.4 Lemma. *The $\mathbb{Z}_2$-homology of ΓS^n mod $\Gamma^0 S^n$ possesses as a basis the homology classes represented by the $n(n+1)/2$ cycles $\{a_1,a_2\}$.*

There exist $g(n)=2n-s-1$ pairwise subordinated homology classes in $H_*(\Gamma S^n,\Gamma^0 S^n)$ which are obtained by forming the cap product between the fundamental cycle $\{n-1\,.\,n-1\}$ and an increasing number of factors

$$(0,1),(0,1)^2,\ldots,(0,1)^{2n-2s-2},\ldots,(0,1)^{2n-2s-2}\cup(1,1)^s$$

of the product

$$(0,1)^{2n-2s-2}\cup(1,1)^s=(n-1,n-1).$$

It remains to check whether, under the canonical inclusion of ΓS^n into ΠS^n, these $g(n)$ pairwise subordinated homology classes survive. That this is indeed the case is the content of the following theorem, due to Alber [Al 1].

2.3.5 Theorem. *The inclusion*

$$i:(\Gamma S^n, \Gamma^0 S^n)\to(\Pi S^n, \Pi^0 S^n)$$

induces a surjective map

$$i^*: H^*(\Pi S^n - \Pi^0 S^n)\to H^*(\Gamma S^n - \Gamma^0 S^n).$$

As a consequence,

$$i_*: H_*(\Gamma S^n, \Gamma^0 S^n)\to H_*(\Pi S^n, \Pi^0 S^n)$$

is injective and carries subordinated classes into subordinated classes.

Remark. Alber [l.c.] considered the homology groups of $\Pi S^n \bmod \Pi^0 S^n$ which are derived from the projected singular complex of ΛS^n under the map $\pi:\Lambda S^n\to\Pi S^n$. From a general result of Bredon [Br] it follows, however, that this projected complex yields the same homology as the singular complex of ΠS^n.

Proof. We define, for $0\leqslant q\leqslant n-1$, cocycles $(0, q)'$ of $\Pi S^n - \Pi^0 S^n$ satisfying $i^*(0, q)' = (0,q)\in H^q(\Gamma S^n - \Gamma^0 S^n)$.

We can assume that $q>0$. For a fixed q we call a singular m-simplex of $\Pi S^n - \Pi S^0$ q-distinguished if its $(q-1)$-dimensional faces do not meet the sphere

$$S'^{n-q-1} = \left\{\sum_{q+1}^{n} x_i^2 = 1\right\}.$$

One can show, using subdivisions and deformations, that the q-dimensional homology group derived from the complex of q-distinguished singular simplices, coincides with the q-dimensional homology group derived from the complex of all singular simplices.

We now define the cocycle $(0, q)'$ as follows. Let

$$t^q: \Delta^q \to \Pi S^n$$

be a q-distinguished singular simplex. Then

$$\hat{t}^q: \partial\Delta^q \times S \to S^n; \qquad \hat{t}^q(d, t) = t^q(d)(t)$$

is a cycle on S^n. (For $q=1$ it consists of the image of two circles.) Define (0, q)′ on t^q to be the linking number mod 2 of $\hat{t}^q$ and S'^{n-q-1}.

From standard properties of the linking number one has $(0, q)' = 0$ on ∂t^{q+1}; hence, $(0, q)'$ is a cocycle. Finally, $(0, q)'$ is equal to 1 on the cycle $[0, q]\subset \Gamma S^n - \Gamma^0 S^n$, consisting of all great circles on S^{q+1} passing through the sphere S^0.

Observe now that the cocycles $(0, q)$, $0\leqslant q\leqslant n-1$, are generators of the ring $H^*(\Gamma S^n - \Gamma^0 S^n) = H^*(G(2, n-1))$, cf. (2.3.2). Thus we have proved the first part of the theorem.

To prove the remainder we recall that $\{a_1, a_2\} \cap (a_1, a_2) = \{0, 0\}$, the basic (or Thom) $(n-1)$-dimensional homology class of ΓS^n mod $\Gamma^0 S^n$, cf. (2.3.4).

Observe now that the cycle $[0, n-1]$ represents the Euler class mod 2 of the bundle γ^{n-1}. It is well known that $[0, n-1]$ is homologous mod 2 to the Thom homology class $\{0, 0\}$. Hence, since $i_*[0, n-1] \nsim 0$ also $i_*\{0, 0\} \nsim 0$. Therefore

$$i_*\{0, 0\} = i_*(\{a_1, a_2\} \cap i^*(a_1, a_2)') = i_*\{a_1, a_2\}' \nsim 0,$$

i.e., i_* is injective. □

As a corollary we have, using (2.2.10):

2.3.6 Theorem. *Let M be a Riemannian manifold homeomorphic to the n-sphere S^n. Let*

$$h: S^n \to M$$

be a differentiable homotopy equivalence carrying non-trivial circles into non-trivial curves. Choose $\kappa^* > 0$ such that $h_\Gamma(\Gamma S^n) \subset \Pi^{\kappa^*-} M$.

Claim. *There exist $g(n) = 2n - s - 1$, $0 \leqslant s = n - 2^k < 2^k$, critical values*

$$0 < \kappa_1 \leqslant \ldots \leqslant \kappa_{g(n)} < \kappa^*$$

in ΠS^n, which are the critical values of $g(n)$ pairwise subordinated homology classes of ΠM mod *$\Pi^0 M$.*

In particular, if two of these values coincide, then there exists an infinite number of closed geodesics in ΠM.

Note. This theorem is due to Alber [Al 1]. We do not claim, however, that the $g(n)$ closed geodesics constructed here are all prime or that the underlying prime closed geodesics all differ from one another. This will be proved much later using completely different methods, cf. (5.1).

In the case $n=2$, Lusternik and Schnirelmann [LS 1] were able to show that, by choosing $h: S^2 \to M$ as diffeomorphism and applying special E-decreasing deformations which carry the image of a circle into curves without self-intersections, the images $h_\Gamma\{0, 0\}$, $h_\Gamma\{0, 1\}$, $h_\Gamma\{1, 1\}$ of the $g(2) = 3$ pairwise subordinated cycles of ΓS^2 mod $\Gamma^0 S^2$ remain hanging at three closed geodesics which are not only prime but even have no self-intersections; see the Appendix for a proof.

For $n > 2$, the existence of closed geodesics without self-intersections has been proved only under certain restrictions on the Riemannian metric of M, cf. [Kl 7], [Al 2] for a listing of results in this direction; they all are based on the following result, cf. [Kl 1], [GKM]:

2.3.7 Theorem. *Let M be a simply connected, compact manifold for which the sectional curvature K satisfies the relation*

$$K_0/4 < K \leqslant K_0$$

for some $K_0 > 0$. Then a geodesic loop on M has length $\geqslant 2\pi/\sqrt{K_0}$, i.e. a closed geodesic of length $< 4\pi/\sqrt{K_0}$ has no self-intersections.

Moreover, there exists a homeomorphism

$$h: S^n \to M$$

having the following properties. Denote by $p_\pm$ the points $(\pm 1, 0, \ldots, 0) \in S^n$. Let $H^n_\pm$ be the half-sphere around $p\pm$. Then $h|H^n_\pm$ is differentiable. The image $h(S^{*n-1})$ of the "equator" $S^{*n-1} = \partial H^n_\pm$ is a submanifold M^* of codimension 1 formed by the points $q \in M$ with $d(h(p_+), q) = d(h(p_-), q) < \pi/\sqrt{K_0}$.

With the help of (2.3.7) we can easily prove, cf. [Kl 7]:

2.3.8 Theorem. *Let M be homeomorphic to S^n and assume that the sectional curvature of M satisfies*

$$(\dagger) \qquad K_0/4 < K \leqslant K_0$$

for some $K_0 > 0$. Then there exist on M n closed geodesics without self-intersections having length in the interval $[2\pi/\sqrt{K_0}, 4\pi/\sqrt{K_0}[$.

Proof. We use the map h of (2.3.7) to show that there exist n pairwise subordinated homology classes in ΠM mod $\Pi^0 M$ which can be realized by cycles having their carrier in Π^{κ^*-}, $\kappa^* = 8\pi^2/K_0$. It follows that there are n different closed geodesics on M of length $< 4\pi/\sqrt{K_0}$. (2.3.7) then implies the theorem.

It remains to construct these cycles. From (2.3.7) we have that the great circles passing through p_- and p_+ go into curves of length $< 4\pi/\sqrt{K_0}$. Hence, we can assume (following an appropriate parameterization) that $h_\Gamma([0, n-1]) \subset \Pi^{\kappa^*-}$.

Since it is not obvious that the images under h of all the circles parallel to one of the great circles in $[0, n-1]$ have length $< 4\pi/\sqrt{K_0}$, we replace the cycle $\{0, n-1\}$ by the following homologous cycle. Each non-constant circle in $\{0, n-1\}$ meets the equator S^{*n-1} in exactly two points; replace the circle by the closed curve composed of the two half great circles going from p_- to p_+ through the two points on the equator. Replace the constant circles on the equator by the closed curve composed of the two half great circles going from p_- to p_+ through the point circle and back. Finally, shrink each of these latter closed curves to the point curve on the equator.

Thus we have constructed for each of the n subordinated cycles $\{0, p\}$, $0 \leqslant p \leqslant n-1$, homologous cycles which go into Π^{κ^*-}.

This proves (2.3.8). □

Remark. In this section we used the subspace ΓS^n of circles of the space ΠS^n to construct subordinated homology classes. Similarly, one can use for the other symmetric spaces of rank 1 (i.e. the projective space over the complex numbers, the quaternions or the Cayley numbers) certain subspaces of circles in Π to construct subordinated homology classes. For details, we refer to [Kl 2], [Kl 6].

Eliasson [El 1] has, in a similar fashion, constructed subordinated homology classes on the Grassmann manifolds $G(2, n-1)$.

2.4 Morse Theory on ΛM

In this section we shall investigate the behavior of the flow $\phi_s: \Lambda M \to \Lambda M$ in the neighborhood of critical elements in more detail.

To obtain a complete picture of what is going on and what the topology of $\Lambda^\kappa \bmod \Lambda^{\kappa-}$ looks like at a critical value κ of E, one must assume that the critical points in $E^{-1}(\kappa)$ form a non-degenerate critical submanifold. Similarly, as in the case of a function on a Euclidean manifold having non-degenerate critical submanifolds, one can then develop the Morse theory on ΛM.

Note that we have, as a special feature, the group $O(2)$ acting on ΛM. This leads to an equivariant version of Morse theory.

We begin by defining the *index form at a critical point* $c \in \Lambda M$ of E as being the Hessian $D^2E(c)$. We wish to compute $D^2E(c)$. In order to do so, let ξ, ξ' be elements of $T_c\Lambda M$. Consider a differentiable map

$$\kappa: [0,1]^2 \to \Lambda M;\ (s,s') \mapsto \kappa(s,s')$$

satisfying $\kappa(0,0) = c$,

$$\frac{\partial \kappa}{\partial s}(0,0) = \xi,\ \frac{\partial \kappa}{\partial s'}(0,0) = \xi'.$$

Then

$$D^2E(c)\,(\xi, \xi') = \frac{\partial^2}{\partial s \partial s'}\, E(\kappa(0,0)).$$

In order to derive an explicit expression for $D^2E(c)$, we recall the following formulas of Riemannian geometry on M, cf. [GKM].

First, with κ as above, define

$$\kappa^{\sim}: (s,s',t) \in [0,1]^2 \times S \mapsto \kappa^{\sim}(s,s',t) \in M$$

by

$$\kappa^{\sim}(s,s',t) = \kappa(s,s')\,(t).$$

Then $\partial\kappa^{\sim}/\partial t(0,0,t) = \dot{c}(t)$. We put

$$\partial\kappa^{\sim}/\partial s(s,s',t) = \xi(t;s,s');\ \partial\kappa^{\sim}/\partial s'(s,s',t) = \xi'(t;s,s').$$

If $\kappa^{\sim}$ is differentiable in all arguments we have

$$\text{(i)} \quad \begin{aligned} &\nabla(\nabla\xi'(t)\,.\,\dot{c}(t))\,.\,\xi(t) = R(\xi(t), \dot{c}(t), \xi'(t)) \\ &+\nabla(\nabla\xi'(t)\,.\,\xi(t))\,.\,\dot{c}(t). \end{aligned}$$

Moreover

$$\langle R(\xi(t), \partial c(t), \xi'(t)), \partial c(t)\rangle = -\langle R(\xi(t), \partial c(t), \partial c(t)), \xi'(t)\rangle.$$

If $c \in Cr\Lambda$, we define, for $\xi, \xi' \in T_c\Lambda$,

$$\text{(ii)} \quad \begin{aligned} &R^{\sim}(\xi, \partial c, \xi')(t) := R(\xi(t), \partial c(t), \xi'(t)) \\ &K_c^{\sim}(\xi)(t) := R(\xi(t), \partial c(t), \partial c(t)). \end{aligned}$$

Using the covariant derivative ∇_{α^0} in the bundle $\alpha^0: H^0(H^1(S,M)^*TM) \to H^1(S,M)$ we obtain from (i) and (ii),

$$\text{(i)}' \quad \nabla_{\alpha^0}(\nabla\xi')\,.\,\xi = R^{\sim}(\xi, \partial c, \xi') + \nabla(\nabla_{\alpha^0}\xi'\,.\,\xi)$$

(here we have abbreviated

$\nabla_{\alpha^0}\partial c\,.\,\xi$ by $\nabla\xi$, cf. (1.3.5); and

$$\text{(ii)}' \quad \langle R^{\sim}(\xi, \partial c, \xi'), \partial c\rangle_0 = -\langle K_c^{\sim}(\xi), \xi'\rangle_0).$$

2.4.1 Lemma. *Let c be a critical point of ΛM. Then the index form $D^2E(c)$ is given by*

$$D^2E(c)(\xi, \xi') = \langle \xi, \xi'\rangle_1 - \langle (K_c^{\sim} + \mathrm{id})\xi, \xi'\rangle_0.$$

Proof. $DE(c')\,.\,\xi' = \langle \partial c', \nabla\xi'\rangle_0$. Hence, using (i)′ and (ii)′,

$$\begin{aligned} D\langle \partial c', \nabla\xi'\rangle_0\,.\,\xi|_{c'=c} &= \langle \nabla\xi, \nabla\xi'\rangle_0 + \langle \partial c, \nabla_{\alpha^0}(\nabla\xi')\,.\,\xi\rangle_0 \\ &= \langle \xi, \xi'\rangle_1 + \langle \partial c, R^{\sim}(\xi, \partial c, \xi')\rangle_0 - \langle \xi, \xi'\rangle_0 + \langle \partial c, \nabla(\nabla_{\alpha^0}\xi'\,.\,\xi)\rangle_0. \end{aligned}$$

But the last term vanishes, since $DE(c) = 0$. □

Let c be a critical point of E in ΛM. From the index form $D^2E(c)$ we obtain in the usual way a self-adjoint operator $A_c: T_c\Lambda \to T_c\Lambda$ by the identity

$$\langle A_c\xi, \xi'\rangle_1 = \langle \xi, {}^tA_c\,\xi'\rangle_1 = D^2E(c)(\xi, \xi').$$

2.4.2 Theorem. *The operator A_c is of the form*

$$A_c = \mathrm{id} + k_c$$

where $k_c = -(1-\nabla^2)^{-1} \circ (K_c^{\sim}+1)$

is a compact operator characterized by the identity

$$\langle k_c\xi, \xi'\rangle_1 = -\langle (K_c^{\sim}+1)\xi, \xi'\rangle_0 .$$

2.4.3 Corollary 1. *Either A_c has only finitely many eigenvalues including 1, or the eigenvalues form an infinite discrete bounded set with 1 as accumulation point. Let*

$$T_c\Lambda M = T_c^{-}\Lambda M + T_c^{0}\Lambda M + T_c^{+}\Lambda M$$

be the orthogonal decomposition of the tangent space $T_c\Lambda$ into subspaces spanned by the eigenvectors of A_c having eigenvalue $<0, =0$ and >0 respectively. Then $\dim T_c^{-}\Lambda M$ *and* $\dim T_c^{0}\Lambda M$ *are finite.*

We call $\dim T_c^{-} M$ and $\dim T_c^{0}\Lambda M - 1$ the *index* and the *nullity* of c, respectively.

Proof. It suffices to consider $\xi \in T_c\Lambda M$ differentiable, since these elements are dense in $T_c\Lambda M$. We then have

$$\langle \nabla\xi, \nabla\xi'\rangle_0 = -\langle \nabla^2\xi, \xi'\rangle_0 ,$$

hence

$$-\langle (K_c^{\sim}+1)\xi, \xi'\rangle_0 = -\langle (1-\nabla^2)^{-1} \circ (K_c^{\sim}+1)\xi, \xi'\rangle_1 = \langle k_c\xi, \xi'\rangle_1 .$$

Therefore

$$\langle A_c\xi, \xi'\rangle_1 = \langle \xi, \xi'\rangle_1 - \langle (K_c^{\sim}+1)\xi, \xi'\rangle_0 = \langle (\mathrm{id}+k_c)\xi, \xi'\rangle_1 .$$

It remains to show that $k_c = -(1-\nabla^2)^{-1} \circ (K_c^{\sim}+1)$ is compact. This follows from the well-known fact that the inverse of the elliptic differential operator $(1-\nabla^2)$ is compact.

A direct proof would start with the identity

$$\langle k_c\xi, k_c\xi\rangle_1 = -\langle (K_c^{\sim}+1)\xi, k_c\xi\rangle_0 ,$$

hence

$$\begin{aligned}\|k_c\xi\|_1^2 &\le \|K_c^{\sim}+1\|_\infty \|k_c\xi\|_\infty \|\xi\|_0 \\ &\le \text{const } \|k_c\xi\|_1 \|\xi\|_0 .\end{aligned}$$

If $\{\xi_m\}$ is bounded in $H^1(c^*TM) = T_c\Lambda M$ then we know that it has accumulation points in $H^0(c^*TM)$. Hence, using (1.2.1),

$$\|k_c\xi_m\|_1 \le \text{const } \|\xi_m\|_0 ,$$

thus, $\{k_c\xi_m\}$ has accumulation points in $H^1(c^*TM)$. □

The corollary is a well-known statement about self-adjoint operators of the form "identity plus compact operator". □

2.4.4 Corollary 2. *The eigenvectors of A_c belonging to the eigenvalue $\lambda \in \mathbb{R}$, are the periodic solutions of the differential equation*

$$(\lambda-1)(\nabla^2-1)\xi-(K_c^{\sim}+1)\xi=0.$$

Proof. Observe that the eigenvectors are differentiable. Then this is just another version of $A_c\xi=\lambda\xi$. □

We call a closed submanifold B of ΛM *critical*, if $E|B=\text{const}$, if it is closed under the S-action $\chi^{\sim}$ and if B consists entirely of critical points of E.

Note that critical submanifolds are compact.

Examples of critical submanifolds are

(i) the orbit of a critical point under the $O(2)$- action χ;

(ii) the set $\Lambda^0 M$ of point curves, cf. (1.4.6).

2.4.5 Proposition. *Let B be a critical submanifold of ΛM. Then the tangent space $T_c B$ of B at $c \in B$ belongs to the null space $T_c^0 \Lambda$ of c.*

Proof. Let $\xi \in T_c B$, $\eta \in T_c \Lambda$. We have to show that $D^2E(c)(\xi,\eta)=0$.

To see this we consider a differentiable map

$$\kappa:(r,s)\in[0,1]^2\to\kappa(r,s)\in\Lambda M$$

satisfying

$$\kappa(0,0)=c,\ \kappa(r,0)\subset B,\ \frac{\partial\kappa}{\partial r}(0,0)=\xi,\ \frac{\partial\kappa}{\partial s}(0,0)=\eta.$$

Then

$$\frac{\partial}{\partial s}E(\kappa(r,0))=DE(\kappa(r,0))\cdot\frac{\partial\kappa}{\partial s}(r,0)=0.$$

Hence,

$$D^2E(c)(\xi,\eta)=\partial^2E(\kappa(0,0))/\partial r\partial s=0. \quad \square$$

Note that every closed geodesic, being part of the 1-dimensional critical submanifold $\chi^{\sim}(S,c)=S.c$, has nullity $\geqslant 1$, $\partial c \neq 0$ and $\partial c \in$ nullspace of A_c.

If, for all elements c of a critical submanifold B, index c is constant, then we call this number the *index of B*. Similarly, if the nullity of $D^2E(c)$ is constant for $c\in B$, then we call this number the *nullity of B*.

If the critical submanifold B has an index and if the nullity of B coincides with dim B, i.e. if for all $c\in B$ we have $T_cB=T_c^0\Lambda$, then we call B a *non-degenerate critical submanifold*.

If c is a closed geodesic and if the orbit $S.c$ is non-degenerate, then we call c a *non-degenerate closed geodesic*. This is equivalent to saying that the nullity of $c=0$ or the nullity of $D^2E(c)=1$.

2.4.6 Proposition. *$\Lambda^0 M$ is a non-degenerate critical submanifold of index* 0.

Proof. We have from (2.4.4) with $c = \text{const}$, $K_c^{\sim} = 0$, hence

$$(\lambda - 1)\nabla^2 \xi - \lambda \xi = 0.$$

There are no periodic solutions for $\lambda < 0$, index $\Lambda^0 M = 0$. For $\lambda = 0$, only $\xi = \text{const}$ are periodic solutions, hence, $T_c^0 \Lambda M = T_c \Lambda^0 M$. □

As a preliminary step for the main theorem of Morse theory we consider a *Riemannian $O(2)$-vector bundle*

$$\mu : N \to B$$

where B is a compact manifold and the fibres are Hilbert spaces with a Riemannian metric $\langle\ ,\ \rangle_1$, cf. (1.1). Moreover, we shall give a continuous $O(2)$-action

$$\chi : O(2) \times N \to N$$

such that, for every $\alpha \in O(2)$, the map $\chi_\alpha := \chi(\alpha,\ ) : N \to N$ is an isometric bundle map.

This implies that the projection μ is equivariant, as is the zero section $c \in B \mapsto 0_c \in N$.

Denote by $D_\varepsilon \mu$ the ε-disc bundle of μ, for some $\varepsilon > 0$. $D\mu$ will denote the (open) unit disc bundle. Assume that there is given a differentiable invariant function

$$E : D_\varepsilon \mu \to \mathbb{R}.$$

We call E a *Morse function* if E has the 0-section B as non-degenerate critical submanifold of index $k \geqslant 0$ and if the index form $D^2 E(c)$ at $c \in B$ is represented by an operator A_c of the form "identity plus compact operator".

Clearly, we have modeled this definition on the following example. Let the critical set in ΛM at the E-value κ be a non-degenerate critical submanifold B of index k. Let $\mu : N \to B$ be the normal bundle of this manifold. μ then is a Riemannian $O(2)$-vector bundle.

Consider, for some $\varepsilon > 0$, the ε-disc bundle $D_\varepsilon \mu$ of μ. The fibre over $c \in B$ consists of the elements $\xi \in N_c$ with $\|\xi\|_1 < \varepsilon$. Using the exponential map, exp, of the Levi-Cività connection, we can map $D_\varepsilon \mu$, for ε sufficiently small, onto an open neighborhood of B in ΛM. We could also use the map, $\exp^{\sim}$, of (1.3.7.), since the $\|\ \|_1$-norm dominates the $\|\ \|_\infty$-norm, cf. (1.2.1).

$E \circ \exp : D_\varepsilon \mu \to \mathbb{R}$ is then a Morse function on μ.

We now prove the *generalized Morse Lemma:*

2.4.7 Lemma. *Let μ be a Riemannian $O(2)$-bundle over B and E an invariant Morse function, $E|B = 0$.*

Claim. *There exists $\varepsilon > 0$ and an equivariant fibre preserving diffeomorphism*

$$\psi : D_\varepsilon \mu \to \psi D_\varepsilon \mu \subset N$$

and an equivariant differentiable section: $c \in B \mapsto P_c \in L(N)$ with P_c an orthogonal bundle projection, such that

$$E \circ \psi(\xi) = \|P_c \xi\|_1^2 - \|(\mathrm{id} - P_c)\xi\|_1^2, \; c = \mu(\xi).$$

Note. For the case that B is a point, this extension of the Morse lemma to Hilbert manifolds was proved by Palais [Pa 2]; see also [Sch]. If we have the trivial $O(2)$-action (that is to say, no $O(2)$-action), this lemma was proved by Meyer [Me]. If we have a differentiable $O(2)$-action, the lemma is contained in Wasserman [Wa]. Here we need the case of a continuous $O(2)$-action. We will see that the proofs of the special cases mentioned previously can be carried over to our case in a rather straightforward manner.

Proof. To simplify our notation, we neglect to indicate the fact that E might be defined only in a small neighborhood of the zero section B of μ.

We write $E|N_c$ in the form

$$E(\eta) = \int_0^1 (1-t) D^2 E(t\eta) \,.\, (\eta, \eta) dt.$$

Then

$$h(\eta) \,.\, (\xi, \xi') = \int_0^1 (1-t) D^2 E(t\eta) \,.\, (\xi, \xi') dt$$

is an equivariant symmetric bilinear form.

Since $D^2(E|N_c)(0_c)$ is non-degenerate, $h(\eta)$ is non-degenerate in some ε-neighborhood of the zero section N. We represent $h(\eta)$ by a self-adjoint operator $k(\eta)$

$$\langle k(\eta)\xi, \xi' \rangle_1 = h(\eta) \,.\, (\xi, \xi').$$

Our first goal is to find on $D_\varepsilon(\mu)$, $\varepsilon > 0$ sufficiently small, an equivariant differential section m of invertible operators such that

$$\langle k(\eta)\eta, \eta \rangle_1 = \langle k(0_c) m(\eta)\eta, \, m(\eta)\eta \rangle_1 .$$

That is to say, we are looking for a family $m(\eta)$ of operators satisfying

$$(*) \qquad m(\eta)^* k(0) m(\eta) = k(\eta)$$

where * denotes the adjoint map.

If we consider $l(\eta) = k(0)^{-1} k(\eta)$ for $\|\eta\|_1$ small, then it is close to the identity and therefore possesses a square root which we will denote by $m(\eta)$.

To show that $m(\eta)$ satisfies (*), we first observe that, from $l(\eta)^* = k(\eta) k(0)^{-1}$, we have

$$l(\eta)^* k(0) = k(0) l(\eta).$$

But $m(\eta)$ can be represented locally by a power series in $(1-l(\eta))$, i.e. the series determined by $(1-(1-l(\eta)))^{1/2}$, and $m(\eta)^*$ is then the same power series in $(1-l(\eta)^*)$. Hence, the preceding relation holds if we replace $l(\eta)$ and $l(\eta)^*$ by $m(\eta)$ and $m(\eta)^*$, and thus

$$m(\eta)^* k(0) m(\eta) = k(0) m(\eta) m(\eta) = k(0) l(\eta) = k(\eta).$$

i.e. we have proved (*).

If we define, for $\|\eta\|_1$ small, $\phi'^{-1}: \eta \mapsto \xi' = m(\eta)\eta$, this is an invertible function. For the differentiability see Palais [Pa 2]. We can write

$$E(\phi'(\xi')) = \langle k(0_c)\xi', \xi' \rangle_1, \quad c = \mu(\xi').$$

Since we supposed that $k(0_c)$ (the self-adjoint operator representing the Hessian $D^2E(c)$ at $0_c = c \in B$) is of the form "identity plus compact operator", we can equivariantly introduce new coordinates ξ by $\xi' = \phi(\xi)$ and an orthogonal projection P_c such that

$$\langle k(0_c)\xi', \xi' \rangle_1 = \|P_c\xi\|_1^2 - \|(\text{id-}P_c)\xi\|_1^2.$$

By putting $\phi' \circ \phi = \psi$, we thus get the lemma, except for the differentiability of P_c. For this see Meyer [Me]. □

Remark. The last coordinate transformation $\xi' = \phi(\xi)$ and the orthogonal projection P_c can be read off from the eigenvectors of $D^2(E(c)|N_c)$. If ξ'_λ is the component of ξ' at the eigenvector e_λ belonging to the eigenvalue $\lambda \neq 0$, then we define the component ξ_λ of $\xi = \phi^{-1}(\xi')$ by $\xi_\lambda = \sqrt{|\lambda|}\, \xi'_\lambda$. P_c is the projection of the fibre N_c onto the positive eigenspace.

2.4.8 Corollary. *Assume that the critical set B in ΛM of E at $E^{-1}(\kappa)$ is a non-degenerate critical submanifold of index k. Let $\mu: N \to B$ be the normal bundle. Then there exists, for $\varepsilon > 0$ sufficiently small, an equivariant diffeomorphism ϕ of the ε-disc bundle $D_\varepsilon\mu$ of μ onto an open neighborhood of B such that*

$$E \circ \phi(\xi) = \kappa + \|P_c\xi\|_1^2 - \|(\text{id} - P_c)\xi\|_1^2, \quad c = \mu(\xi),$$

where the P_c are equivariant orthonormal bundle projections.

In particular, κ is an isolated critical value.

Proof. Recall that we constructed above an equivariant diffeomorphism of $D_\varepsilon(\mu)$ onto a neighborhood of B using the exponential map. Thus we obtain (2.4.8). □

Following Palais [Pa 2] we consider a differentiable monotonic decreasing function $\lambda: \mathbb{R} \to \mathbb{R}$ satisfying $\lambda(x) = 0$ for $x \geqslant 1$, $\lambda(x) > 0$ for $x < 1$, $\lambda(x) = 1$ for $x \leqslant 1/2$. Using λ, we define $\sigma(s)$ for $s \in [0,1]$ to be the unique solution of $\lambda(\sigma)/(1+\sigma) = 2(1-s)/3$.

Claim. *σ is continuous, strictly monotonic increasing and differentiable in* $[0,1]$. *Moreover,* $\sigma(0)=1/2$, $\sigma(1)=1$, *and, if* $\varepsilon>0$, *then*

$$-\varepsilon \leqslant u^2 - v^2 \leqslant -\varepsilon + \frac{3\varepsilon}{2}\lambda(u^2/3)$$

implies

$$u^2 \leqslant \varepsilon\sigma\big(v^2/(\varepsilon+u^2)\big).$$

For the proof we refer to [Pa 2].

2.4.9 Proposition. *Let $\mu: N \to B$ be a Riemannian $O(2)$-vector bundle endowed with an equivariant orthogonal bundle projection $c \in B \mapsto P_c \in L(N_c; N_c)$ with co-kernel of dimension k. Consider, for some ε, on the 2ε-disc bundle $D_{2\varepsilon}\mu$ the equivariant functions*

$$E(\xi) = \|P_c\xi\|_1^2 - \|(\mathrm{id}-P_c)\xi\|_1^2, \quad c=\mu(\xi),$$

$$F(\xi) = E(\xi) - \frac{3\varepsilon}{2}\lambda(\|P_c\xi\|_1^2/\varepsilon), \quad c=\mu(\xi).$$

Claim. *The space $N_F := \{\xi \in D_{2\varepsilon}\mu,\ F(\xi) \leqslant -\varepsilon\}$ arises from the space $N_E = \{\xi \in D_{2\varepsilon}\mu,\ E(\xi) \leqslant -\varepsilon\}$ by attaching equivariantly the handle bundle $\bar{D}\mu^- \oplus \bar{D}\mu^+$, which is formed by the direct sum of the closed unit disc bundles of the subbundles μ^- and μ^+ of μ, consisting of the kernel and the range of P, respectively.*

Note. Attaching the bundle $\bar{D}\mu^- \oplus \bar{D}\mu^+$ equivariantly to N_E means that: we have an equivariant homeomorphism h of $\bar{D}\mu^- \oplus \bar{D}\mu^+$ onto a closed subset H of N_F such that

(i) $N_F = N_E \cup H$;

(ii) $h|\partial\bar{D}\mu^- \oplus \bar{D}\mu^+$ is an equivariant diffeomorphism onto $\partial N_E \cap H$;

(iii) $h|D\mu^- \oplus \bar{D}\mu^+$ is an equivariant diffeomorphism onto $N_F - N_E$.

Compare the note after (2.4.7).

Proof. Define $h: D\mu^- \oplus \bar{D}\mu^+ \to N_F$ by

$$h(\xi_- + \xi_+) = \big(\varepsilon\sigma(\|\xi_-\|_1^2)\|\xi_+\|_1^2 + \varepsilon\big)^{1/2}\xi_- + \big(\varepsilon\sigma(\|\xi_-\|_1^2)\big)^{1/2}\xi_+$$

with the function σ introduced above and $\xi_- = (\mathrm{id}-P)\xi$, $\xi_+ = P\xi$.
Then we have

$$E(h(\xi_- + \xi_+)) = \varepsilon[\sigma(\|\xi_-\|_1^2)\|\xi_+\|_1^2(1-\|\xi_-\|_1^2) - \|\xi_-\|_1^2] \geqslant -\varepsilon$$

$$F(h(\xi_- + \xi_+)) = \varepsilon[\sigma(\|\xi_-\|_1^2)\|\xi_+\|_1^2(1-\|\xi_-\|_1^2) - \|\xi_-\|_1^2$$

$$-\tfrac{3}{2}\lambda(\sigma(\|\xi_-\|_1^2)\|\xi_+\|_1^2)]$$

$$\leqslant\varepsilon[\sigma(\|\xi_-\|_1^2)(1-\|\xi_-\|_1^2)-\|\xi_-\|_1^2-\tfrac{3}{2}\lambda(\sigma(\|\xi_-\|_1^2))]\leqslant-\varepsilon.$$

Hence, im h belongs to the set

$$H:=\{\xi\in D_{2\varepsilon}\mu;\ E(\xi)\geqslant-\varepsilon,\ F(\xi)\leqslant-\varepsilon\}.$$

Actually, h is a homeomorphism onto H with h^{-1} being given by

$$h^{-1}(\eta)=(\varepsilon+\|P\eta\|_1^2)^{-1/2}(\mathrm{id}-P)\eta+$$

$$(\varepsilon\sigma(\|(\mathrm{id}-P)\eta\|_1^2/(\varepsilon+\|P\eta\|_1^2))^{-1/2}P\eta\in\bar{D}\mu^-\oplus\bar{D}\mu^+.$$

This is verified by using the properties of σ.

Since $\sigma|[0,1[$ is differentiable with $\dot{\sigma}\neq 0$, one sees that $h|D\mu^-\oplus\bar{D}\mu^+$ is differentiable of maximal rank. Finally, for $\xi=\xi_-+\xi_+\in\partial\bar{D}\mu^-\oplus\bar{D}\mu^+$, $h(\xi)=(\varepsilon\|\xi_+\|_1^2+\varepsilon)^{1/2}\xi_-+\varepsilon^{1/2}\xi_+$, i.e. $h|\partial\bar{D}\mu^-\oplus\bar{D}\mu^+$ is a diffeomorphism onto $\partial N_E\cap H$.

Note that all the previous definitions and constructions were made equivariently. Thus we have proved (2.4.9). □

We can now state the main theorem of Morse theory of the function E on ΛM. Here again the remarks hold which we made following (2.4.7).

2.4.10 Theorem. *Assume that the set of critical points in $\Lambda=\Lambda M$ at the E-level $\kappa>0$ is a non-degenerate critical submanifold B.*

Then there exists an $\varepsilon>0$ such that $\Lambda^{\kappa+\varepsilon}$ is equivariantly diffeomorphic to $\Lambda^{\kappa-\varepsilon}$ with a handle bundle of type $\bar{D}\mu^-\oplus\bar{D}\mu^+$ attached. Here μ is the normal bundle of B and

$$\mu=\mu^-\oplus\mu^+$$

the decomposition into the negative and the positive sub-bundles, respectively.

Proof. From (2.4.8) we have of the existence of an equivariant diffeomorphism ϕ of some disc bundle $D_{2\delta}\mu$ of the normal bundle μ of B such that

$$E\circ\phi(\xi)=\kappa+\|P\xi\|_1^2-\|(\mathrm{id}-P)\xi\|_1^2.$$

Choose ε such that $0<\varepsilon<\delta^2$ and such that κ is the only critical value in $]-3\varepsilon+\kappa,\ \kappa+3\varepsilon[$. On $W:=\{e\in\Lambda M,\ E(e)\geqslant\kappa-2\varepsilon\}$ we define a function F by

$$F(e)=\begin{cases}E(e)-\dfrac{3\varepsilon}{2}\lambda(\|P\circ\phi^{-1}(e)\|_1^2/\varepsilon), & \text{for } e\in\phi D_{2\delta}\mu,\\ E(e)\ , & \text{otherwise.}\end{cases}$$

If $e=\phi(\xi)$ is in W and $F(e)\neq E(e)$, then $\lambda(\|P\xi\|_1^2/\varepsilon)\neq 0$, so $\|P\xi\|_1^2<\varepsilon$, $E(e)<\kappa+\varepsilon$ and $E(e)=\kappa+\|P\xi\|_1^2-\|(\mathrm{id}-P)\xi\|_1^2\geqslant\kappa-2\varepsilon$, hence $\|\xi\|_1^2<4\varepsilon<4\delta^2$. That is to say, the closure of the set $\{e\in W\cap\phi D_{2\delta}\mu;\ E(e)\neq F(e)\}$ belongs to the interior of $\phi D_{2\delta}\mu$, which proves that F is differentiable.

The above also shows that $\{e\in W,\ E(e)\leqslant\kappa+\varepsilon\}=\{e\in W,\ F(e)\leqslant\kappa+\varepsilon\}$.

We therefore obtain from (2.4.9) that the manifold with boundary

$$N_F:=\{e\in W\cap\phi D_{2\delta}\mu;\ F(e)\leqslant\kappa-\varepsilon\}$$

arises from the manifold with boundary

$$N_E:=\{e\in W\cap\phi D_{2\delta}\mu,\ E(e)\leqslant\kappa-\varepsilon\}$$

by attaching equivariantly the handle bundle $\bar{D}\mu^-\oplus\bar{D}\mu^+$.

Note that $N_E=\Lambda^{\kappa-\varepsilon}\cap\phi D_{2\delta}\mu$ and that $N_F\subset\Lambda^{\kappa+\varepsilon}$ contains the critical submanifold B in its interior, since for $c\in B$ we have $F(c)=\kappa-3\varepsilon/2<\kappa-\varepsilon$. That is to say, $\mathscr{U}=\mathrm{int}\ N_F$ is an open neighborhood of B and we have from (1.4.8) that $\|\mathrm{grad}\ E\|_1$ is bounded away from zero on $\mathscr{U}\cap(\Lambda^{\kappa+\varepsilon}-\Lambda^{\kappa-\varepsilon})$. Therefore, the deformations ϕ_s can be used to construct an equivariant deformation of $\Lambda^{\kappa+\varepsilon}$ onto $\Lambda^{\kappa-\varepsilon}\cup N_F$, cf. Palais [Pa 2].

This proves the theorem. □

Remark. In general it may happen that the critical set at E-level κ decomposes into several $O(2)$-invariant, non-degenerate, critical submanifolds, say $B_1,\dots,B_k$. The B_i will then be disjoint. In the same manner, one can prove that $\Lambda^{\kappa+\varepsilon}$ is equivariantly diffeomorphic to $\Lambda^{\kappa-\varepsilon}$ with handle bundles attached corresponding to each of the submanifolds B_i.

2.4.11 Corollary. *With the hypotheses of* (2.4.10) $H_*(\Lambda^\kappa,\Lambda^{\kappa-})=H_{*-k}(B)$, *where* $k=$ *index* B. *Here one has to take homology with* $\mathbb{Z}_2$*-coefficients, unless the negative bundle* μ^- *over* B *is orientable. In particular, if* B *is the* $O(2)$*-orbit of a non-degenerate closed geodesic* c *of index* k, $H_i(\Lambda^\kappa,\Lambda^{\kappa-})=\mathbb{Z}_2\oplus\mathbb{Z}_2$ *for* $i=k,k+1$, *and* $=0$ *otherwise. The same is true also for integer coefficients, if* μ^- *is orientable; otherwise,* $H_*(\Lambda^\kappa,\Lambda^{\kappa-})=0$.

Proof. Since the zero section of $\mu^+=\mu^+B$ is a strong equivariant deformation retract of $D\mu^+$, $\Lambda^{\kappa+\varepsilon}$ can be deformed into $\Lambda^{\kappa-\varepsilon}\bigcup_h(\bar{D}\mu^-\oplus\bar{D}\mu^+)$ and then into $\Lambda^{\kappa-\varepsilon}\bigcup_h\bar{D}\mu^-$. Consequently, $\Lambda^{\kappa+\varepsilon}$ mod $\Lambda^{\kappa-\varepsilon}$ is topologically equivalent to $\bar{D}\mu^-$ mod $\partial\bar{D}\mu^-$, i.e. the Thom space of the negative bundle μ^-B. Thus the Thom isomorphism yields

$$H_*(\Lambda^{\kappa+\varepsilon},\ \Lambda^{\kappa-\varepsilon})=H_{*-k}(B).$$

Finally, since $H_*(\Lambda^{\kappa+\varepsilon},\ \Lambda^\kappa)=0$ and $H_*(\Lambda^{\kappa-},\ \Lambda^{\kappa-\varepsilon})=0$, the exact homology sequence of the triples $(\Lambda^{\kappa+\varepsilon},\ \Lambda^\kappa,\ \Lambda^{\kappa-})$ and $(\Lambda^\kappa,\ \Lambda^{\kappa-},\ \Lambda^{\kappa-\varepsilon})$ yields the corollary. □

A consequence of (2.4.10) and (2.4.11) are the generalized Morse inequalities:

2.4.12 Theorem. *Let $\kappa_0, \kappa_1, \kappa_0 < \kappa_1$, be non-critical values of E. Assume that, for all critical values κ in the interval $[\kappa_0, \kappa_1]$ the critical set at $E^{-1}(\kappa)$ decomposes into non-degenerate $O(2)$-invariant critical submanifolds B with index $k(B)$. Denote by b_l the l-th Betti number in $\mathbb{Z}_2$-coefficients.*

Claim. *For every $m \geqslant 0$*

$$\sum_{l=0}^{m} (-1)^{m-l} b_l(\Lambda^{\kappa_1}, \Lambda^{\kappa_0}) \leqslant \sum_{B} \sum_{l=0}^{m} (-1)^{m-l} b_{l-k(B)}(B)$$

hence

$$b_m(\Lambda^{\kappa_1}, \Lambda^{\kappa_0}) \leqslant \sum_{B} b_{m-k(B)}(B).$$

Here, the sums on the right-hand side are taken over all the non-degenerate critical submanifolds B having an E-value in the interval $[\kappa_0, \kappa_1]$. For m sufficiently large, equality holds.

Proof. If there is only one critical level in $[\kappa_0, \kappa_1]$, say κ, then the theorem follows immediately from (2.4.11) since in this case

$$H_*(\Lambda^{\kappa_1}, \Lambda^{\kappa_0}) = H_*(\Lambda^{\kappa}, \Lambda^{\kappa-}).$$

Assume now that there are exactly two critical levels, say $\kappa, \kappa', \kappa < \kappa'$, in $[\kappa_0, \kappa_1]$. We will prove our theorem under this hypothesis. The general case then follows by induction on the number of critical levels.

From the exact homology sequence of the triple $(\Lambda^{\kappa_1}, \Lambda^{\kappa_{1/2}}, \Lambda^{\kappa_0})$, $\kappa_0 < \kappa < \kappa_{1/2} < \kappa' < \kappa_1$ we have, since $H_*(\Lambda^{\kappa_1}, \Lambda^{\kappa_{1/2}}) = H_*(\Lambda^{\kappa'}, \Lambda^{\kappa'-})$ and $H_*(\Lambda^{\kappa_{1/2}}, \Lambda^{\kappa_0}) = H_*(\Lambda^{\kappa}, \Lambda^{\kappa-})$,

$$\sum_{l=0}^{m} (-1)^{m-l} b_l(\Lambda^{\kappa_1}, \Lambda^{\kappa_0}) \leqslant \sum_{l=0}^{m} (-1)^{m-l} b_l(\Lambda^{\kappa'}, \Lambda^{\kappa'-}) + \sum_{l=0}^{m} (-1)^{m-l} b_l(\Lambda^{\kappa}, \Lambda^{\kappa-})$$

and we have equality as soon as m becomes $\geqslant$ the largest l for which one of the l-dimensional Betti numbers appearing in the formula does not vanish.

Now apply (2.4.11) to obtain (2.4.12). □

2.5 The Morse Complex

In this section we shall present an important refinement of the usual Morse theory as developed in (2.4) for $(\Lambda M, \langle\,,\rangle_1, E)$. The general hypothesis is that all closed geodesics on M are non-degenerate, i.e. the critical orbits S.c, $E(c)>0$, are non-degenerate critical submanifolds. It seems likely, though, that we need only the assumption that the critical set on ΛM decomposes into non-degenerate critical submanifolds. Certainly, this is true if M is a symmetric space; we shall show this in particular for the case in which M is the sphere with the standard metric.

The basic new elements in the theory are the unstable and local stable manifolds of a critical S-orbit. For each $\kappa>0$, the closure of the unstable manifolds in $\Lambda^\kappa M$ defines the Morse complex $\mathscr{M}^\kappa M$; $\mathscr{M}^\kappa M$ is the canonical finite-dimensional representation of $\Lambda^\kappa M$ insofar as $\Lambda^\kappa M$ can be retracted onto $\mathscr{M}^\kappa M$. It is invariant under the S-action $\tilde{\chi}$ and the $\mathbb{Z}_2$-action θ and under the induced action of the isometry group of M. Most important, $\mathscr{M}^\kappa M$ is tangential to the vector field grad E and grad $E|\mathscr{M}^\kappa M$ is complete.

We conclude the section with a detailed study of the Morse complex of ΛS^n.

Let c be a non-degenerate closed geodesic of multiplicity m. Then

$$z\in S\mapsto z^{\frac{1}{m}}\,.\,c\in\Lambda M$$

is an embedding of the circle as non-degenerate critical submanifold $S\,.\,c$ in the sense of (2.4). Let

$$\mu=\mu(S\,.\,c):N=N(S\,.\,c)\to S$$

be the *normal bundle of this submanifold.* μ splits into the positive and negative subbundle, $\mu=\mu^-\oplus\mu^+$, with

$$\mu^\pm=\mu^\pm(S\,.\,c):N^\pm=N^\pm(S\,.\,c)\to S.$$

Recall from (2.4) that the S-action $\tilde{\chi}$ on ΛM induces S-actions on $\mu,\mu^\pm$, which respect the splitting. Also, the map $\theta:\Lambda M\to\Lambda M$ defines a canonical identification of $\mu(S\,.\,c)$ with $\mu(S\,.\,\theta c)$.

Let $E|S\,.\,c=\kappa$, index $c=k$.

2.5.1 Theorem. *Let c be a non-degenerate closed geodesic. Then there exists an embedding*

$$W_u=W_u(S\,.\,c):(N^-(S\,.\,c),S)\to(\Lambda^\kappa M,S\,.\,c)$$

and, at least for a restriction to a sufficiently small disc bundle associated to μ^+, an embedding

$$W_s=W_s(S\,.\,c):(N^+(S\,.\,c),S)\to(\Lambda M-\Lambda^{\kappa-}M,S\,.\,c)$$

having the following properties:

(i) *W_u and W_s commute with the S-action and $\mathbb{Z}_2$ action;*

(ii) $TW_u|N_z^-: N_z^- \to T_{z.c}^- \Lambda M$ and $TW_s|N_z^+: N_z^+ \to T_{z.c}^+ \Lambda M$ *are linear isomorphisms;*

(iii) *set* $W_u|N_z^- = W_{uu}(z.c)$ *and* $W_s|N_z^+ = W_{ss}(z.c)$. *The elements* $c^* \in$ *image* $W_{uu}(z.c)$ *are characterized as those* $c^* \in \Lambda M$ *for which* $\lim_{s\to-\infty} \phi_s c^* = z.c$; *the elements* $c^* \in$ *image* $W_{ss}(z.c)$, c^* *sufficiently close to* $S.c$, *are characterized by the property* $\lim_{s\to\infty} \phi_s c^* = z.c$.

We call $W_u(S.c)$ and $W_{uu}(c)$ the *unstable* and the *strong unstable* manifold of $S.c$ and c, respectively. Similarly, $W_s(S.c)$ and $W_{ss}(c)$ are called the *stable* and the *strong stable* manifolds of $S.c$ and c, respectively.

Note. $z.W_{uu}(c) = W_{uu}(z.c)$, $z.W_{ss}(c) = W_{ss}(z.c)$, $\theta W_{uu}(c) = W_{uu}(\theta c)$, $\theta W_{ss}(c) = W_{ss}(\theta c)$.

Proof. As in (2.4), we identify, via the exponential map, the total space $D = D(S.c)$ of a sufficiently small disc bundle associated to $\mu(S.c)$ with a tubular neighborhood of $S.c$. In (1.4.13) we defined an equivariant vector field u on ΛM which is $\neq 0$ outside $\Lambda^0 M$ and is orthogonal to grad E. Moreover, $u = -\partial c$ if c is a critical point of E.

It follows that u is orthogonal to the fibre D_z of D over z at the base point z. Hence, we only need choose the diameter of D sufficiently small, to obtain n transversal to the fibres of D everywhere.

Using $u(\xi)$ as generator of a horizontal space $T_\xi^0 N$ at $\xi \in D_z(S.c)$, we have the decomposition

$$T_\xi D = T_\xi^- N \oplus T_\xi^0 N \oplus T_\xi^+ N$$

where $T_\xi^\pm N$ are the subspaces of the vertical space (= tangent space to the fibre at ξ) which are parallel to $N_z^\pm$, respectively, $z = \mu(\xi)$.

At the same time we use the vector field u to introduce local coordinates at $z_0 \in S$, a neighborhood of $z_0 \in D$ being identified with a neighborhood of the origin of $N_{z_0}^- \oplus \mathbb{R} \oplus N_{z_0}^+$.

Since $\langle \text{grad } E, u\rangle_1 = 0$, grad $E(\xi)$ at $\xi = (\xi_-, \xi_+) \in (N_z^- \oplus N_z^+) \cap D$ has the form grad $E(\xi) = (A^- . \xi_- + C^-(\xi), C^0(\xi), A^+ . \xi_+ + C^+(\xi)) \in T_\xi^- N \oplus T_\xi^0 N \oplus T_\xi^+ N$. Here, $A = A_{z.c}: T_{z.c}\Lambda \to T_{z.c}\Lambda$ is the self-adjoint operator representing $D^2 E(z.c)$, cf. (2.4.2). $A^\mp = A|T_{z.c}^\mp \Lambda$. Thus, A^- is an operator on a k-dimensional space with strictly negative eigenvalues, whereas A^+ is a positive definite operator on a proper Hilbert space.

Moreover, $\|C^*(\xi)\|_1 = o(\|\xi\|_1)$,

and

$$(*) \qquad \|C^*(\xi) - C^*(\xi')\|_1 \leq \varepsilon \|\xi - \xi'\|_1, * \in \{-, 0, +\},$$

where $\varepsilon > 0$ can be made arbitrarily small, if only $\|\xi\|_1$ and $\|\xi'\|_1$ are sufficiently small.

Now consider, for a given $\xi_+ \in N_z^+ \cap D$, the following integral equation:

$$(\dagger) \quad u(s;\xi_+) = (u_-(s;\xi_+),\, u_0(s;\xi_+),\, u_+(s;\xi_+))$$

$$= \left(\int_s^\infty e^{(t-s)A^-} \,.\, C^-(u(t;\xi_+))dt,\ \int_s^\infty C^0(u(t;\xi_+))dt,\right.$$

$$\left. e^{-sA^+} \,.\, \xi_+ + \int_s^0 e^{-(s-t)A^+} \,.\, C^+(u(t;\xi_+))dt\right).$$

Assume, for the moment, that we have a solution of (†). Then,

$$du(s;\xi_+)/ds = -\operatorname{grad} E(u(s;\xi_+))$$

$$u(0;\xi_+) = \left(-\int_0^\infty e^{tA^-} \,.\, C^-(u(t;\xi_+))dt,\ \int_0^\infty C^0(u(t;\xi_+))dt,\ \xi_+\right)$$

and

$$\lim_{s\to\infty} u(s;\xi_+) = (0,0,0).$$

That is to say, $u(s;\xi_+)$, $s \geqslant 0$, describes a trajectory of $-\operatorname{grad} E$ which has, for $s\to\infty$, the base point z of $\xi_+ \in N^+$ as limit point.

To prove the existence of a solution of (†), we generalize the proof in Coddington-Levinson [CL] of the existence of a stable manifold. We first observe that there exist $a>0$ and $K\geqslant 1$ such that $\|e^{tA^-}\|_1 \leqslant Ke^{-at}$, $\|e^{-tA^+}\|_1 \leqslant Ke^{-at}$. Choose b satisfying $0<b<a<2b$. Set $(a-b)/6K=\varepsilon$. Choose $\delta>0$ so small that (*) holds, for $\|\xi\|_1<\delta$, $\|\xi'\|_1<\delta$.

We start an iteration process for the solution of (†) with $u^0(s;\xi_+)=0$. We claim, with $u^k(s;\xi_+)$ denoting the $k-th$ iteration, that

$$(\dagger\,\dagger) \quad \|u^{n+1}(s;\xi_+) - u^n(s;\xi_+)\|_1 \leqslant K\|\xi_+\|_1 2^{-n} e^{-sb}.$$

This will clearly imply the existence of a solution $u(s;\xi_+) = \lim_{n\to\infty} u^n(s;\xi_+)$ satisfying

$$(\dagger\,\dagger\,\dagger) \quad \|u(s;\xi_+)\|_1 \leqslant 2K\|\xi_1\| e^{-sb}.$$

To prove († †), we note that it is true for $n=0$, since

$$u^1(s;\xi_+) = (0,0,e^{-sA^+}\xi^+).$$

For $n>0$ we have from (*) and the induction hypothesis that

$$\|u_-^{n+1} - u_-^n\|_1 \leqslant \int_s^\infty Ke^{-(t-s)a} \varepsilon K\|\xi_+\|_1 2^{-(n-1)} e^{-tb} dt$$

$$\leqslant \varepsilon K^2 \|\xi_+\|_1 2^{-(n-1)} e^{-sb}(a+b)^{-1} < K\|\xi_+\|_1 2^{-n} e^{-sb}/3,$$

where we use $\varepsilon K = (a-b)\,(a+b)^{-1}/6 < 1/6$.

Similarly,

$$\|u_0^{n+1}-u_0^n\|_1 \leqslant \int_s^\infty \varepsilon K\|\xi_+\|_1 3^{-(n-1)}e^{-tb}dt$$

$$=(a-b)\|\xi_+\|_1 2^{-n}e^{-sb}/3b < K\|\xi_+\|_1 2^{-n}e^{-sb}/3$$

and

$$\|u_+^{n+1}-u_+^n\|_1 \leqslant \int_0^s Ke^{-(s-t)a}\varepsilon K\|\xi_+\|_1 2^{-(n-1)}e^{-tb}dt$$

$$\leqslant \varepsilon K^2\|\xi_+\|_1 2^{-(n-1)}e^{-sa}\big(e^{s(a-b)}-1/(a-b)\big) < K\|\xi_+\|_1 2^{-n}e^{-sb}/3.$$

By the same procedure we can prove, for a given $\xi_-\in N_z^-\cap D$, the existence of a solution of the integral equation

$$v(s;\xi_-)=\big(v_-(s;\xi_-),\, v_0(s;\xi_-),\, v_+(s;\xi_-)\big)=(e^{-sA^-}\,.\,\xi_-$$

$$-\int_0^s e^{-(s-t)A^-}\,.\,C^-\big(v(t;\xi_-)\big)dt,\ -\int_{-\infty}^s C^0\big(v(t;\xi_-)\big)dt,$$

$$-\int_{-\infty}^s e^{(t-s)A^+}\,.\,C^+\big(v(t;\xi_-)\big)dt).$$

Then,

$$dv(s;\xi_-)/ds = -\operatorname{grad} E\big(v(s;\xi_-)\big)$$

$$v(0;\xi_-)=(\xi_-, -\int_{-\infty}^0 C^0\big(v(t;\xi_-)\big)dt, -\int_{-\infty}^0 e^{tA^+}\,.\,C^+\big(v(t;\xi_-)\big)dt)$$

and

$$\lim_{s\to-\infty} v(s;\xi_-)=(0,0,0)$$

We now define, for $\xi^\pm\in D\cap N_z^\pm$,

$$W_{ss}(z\,.\,c)\,(\xi_+)=u(0;\xi_+);\ W_{uu}(z\,.\,c)\,(\xi_-)=v(0;\xi_-).$$

These are the strong stable and strong unstable manifolds in a neighborhood of the origin of the fibre.

Clearly, $z\,.\,W_{ss}(c)=W_{ss}(z\,.\,c)$ and $z\,.\,W_{uu}(c)=W_{uu}(z\,.\,c)$. Thus, we define $W_s(S\,.\,c)$ and $W_u(S\,.\,c)$ as the union of the $W_{ss}(z\,.\,c)$, $z\in S$, and $W_{uu}(z\,.\,c)$, $z\in S$, respectively. By taking the complete ϕ_s trajectories on $W_u(S\,.\,c)$, we get the global unstable manifold.

It remains to show that $W_s(S\,.\,c)$ and $W_u(S\,.\,c)$ are differentiable immersions. This is clear from our definitions, cf. [CL].

In order to characterize the $c^*\in W_s(S\,.\,c)$ by the property $\lim_{s\to\infty}\phi_s c\in S\,.\,c$, we proceed as follows. For c^* sufficiently close to $S\,.\,c$, c^* can be represented as $u(0;\xi_+)+v(0;\xi_-)\in D(S\,.\,c)$, where $\xi_+\in N_z^+$ and $\xi_-\in N_z^-$, need not have the same base point. From the integral equation for $v(s;\xi_-)$ we get constants $K'>0$, $b'>0$ such that

$$\|v(s;\xi_-)\|_1 \geqslant K'\|\xi_-\|_1 e^{sb'}.$$

Thus, $\|v(s;\xi_-)\|_1$ can become arbitrarily small for a sequence $\{s_i\}$ going to ∞ only if $\xi_-=0$, i.e. $v(s;\xi_-)\equiv 0$, i.e. $c^*\in W_s(S\,.\,c)$.

This concludes the proof of (2.5.1). □

Remark. A particular consequence of (2.5.1) is the following. If a trajectory $\phi_s c^*$, $s\geqslant 0$, has among its limit points a non-degenerate closed geodesic c, then $\lim_{s\to\infty}\phi_s c^*=c$. This was proved already by Eliasson [El 4].

Now assume that all closed geodesics on M are non-degenerate. Then we have, for each critical orbit $S\,.\,c$, $E(c)=\kappa>0$, the unstable manifold

$$W_u(S\,.\,c)\colon(N^-(S\,.\,c),\,S)\to(\Lambda^\kappa M,\,S\,.\,c)$$

where $N^-(S\,.\,c)$ is the total space of a k-dimensional vector bundle over S.

We already know from (2.4) that $\Lambda^\kappa M$ is obtained from $\Lambda^{\kappa-\varepsilon}M$, $\varepsilon>0$ sufficiently small, by attaching the negative bundles over each of the critical orbits $S\,.\,c_j$, $S\,.\,\theta c_j$, $1\leqslant j\leqslant k$, at the critical level κ.

We now elaborate on this by introducing, for $\kappa^*\geqslant 0$, the *Morse complex* $\mathcal{M}^{\kappa^*}M$ of ΛM. $\mathcal{M}^{\kappa^*}M$ will be the closure of the unstable manifolds $W_u(S\,.\,c)$ of the critical orbits $S\,.\,c$, $E(c)\leqslant\kappa^*$.

Here, $\mathcal{M}^0M$ is equal to M since, for $E(c)=0$, $W_u(S\,.\,c)=c$. If κ_0 is the smallest positive critical value, $\mathcal{M}^{\kappa_0}M$ consists of M to which the finitely many pairs $W_u(S\,.\,c_0)$ and $W_u(S\,.\,\theta c_0)$ of disc bundles with $E(c_0)=\kappa_0$ are attached with their boundaries. In the case that index $c_0=0$, $W_u(S\,.\,c_0)\cong S\cdot c_0$; thus, $\mathcal{M}^{\kappa_0}M$ need not be connected even if ΛM is connected, and the same is true for $\mathcal{M}^{\kappa}M$ with arbitrary κ.

In general, if κ' is a positive critical value, $\mathcal{M}^{\kappa'}M$ is obtained from $\mathcal{M}^{\kappa}M$, where $\kappa<\kappa'$ is the maximal critical value $<\kappa'$, by attaching the finitely many disc bundles $W_u(S\,.\,c')$ and $W_u(S\,.\,\theta c')$ with $E|S\,.\,c'\cup S\,.\,\theta c'=\kappa'$. If κ^* is arbitrary, we define $\mathcal{M}^{\kappa^*}M$ to be equal to $\mathcal{M}^{\kappa}M$ where κ is the greatest critical value $\leqslant\kappa^*$.

The union of the $\mathcal{M}^{\kappa^*}M$, $\kappa^*\to\infty$, is also called the *full Morse complex*, $\mathcal{M}M$, of ΛM.

2.5.2 Lemma. *The Morse complex $\mathcal{M}^{\kappa}M$ contains only complete ϕ-trajectories $\phi_s c^*$, $s\in\mathbb{R}$. These trajectories have well-defined limit points $\phi_\infty c^*$ and $\phi_{-\infty}c^*$ which are critical points and also belong to $\mathcal{M}^{\kappa}M$. Thus, we have a $\mathbb{R}$-action on $\mathcal{M}^{\kappa}M$ given by*

$$\mathbb{R}\times\mathcal{M}^{\kappa}M\to\mathcal{M}^{\kappa}M,\ (s,\,c^*)\mapsto\phi_s c^*.$$

In addition, $\mathcal{M}^{\kappa}M$ is closed under the S-action $\tilde{\chi}$ and the $\mathbb{Z}_2$-action θ as well as under the induced action of the isometry group of M.

The proof follows immediately from the definitions and constructions which were all made equivariantly. □

The previous arguments apply only to manifolds M for which the closed geodesics all are non-degenerate. There are interesting examples, however, for which this hypothesis is not satisfied, in particular, the irreducible symmetric spaces. Ziller [Zi] has shown that for such a manifold M the critical set on ΛM decomposes into non-degenerate critical submanifolds.

Here we want to consider only the sphere S^n with the standard metric of constant curvature equal to 1. The homology of ΛS^n has been studied by various authors, see in particular, Morse [Mor 2], Eells [Ee 1], Švarc [Šv], Eliasson [El 2].

2.5.3 Proposition. *The critical set $Cr\Lambda S^n$ decomposes into the non-degenerate critical submanifolds $\Lambda^0 S^n \cong S^n$ and $B_q = B_q S^n$, consisting of the q-fold covered great circles, $q=1,2,\dots$. B_q is isomorphic to the Stiefel manifold $V(2,n-1)$ of orthonormal 2-frames in $\mathbb{R}^{n+1}$; in particular,* $\dim B_q = 2n-1$. *$E|B_q = 2\pi^2 q^2$, index $B_q = (2q-1)(n-1)$.*

Proof. The special form of the curvature tensor of S^n yields, for $c \in B_q$, i.e. $|\dot c| = 2\pi q$,

$$K_c^{\sim}(\xi)(t) = R(\xi(t), \dot c(t), \dot c(t)) =$$

$$-\langle \dot c(t), \xi(t)\rangle \dot c(t) + \langle \dot c(t), \dot c(t)\rangle \xi(t) =$$

$$-\langle \dot c(t), \xi(t)\rangle \dot c(t) + 4\pi^2 q^2 \xi(t).$$

Thus, $\lambda = 1$ is not eigenvalue. With this, formula (2.4.4) is

$$(*) \qquad \nabla^2 \xi + \frac{4\pi^2 q^2 + \lambda}{1-\lambda}\xi + \frac{1}{\lambda - 1}\langle \dot c, \xi\rangle \dot c = 0.$$

We decompose ξ into the subset of tangential vectors, $\xi(t) = \alpha(t)\dot c(t)$, and vertical vectors, $\xi(t) \perp \dot c(t)$. Then (*) decomposes into

$$\text{(tan)} \quad \ddot\alpha(t) + \lambda\alpha(t)/(1-\lambda) = 0,$$

$$\text{(ver)} \quad \nabla^2\xi + (4\pi^2 q^2 + \lambda)\xi/(1-\lambda) = 0.$$

For $\lambda < 0$, (tan) has no periodic solutions. The non-trivial solutions of (ver) occur for

$$\lambda = 4\pi^2(p^2 - q^2)/(1 + 4\pi^2 p^2), \quad p = 0,1,\dots, q-1.$$

For $p=0$, $\xi(t) = \xi_0$, $\xi_0 \perp \dot c$ are the solutions. For $0 < p < q$,

$$\xi(t) = \xi_0 \cos 2\pi p t + \xi_1 \sin 2\pi p t,$$

$\xi_0, \xi_1 \perp \dot c$, are the solutions. The dimension of $T_c^- \Lambda S^n$ for $c \in B_q$ therefore becomes

$$(n-1) + (q-1)2(n-1) = (2q-1)(n-1).$$

In the case $\lambda=0$, (tan) has the solution $\alpha=\alpha_0$ and (ver) has the solution

$$\xi(t)=\xi_0 \cos 2\pi qt+\xi_1 \sin 2\pi qt$$

$\xi_0, \xi_1 \perp \dot{c}$. Hence, nullity $B_q=2n-1=\dim B_q$, i.e., B_q is non-degenerate. □

We can now easily determine the $\mathbb{Z}_2$-homology of ΛS^n, cf. [Kl 4].

2.5.4 Theorem. *Let $n>2$. Then the inclusion homomorphism in $\mathbb{Z}_2$-homology*

$$i_*: H_*(\Lambda^{2q^2\pi^2-} S^n, \Lambda^0 S^n) \to H_*(\Lambda^{2q^2\pi^2} S^n, \Lambda^0 S^n)$$

is injective. Hence

$$H_*(\Lambda S^n, \Lambda^0 S^n) = \bigoplus_{q\geq 1} H_{*-(2q-1)(n-1)}(V(2,n-1)).$$

Note. Ziller [Zi] proved the corresponding theorem for arbitrary global symmetric spaces. We restrict ourselves to the case $n\geqslant 4$; the cases $n=2$ and $n=3$ must be handled separately, see Švarc [Šv], Eliasson [El 2] and Klein [Kle].

Proof. We claim that none of the cycles of $\Lambda^{2q\pi-} S^n$ is killed by attaching the negative bundle μ_q^- over B_q.

To see this, first observe that the $\mathbb{Z}_2$-homology of $V(2,n-1) \cong B_q$ can be represented by the following cycles u_i, $\dim u_i = i$,
u_0 = a single parameterized great circle S^1;
u_{n-1} = all S^1 which start at a fixed point $p_0 \in S^n$;
u_n = all S^1 which pass through a fixed point $p_0 \in S^n$;
u_{2n-1} = all S^1 on S^n.

Now consider the negative bundle $\mu_{q'}^-$, over $B_{q'}$, $q'<q$. The difference between the fibre dimensions of μ_q^- and $\mu_{q'}^-$ is $2(q-q')(n-1)$. If $\mu_q^-|u_j$ kills $\mu_{q'}^-|u_i$, this implies that their dimensions differ by 1,

$$2(q-q')(n-1)+j-i=1;\ i,j\in\{0,n-1,n,2n-1\}.$$

One easily verifies that this implies that $i>0$ and

$$2(q-q')(n-1)\in\{2,n,n+1,2n-1\}.$$

But this cannot hold if $n>3$.

Thus we have proved the theorem. □

For the construction of a Morse complex of ΛS^n we consider the canonical embedding $i_q: B_q \to B_q \subset \Lambda S^n$ of the critical submanifold B_q. This induces the normal bundle over B_q

$$\mu_q = \mu(B_q): N(B_q) \to B_q.$$

The decomposition of the fibres $N_{c_q}(B_q)$ over c_q into the positive and negative eigenspaces $N^+_{c_q}(B_q)$ and $N^-_q(B_q)$ gives the subbundles

$$\mu_q^\pm = \mu^\pm(B_q): N^\pm(B_q) \to B_q.$$

Here, $\mu^-(B_q)$ has fibre dimension equal to index B_q, i.e. $(2q-1)(n-1)$, whereas the fibre of $\mu^+(B_q)$ is a proper Hilbert space.

Recall that B_q is isomorphic to the Stiefel manifold $V(2,n-1) = SO(n+1)/SO(n-1)$. The isometric action of $SO(n+1)$ on S^n extends to an isometric action on ΛS^n, leaving the function E invariant. This is a consequence of the intrinsic definition of ΛM with its Riemannian metric and E.

In particular, the bundles $\mu(B_q)$, $\mu^\pm(B_q)$ are carried into themselves by the action of $SO(n+1)$.

We use this to prove for each of the non-degenerate critical submanifolds the existence of stable and unstable manifolds; set $E|B_q = 2\pi^2 q^2 = \kappa(q)$.

2.5.5 Theorem. *For every $q = 1,2,\dots$, there exists an injective immersion*

$$W_u(B_q): (N^-(B_q), B_q) \to (\Lambda^{\kappa(q)}, B_q)$$

and, at least on a certain neighborhood of the base space, an injective immersion

$$W_s(B_q): (N^+(B_q), B_q) \to (\Lambda - \Lambda^{\kappa(q)-}, B_q)$$

having the following properties:

(i) *the immersions are equivariant with respect to the S-action $\chi^\sim$ and the $\mathbb{Z}_2$-action θ as well as with respect to the $SO(n+1)$-action;*

(ii) *when restricted to the base manifold B_q, these immersions are the canonical embeddings;*

(iii) *for $c_q \in B_q$, the differential $DW_u(B_q)(c_q)|N^-_{c_q}(B_q)$ is the canonical map from $N^-_{c_q}(B_q)$ onto the subspace $T^-_{c_q}\Lambda S^n$, spanned by the eigenvectors of $D^2E(c_q)$ with negative eigenvalue. The corresponding statement is true for $DW_s(B_q)(c_q)|N^+_{c_q}(B_q)$;*

(iv) *the elements $c^* \in$ image $W_u(B_q)$ are characterized by the property $\lim_{s\to-\infty} \phi_s c^* \in B_q$. The elements $c^* \in$ image $W_s(B_q)$ are characterized by the property $\lim_{s\to\infty} \phi_s c^* \in B_q$.*

Remark. We call $W_s(B_q)$ and $W_u(B_q)$ the *stable* and *unstable* manifolds of the non-degenerate critical submanifold B_q, respectively.

$$W_{ss}(c_q) := W_s(B_q)|N^+_{c_q}(B_q), \quad \text{and} \quad W_{uu}(c_q) := W_u(B_q)|N^-_{c_q}(B_q)$$

are called *strong stable* and *strong unstable* manifolds at $c_q \in B_q$, respectively.

Proof. Let $\delta_q: D(B_q) \to B_q$ be a disc bundle over B_q, associated with μ_q. For sufficiently small discs, we can identify $D(B_q)$ with a tubular neighborhood of B_q in ΛM, via the exponential map. This identification gives a Riemannian metric $\langle\ ,\ \rangle_1$ on $D(B_q)$ as well as the function E and the corresponding gradient field, grad E.

The group $SO(n+1)$ carries the vector field grad E into itself. In particular, the isotropy group $SO(n-1)$ of a point c_q on the base B_q leaves the fibre D_{c_q} over c_q in $D(B_q)$ invariant. Since $SO(n+1)$ is operating transitively on B_q, the tangent space of the $SO(n+1)$-orbit through $\xi \in D_{c_q}$ is transversal to $T_\xi D_{c_q} \cong N_{c_q}$.

Thus, the tangent space $T_\xi SO(n+1)$ to the $SO(n+1)$-orbit through $\xi \in D_{c_q}$ contains a "horizontal" subspace $T_\xi^0 D(B_q)$ which is complementary to the "vertical" space $T_\xi D_{c_q} \cong N_{c_q}$. We fix $T_\xi^0 D(B_q)$ by the condition that it should be orthogonal to $T_\xi SO(n+1) \cap T_\xi D_{c_q}$. By $T_\xi^\pm N(B_q)$, we denote the subspaces of $T_\xi D_{c_q}$ which are parallel to $N_{c_q}^\pm$. Thus, we can identify

$$T_\xi N(B_q) = T_\xi^- N(B_q) \oplus T_\xi^0 N(B_q) \oplus T_\xi^+ (B_q)$$

componentwise with

$$T_{c_q} N(B_q) = N_{c_q}^- (B_q) \oplus T_{c_q} B_q \oplus N_{c_q}^\pm (B_q),$$

where, for the identification of $T_\xi^0 N(B_q)$ with $T_{c_q} B_q$, we use the bundle projection $\mu(B_q)$.

We also use this decomposition for local coordinates near c_q.

The operation of $SO(n+1)$ on $D(B_q)$ is isometric and preserves E. Thus, grad E is orthogonal to $T_\xi^0 N(B_q)$. For $\xi = 0$, $T_\xi^0 N(B_q)$ coincides with $T_{c_q} B_q$. Therefore, grad $E(\xi)$ possesses a decomposition of the following form, with

$$\xi = (\xi_-, \xi_0, \xi_+) \in N_{c_q}^- (B_q) \oplus T_{c_q} B_q \oplus N_{c_q}^+ (B_q):$$

$$\text{grad } E(\xi) = (A^- . \xi_- + C^-(\xi), C^0(\xi), A^+ . \xi_+ + C^+(\xi)) \in$$

$$N_{c_q}^- (B_q) \oplus T_{c_q} B_q \oplus N_{c_q}^+ (B_q).$$

Here, $A^\pm$ is the restriction of the self-adjoint operator A_{c_q} (2.4.2) to $T_{c_q}^\pm \Lambda M \cong N_{c_q}^\pm (B_q)$. The $C^*(\xi)$, $* \in \{-, 0, +\}$, satisfy

$$\|C^*(\xi)\|_1 = o(\|\xi\|_1)$$

$$(*) \qquad \|C^*(\xi) - C^*(\xi')\|_1 \leq \varepsilon \|\xi - \xi'\|_1$$

where $\varepsilon > 0$ can be made arbitrarily small, as long as $\|\xi\|_1$, $\|\xi'\|_1$ are sufficiently small.

These are exactly the same hypotheses as in the proof of (2.5.1). Hence, we can draw the same conclusions. □

Remark. The existence of stable and unstable manifolds can be proved quite generally for any non-degenerate critical submanifold of ΛM. This was shown by Duistermaat, who uses a refinement of a technique of Perron [Pe]. Duistermaat's result is unpublished. There is a manuscript by him entitled "Stable manifolds" from August 1972.

In any case, we now can define the Morse complex $\mathcal{M}^\kappa S^n$ of ΛS^n as above as the closure of the unstable manifolds in $\Lambda^\kappa S^n$.

For later application we include here a few facts about the loop space $(\Omega M,*)$ of a compact Riemannian manifold M with base point $*$, and, in particular, about the stable and unstable manifolds of $(\Omega S^n, *)$. It was Palais [Pa 2] who first employed the theory of Hilbert manifolds with a function satisfying condition (C) for the study of ΩM.

Recall, from (2.1.4), that in the bundle

$$\gamma:\Lambda M\to M; \qquad c\mapsto c(0)=c(1).$$

γ is a differentiable submersion; cf. also Grove [Gro 1]. The fibre $(\Omega M,*)$ over a point $*\in M$ consists of the H^1-maps $c:([0,1],\{0,1\})\to(M,*)$. The function $E^\Omega=E|\Omega M$ is differentiable, and the critical points of E^Ω are precisely the geodesics from $*$ to $*$, parameterized from 0 to 1.

$(\Omega M,*)$ also carries a canonical Riemannian metric $\langle\ ,\ \rangle_1$, induced from the metric on ΛM. The tangent space $T_c\Omega M$ at $c\in\Omega M$ consists of the H^1-vector fields $\xi(t)$, $t\in[0,1]$, along c, vanishing at $t=0$ and $t=1$. The associated gradient field grad E^Ω satisfies condition (C).

Let $c\in(\Omega M,*)$ be a critical point. Denote by $T_c^\pm\Omega M$ the subspace of the tangent space $T_c\Omega M$, spanned by the eigenvectors with positive and negative eigenvalues of $D^2E^\Omega(c)$, respectively. dim $T_c^-\Omega M$ is always finite; it is called the index of c in $(\Omega M,*)$. c is called non-degenerate if $D^2E^\Omega(c)$ has no eigenvalue equal to zero.

Note that, for a closed geodesic c in ΛM with $c(0)=c(1)=*$, the index of c as a critical point of E is $\geqslant$ the index of c as a critical point of E^Ω.

We also have the concept of a non-degenerate critical submanifold of $(\Omega M,*)$: This is a closed connected submannifold B consisting entirely of critical points such that its tangent space T_cB coincides with the null space of $D^2E^\Omega(c)$, for all $c\in B$.

As an important example we consider $M=S^n=$the sphere of constant curvature equal to 1. The critical points of $(\Omega S^n,*)$ are, besides the trivial geodesic $c=*$, the q-fold covered great circles through $*$, $q=1,2,\dots$. The index of a q-fold covered great circle, considered as critical point of $(\Omega S^n,*)$, is $(2q-1)(n-1)$.

The canonical isometric action of $SO(n)$ on $(S^n,*)$ induces an isometric action on $(\Omega S^n,*)$, leaving E^Ω invariant. The $(n-1)$-parameter family of q-fold covered great circles through $*\in S^n$ forms a non-degenerate critical submanifold B_q^Ω, isomorphic to S^{n-1}, of index $(2q-1)(n-1)$. $SO(n)$ operates transitively on B_q^Ω.

Let

$$\mu_q^\Omega=\mu(B_q^\Omega):N(B_q^\Omega)\to B_q^\Omega$$

be the normal bundle of the submanifold B_q^Ω in $(\Omega S^n,*)$. Denote by

$$\mu_q^{\Omega\pm}:N^\pm(B_q^\Omega)\to B_q^\Omega$$

the positive and negative subbundle of μ_q^Ω, spanned by the eigenvectors of D^2E^Ω with positive and negative eigenvalues, respectively. The $SO(n)$-action on $(\Omega S^n, *)$ induces an $SO(n)$-action on

$$\mu_q^\Omega = \mu_q^{\Omega-} \oplus \mu_q^{\Omega+}$$

which leaves this splitting invariant.

Using exactly the same methods as for the proof of (2.5.5) one shows, with $\kappa(q) = E^\Omega | B_q^\Omega = 2\pi^2 q^2$, that:

2.5.6 Theorem. *For every $q = 1, 2, \ldots$, there exists an injective immersion*

$$W_u(B_q^\Omega): (N^-(B_q^\Omega), B_q^\Omega) \to (\Omega^{\kappa(q)} S^n, B_q^\Omega)$$

and, at least on a certain neighborhood of the base space, an injective immersion

$$W_s(B_q^\Omega): (N^+(B_q^\Omega), B_q^\Omega) \to (\Omega S^n - \Omega^{\kappa(q)-} S^n, B_q^\Omega)$$

having the following properties:

(i) *the immersions are equivariant with respect to the $SO(n)$-action;*

(ii) *when restricted to the base manifold B_q^Ω, these are the canonical embeddings;*

(iii) *with $c \in B_q^\Omega$, we have*

$$DW_u(B_q^\Omega) | N_c^-(B_q^\Omega): N_c^-(B_q^\Omega) \to T_c^- \Omega S^n, \quad \textit{and}$$

$$DW_s(B_q^\Omega) | N_c^+(B_q^\Omega): N_c^+(B_q^\Omega) \to T_c^+ \Omega S^n$$

are isomorphisms;

(iv) *the elements of $c^* \in$ image $W_u(B_q^\Omega)$ are characterized by the property that $\lim_{s \to -\infty} \phi_s c^* \in B_q^\Omega$, and the elements $c^* \in$ image $W_s(B_q^\Omega)$, sufficiently near B_q, are characterized by the property $\lim_{s \to +\infty} \phi_s c^* \in B_q$.*

Chapter 3. The Geodesic Flow

In this chapter we are going to introduce a new aspect of a closed geodesic. Whereas in the previous chapters a closed geodesic was considered as a closed curve distinguished in the space of all closed curves by a certain property (i.e. being a critical value of a functional), we are now going to view a closed geodesic (or rather the tangent vector field along a closed geodesic) as a periodic orbit in the geodesic flow on the tangent bundle of a Riemannian manifold.

The geodesic flow is a special case of a Hamiltonian flow. This observation will put at our disposal the extensive theory of Hamiltonian systems, with particular attention being paid to periodic orbits in such a system.

In section one we give a brief account of the theory of Hamiltonian systems with special emphasis on the geodesic flow. The second section is devoted to the index theorem, whereas in section three we study generic properties of the geodesic flow. A major point here is the Birkhoff-Lewis Fixed Point Theorem. In an appendix we present Moser's proof of this theorem for the differentiable case.

3.1 Hamiltonian Systems

In this section we recollect the basic theory of Hamiltonian systems and then introduce the geodesic flow as a particular system of this sort.

A *symplectic manifold* is an even-dimensional manifold N endowed with a closed 2-form α of maximal rank, i.e. $d\alpha=0$, $\alpha^n \neq 0$ if $\dim N=2n$.

Darboux's Lemma implies the existence of a *symplectic atlas* on a symplectic manifold (N, α). That is to say, in the charts (ϕ, U) of such an atlas with $(x^1,\ldots,x^n, y^1,\ldots,y^n)$ as coordinates in $\phi(U)$, α is represented by the canonical symplectic form over $\mathbb{R}^{2n}$:

$$\alpha=\phi^*\left(\sum_1^n dx^i \wedge dy^i\right).$$

It follows that the coordinate transformations $\phi' \circ \phi^{-1}$ are *canonical* or *symplectic* transformations. That is to say, if $\phi' \circ \phi^{-1}$ is represented by

$$x'^i=x'^i(x, y) \qquad y'^i=y'^i(x, y)$$

then $\sum_i dx'^i \wedge dy'^i = \sum_i dx^i \wedge dy^i$, i.e. the functional matrix

$$d(\phi' \circ \phi^{-1}) = \begin{pmatrix} \dfrac{\partial x'^i}{\partial x^j} & \dfrac{\partial x'^i}{\partial y^j} \\ \dfrac{\partial y'^i}{\partial x^j} & \dfrac{\partial y'^i}{\partial y^j} \end{pmatrix}$$

is symplectic:

$$d(\phi' \circ \phi^{-1}) \circ J \circ {}^t d(\phi' \circ \phi^{-1}) = J$$

where

$$J = \begin{pmatrix} 0_n & E_n \\ -E_n & 0_n \end{pmatrix}.$$

The most important example of the natural occurence of a symplectic manifold is the cotangent bundle:

3.1.1 Proposition. *The cotangent bundle T^*M of a differentiable (Euclidean) manifold carries a canonical symplectic structure α which is defined as follows.*

Consider the diagram of bundle maps

$$\begin{array}{ccc} TT^*M & \xrightarrow{\tau_{T^*M}} & T^*M \\ {\scriptstyle T\tau_M^*}\downarrow & & \downarrow{\scriptstyle \tau_M^*} \\ TM & \xrightarrow[\tau_M]{} & M \end{array}$$

Define a 1-form θ on T^*M by $\theta\colon T_\omega T^*M \to \mathbb{R}$; $\xi \mapsto \omega . T\tau_M^* \xi$ and define α by $-d\theta$.

Proof. In a local representation this diagram has the form

$$\begin{array}{ccc} (x, y, \xi, \eta) & \longrightarrow & (x, y) \\ \downarrow & & \downarrow \\ (x, \xi) & \longrightarrow & (x) \end{array}$$

Here we write, as usual, x^i instead of $x^i \circ \tau_M^*$ for the coordinate on T^*M. Now θ is represented by $\sum_i y^i dx^i$. Hence, α is represented by $\sum_i dx^i \wedge dy^i$, i.e. a closed 2-form of maximal rank. $\square$

Note that the natural atlas on T^*M is a symplectic (or canonical) atlas for the symplectic form $\alpha = -d\theta$.

A *Hamiltonian system* (N, α, H) is a symplectic manifold (N, α) endowed with a differentiable function $H: N \to \mathbb{R}$, the so-called *Hamilton function*.

The associated *Hamiltonian vector field* ζ_H is defined by

$$i_{\zeta_H} . \alpha = dH \quad \text{or} \quad \alpha_*(\zeta_H) = \tfrac{1}{2} dH.$$

Here i_ζ denotes the interior multiplication, i.e. $i_\zeta.\alpha(\eta) = 2\alpha(\zeta, \eta)$. On the other hand,

$$\alpha_* : TN \to T^*N$$

denotes the vector bundle isomorphism determined by the non-degenerate 2-form α, i.e. $\alpha_* | T_pN : T_pN \to T_p^*N$ maps a vector ζ into the linear form $\alpha(\zeta, \ )$.

Assume now that α is represented in symplectic coordinates (x^i, y^i). From

$$\sum_i \xi_H^i dy^i - \sum_i \eta_H^i dx^i = \sum_i H_{x^i} dx^i + \sum_i H_{y^i} dy^i$$

we then have ζ_H represented by $(H_{y^i}, -H_{x^i})$. That is to say, we obtain, in these coordinates for the *Hamiltonian flow* determined by ζ_H, the equations

$$\frac{dx^i}{dt} = H_{y^i}, \quad \frac{dy^i}{dt} = -H_{x^i}$$

i.e. the so-called *Hamiltonian equations*.

The integral curves of the Hamiltonian vector field which start at $p \in N$ will be denoted by $\phi_t p$.

3.1.2 Proposition. *Let (N, α, H) be a Hamiltonian system. Then $H = const$ along the flow lines and α is invariant under the flow.*

Proof. Recall the following formula for the Lie derivative, cf. [AM], [La].

$$L_\zeta \beta = d(i_\zeta . \beta) + i_\zeta . d\beta.$$

Since $i_{\zeta_H} . \alpha = dH$ it follows that

$$dH . \zeta_H = 2\alpha(\zeta_H, \zeta_H)$$

$$L_{\zeta_H} . \alpha = ddH + i_{\zeta_H} . d\alpha = 0. \quad \square$$

We now come to the case in which we are primarily interested. Let M be a Riemannian manifold and denote its scalar product by $g(\ ,\ )$ or $\langle\ ,\ \rangle$. Let (T^*M, α) be its cotangent bundle endowed with the canonical 2-form α. We define as Hamilton function the *kinetic energy:*

$H^* : T^*M \to \mathbb{R}$, $\omega \mapsto \frac{1}{2} \langle \omega, \omega \rangle^*$, where $\langle\ ,\ \rangle^*$ denotes the metric dual to $\langle\ ,\ \rangle$.

In the natural coordinates (x, y) of T^*M, coming from coordinates (x) on M, let $g^{ik}(x)$ be the fundamental tensor of g^*. Then

$$H^*(x, y) = \tfrac{1}{2} \sum_{k,l} g^{kl}(x) y^k y^l.$$

Hence, the Hamiltonian equations become

$$\frac{dx^i}{dt} = H^*_{y^i} = \sum_l g^{il}(x)\, y^l; \qquad \frac{dy^i}{dt} = -H^*_{x^i} = -\tfrac{1}{2} \sum_{k,l} \frac{\partial g^{kl}(x)}{\partial x^i}\, y^k y^l.$$

The Riemannian metric $g = \langle\ ,\ \rangle$ on M defines a bundle isomorphism

$$g_* : TM \to T^*M, \ X \to \langle X,\ \rangle.$$

Thus, the symplectic form α on T^*M is carried into a symplectic form on TM which we denote again by α. That is to say, if Z and Z' are elements of $T_X TM$ then we have

$$\alpha(Z, Z') = \alpha(Dg_* Z, Dg_* Z').$$

3.1.3 Proposition. *The symplectic form α on TM can be described by*

$$2\alpha(Z, Z') = \langle Z_h, Z'_v \rangle - \langle Z'_h, Z_v \rangle$$

where $Z = (Z_h, Z_v)$ and $Z' = (Z'_h, Z'_v)$ are the decompositions into the horizontal and vertical components.

Note. Here we have identified $T_{Xh}TM$ with $T_{\tau X}M$ and $T_{Xv}TM$ with $T_{\tau X}M$ via the linear bijections $T\tau | T_{Xh}TM$ and $K | T_{Xv}TM$.

Proof. Both sides are defined in an invariant manner. Thus it suffices to verify (3.1.3) in special coordinates.

As such coordinates we take orthonormal coordinates based at $p = \tau_M X$. Then we have

$$x^i(p) = 0, \ g_{ik}(x(p)) = \delta_{ik}; \qquad \Gamma^l_{ik}(x(p)) = 0.$$

The local representation of the diagram

$$\begin{array}{ccc} TTM & \xrightarrow{Tg_*} & TT^*M \\ \downarrow{\scriptstyle \tau_{TM}} & & \downarrow{\scriptstyle \tau_{T^*M}} \\ TM & \xrightarrow{g_*} & T^*M \end{array}$$

at $X \in T_p X$ is then

$$\begin{array}{ccc} (0, \dot{x}, \xi, \dot{\xi}) & \xrightarrow{\text{id}} & (0, y, \xi, \eta) = (0, \dot{x}, \xi, \dot{\xi}) \\ \downarrow & & \downarrow \\ (0, \dot{x}) & \xrightarrow{\text{id}} & (0, y) = (0, \dot{x}) \end{array}$$

Moreover, the horizontal and vertical subspaces of X are represented by $\{(0, \dot{x})\} \times \mathbb{R}^n \times \{0\}$ and $\{(0, \dot{x})\} \times \{0\} \times \mathbb{R}^n$, respectively.

Since α is represented by $\sum_i dx^i \wedge dy^i$, we have

$$2\alpha(Z, Z') = \sum_i (\xi^i \eta'^i - \xi'^i \eta^i) = \langle Z_h, Z'_v \rangle - \langle Z'_h, Z_v \rangle. \quad \square$$

Note. The natural coordinates $(x, \dot{x})$ on TM, induced from coordinates (x) on M are generally not symplectic coordinates for the induced symplectic form α. Indeed, g_* is represented by

$$(x^i, \dot{x}^i) \mapsto (x^i, y^i) = \left(x^i, \sum_k g_{ik}(x) \dot{x}^k\right).$$

Thus, $\sum_i dx^i \wedge dy^i$ becomes $\sum_i dx^i \wedge d\left(\sum_k g_{ik}(x) \dot{x}^k\right)$ which, in general, will not be equal to $\sum_i dx^i \wedge d\dot{x}^i$.

We now introduce the *geodesic spray* on the tangent bundle TM of a Riemannian manifold M to be the vector field

$$S: TM \to TTM; \qquad X \mapsto (X, 0) \in T_{Xh} TM \oplus T_{Xv} TM.$$

The integral curve of S, starting at $X \in TM$, will be denoted by $\phi_t X$. The map

$$\phi_t: TM \to TM$$

is called the *geodesic flow* on TM.

3.1.4 Proposition. *If $\phi_t X_0$ is an integral curve of the geodesic spray starting at X_0 then $c(t) = \tau_M \circ \phi_t X_0$ is the geodesic determined by $\dot{c}(0) = X_0$.*

Conversely, if $c(t)$ is a geodesic, then $\dot{c}(t) = \phi_t \dot{c}(0)$.

Proof. $\dot{c}(t) = T\tau \dfrac{d}{dt} \phi_t X_0 = T\tau(\phi_t X_0, 0) = \phi_t X_0.$

$$\nabla \dot{c}(t) = K \frac{d}{dt} \dot{c}(t) = K(\phi_t X_0, 0) = 0,$$

i.e. $\quad \tau \circ \phi_t X_0 \quad$ is geodesic. Conversely, assume

$$\nabla \dot{c}(t) = 0; \qquad \text{so} \qquad \frac{d}{dt} \dot{c}(t) = (\dot{c}(t), 0) = S(\dot{c}(t)),$$

i.e. $\dot{c}(t)$ is an integral curve of S. $\square$

3.1.5 Lemma. *Let M be a Riemannian manifold.*

Claim. *Under the bundle isomorphism*

$$g_*: TM \to T^*M$$

the geodesic spray S is transformed into the Hamiltonian vector field ζ_{H^*} of the system (T^*M, α, H^*), H^* = kinetic energy on T^*M.

In particular, S is the Hamiltonian vector field of (TM, α, H) with $H(X, X) = \frac{1}{2}\langle X, X\rangle$ = kinetic energy on TM.

Proof. We choose the same special coordinates as for the proof of (3.1.3). Then S at $(0, \dot{x})$ is represented by $(0, \dot{x}, \dot{x}, 0)$. Since $H^*(x, y) = \frac{1}{2}\sum_i (y^i)^2 = \frac{1}{2}\sum_i (\dot{x}^i)^2$, ζ_{H^*} is represented by $(0, y, H^*_y, -H^*_x) = (0, \dot{x}, \dot{x}, 0)$. □

From now on we set $\dim M = n+1$. M continues to denote a Riemannian manifold.

Let $c(t)$ be a geodesic. A *Jacobi field* along $c(t)$ is defined as a vector field $Y(t)$ along $c(t)$ satisfying

$$\nabla^2 Y(t) + R\big(Y(t), \dot{c}(t), \dot{c}(t)\big) = 0.$$

3.1.6 Lemma. *The Jacobi fields $Y(t)$ along a geodesic $c(t)$ are in $1:1$ correspondence with invariant vector fields $\tilde{Y}(t)$ along the orbit $\phi_t\dot{c}(0) = \dot{c}(t)$, i.e.*

$$D\phi_t\big(\tilde{Y}(0)\big) = \tilde{Y}(t).$$

The correspondence is given by

$$Y(t) \leftrightarrow \tilde{Y}(t) = \big(Y(t), \nabla Y(t)\big) \in T_{\dot{c}(t)h}TM \oplus T_{\dot{c}(t)v}TM.$$

Proof. Put $X_0 = \dot{c}(0)$. Consider

$$A = (A_h, A_v) \in T_{X_0h}TM \oplus T_{X_0v}TM = T_{X_0}TM.$$

Let $\kappa(s)$, $s \geqslant 0$, be a curve in TM, $\kappa(0) = x_0$, $\kappa'(0) = A$. Put $\kappa^\sim(s, t) = \phi_t\kappa(s)$.

Then $D\phi_t\kappa'(0) = D\phi_t A = \big(A_h(t), A_v(t)\big)$ is the flow-invariant vector field along $\phi_t X_0$ generated by A.

From the definitions we have

$$\frac{\partial\kappa^\sim}{\partial t}(s, t) = \big(\kappa^\sim(s, t), 0\big), \quad T\tau\,\frac{\partial\kappa^\sim}{\partial t} = \kappa^\sim(s, t),$$

$$D\phi_t\kappa'(0) = \frac{\partial^2\kappa^\sim}{\partial t\,\partial s}(0, t) = \left(T\tau\,\frac{\partial^2\kappa^\sim}{\partial t\,\partial s}, \; K\frac{\partial^2\kappa^\sim}{\partial t\,\partial s}\right)$$

$$= \left(\frac{\partial\kappa^\sim}{\partial s}(0, t), \nabla\frac{\partial\kappa^\sim}{\partial s}(0, t)\right).$$

That is to say, $A_h(t) = \frac{\partial \kappa^\sim}{\partial s}(0, t)$ and $A_v(t) = \nabla A_h(t)$.

Using formula (i) at the beginning of (2.4) with $(\xi'(t), \dot{c}(t), \xi(t))$ replaced by $(\dot{c}(t), A_h(t), \dot{c}(t))$, we obtain

$$\begin{aligned}\nabla^2 A_h(t) &= \nabla\left(\nabla \frac{\partial \kappa^\sim}{\partial t}(s, t) \cdot \frac{\partial \kappa^\sim}{\partial s}(s, t)\right) \cdot \dot{c}(t)\Big|_{s=0} \\ &= \nabla(\nabla \dot{c}(t) \cdot A_h(t)) \cdot \dot{c}(t) = R(\dot{c}(t), A_h(t), \dot{c}(t)) \\ &\quad + \nabla(\nabla \dot{c}(t) \cdot \dot{c}(t)) \cdot A_h(t) = -R(A_h(t), \dot{c}(t), \dot{c}(t)),\end{aligned}$$

i.e. $A_h(t)$ is a Jacobi field.

Thus we have constructed a linear map from the $(2n+2)$-dimensional space of invariant vector fields $D\phi_t A$ along $\phi_t \dot{c}(0)$ into the $(2n+2)$-dimensional space of Jacobi fields along $c(t)$. The kernel of this map being zero, we have proved (3.1.6). □

3.1.7 Corollary. *Let $Y(t)$, $Z(t)$ be Jacobi fields along the geodesic $c(t)$. Then*

(*) $$2\alpha(\tilde{Y}(t), \tilde{Z}(t)) = \langle Y(t), \nabla Z(t)\rangle - \langle Z(t), \nabla Y(t)\rangle = \text{const.}$$

In particular, if $\dot{c} \neq 0$, the Jacobi fields orthogonal to $\dot{c}$, i.e. the Jacobi fields satisfying

$$\langle \dot{c}(t), Y(t)\rangle = \langle \dot{c}(t), \nabla Y(t)\rangle = 0,$$

form a $2n$-dimensional subspace.

Proof. By differentiating (*) one sees at once, using the Jacobi equation, that one obtains zero. On the other hand, we know from (3.1.2) that α is invariant under the flow. Hence (*) also follows from (3.1.6)

$$2\alpha(D\phi_t \tilde{Y}(0), D\phi_t \tilde{Z}(0)) = 2\alpha(\tilde{Y}(0), \tilde{Z}(0)).$$

If $\dot{c} \neq 0$, the invariant vector fields generated by $(\dot{c}(0), 0)$ and $(0, \dot{c}(0))$ form a 2-dimensional subspace which is non-degenerate with respect to the form α. The Jacobi fields orthogonal to $\dot{c}$ form the orthogonal complement of this subspace. □

We call an orbit $\phi_t X_0$ *periodic* if $X_0 \neq 0$ and if there exists $\omega > 0$ such that $\phi_\omega X_0 = X_0$.

We will consider only the case $|X_0| = 1$; clearly, to every periodic orbit $\phi_t X_0$ there corresponds the well-determined periodic orbit $\phi_t X_0 / |X_0|$ having this property. Moreover, we shall assume, for the time being, that ω is the smallest positive number satisfying $\phi_\omega X_0 = X_0$. ω is then called the (*prime*) *period* of the periodic orbit $\phi_t X_0$.

3.1.8 Proposition. *The periodic orbits are in* 1:1 *correspondence with the prime closed geodesics:*

(i) *if* $\phi_t X_0$ *is a periodic orbit of period* ω, *then* $c(t) := \tau_M \phi_{\omega t} X_0$, $0 \leqslant t \leqslant 1$ *is a prime closed geodesic of E-value* $\kappa = \omega^2/2$.

(ii) *if* $c(t)$ *is a prime closed geodesic of E-value* κ, *then* $\phi_t \dot{c}(0)/|\dot{c}(0)|$ *is a periodic orbit of period* $\omega = \sqrt{2\kappa}$.

The proof follows immediately from (3.1.4). □

Note. The following construction of the Poincaré map associated to a periodic orbit in the geodesic flow is valid for periodic orbits in an arbitrary Hamiltonian system (N, α, H), provided $dH \neq 0$ on the orbit, cf. [AM]. Since we are interested only in the geodesic flow we formulate the results for this special case only.

Let $X_0 \in T_1 M$ = unit tangent bundle of M. Then we denote by $\Sigma = \Sigma(X_0)$, a *local transversal hypersurface* to the flow at X_0.

We can construct a particular example of such a surface in the following manner: Put $\tau X_0 = p_0 \in M$. Let $D^n_\varepsilon(X_0)$ be the ε-disc in $T_{p_0} M$ which is orthogonal to $X_0 \in T_{p_0} M$ – recall that dim M = dim $T_{p_0} M = n+1$. Choose ε so small that the ε-disc around $p_0 \in T_{p_0} M$ belongs to the range of normal coordinates, cf. (1.2.6).

Let $T_1 \exp D^n_\varepsilon(X_0)$ be the restriction of the unit tangent bundle to the n-dimensional local hypersurface $\exp D^n_\varepsilon(X_0)$ passing through p_0. We now define $\Sigma = \Sigma_\varepsilon(X_0)$ to be the intersection of $T_1 \exp D^n_\varepsilon(X_0)$ with the ε-Ball $B^{2n}_\varepsilon(X_0)$ around $X_0 \in TM$. Here we take on TM the Riemannian metric induced by the splitting

$$TTM = T_h TM \oplus T_v TM$$

and the canonical identification

$$T\pi : T_{X_0 h} TM \to T_{p_0} M; \qquad K : T_{X_0 v} TM \to T_{p_0} M,\ p_0 = \tau X_0.$$

3.1.9 Proposition. *For* $\varepsilon > 0$ *sufficiently small,* α *restricted to* $\Sigma = \Sigma_\varepsilon(X_0)$ *induces a symplectic structure.*

Proof. It suffices to show that $\alpha | T_{X_0}\Sigma$ is non-degenerate. Choose local orthonormal coordinates $(x^0, x^1, \dots, x^n)$ at $p_0 = \tau X_0$, with X_0 represented by $(1, 0, \dots, 0)$. Then $T_{X_0}\Sigma$ is represented by the plane $x_0 = 0$, $\dot{x}_0 = 1$, i.e. $\alpha | T_{X_0}\Sigma$ is represented by $\sum_{i>0} dx^i \wedge dy^i$, where $(x^i, \dot{x}^i)$ and (x^i, y^i) are the canonical coordinates in TM, T^*M, induced by (x^i). □

3.1.10 Lemma. *Let* $\phi_t X_0$ *be a periodic orbit of period* ω. *Let* $\Sigma = \Sigma(X_0)$ *be a local transversal hypersurface to the flow at* X_0.

Claim. *There exist open neighborhoods* Σ_0 *and* Σ_ω *of* X_0 *in* Σ *and a differentiable function* $\delta : \Sigma_0 \to \mathbb{R}$ *with* $\delta(X_0) = 0$ *such that*

$$\mathscr{P} : \Sigma_0 \to \Sigma_\omega; \qquad X \mapsto \phi_{\omega + \delta(X)} X$$

is a symplectic diffeomorphism. Moreover, $\phi_t X \notin \Sigma$ *for* $0 < t < \omega + \delta(X)$.

We call $\mathscr{P}$ *the Poincaré map (associated to the periodic orbit* $\phi_t X_0$ *or the corresponding closed geodesic* c).

Proof. We are interested only in the case where M is complete (or even compact), so that

$$\phi_t : T_1 M \to T_1 M$$

is defined for all $t \in \mathbb{R}$ and constitutes a 1-parameter group of diffeomorphisms.

Since $\phi_t X_0 \neq X_0$ for $0 < t < \omega$, there exist a neighborhood U of X_0 in $T_1 M$ and $\varepsilon > 0$ such that

$$\phi_\omega | U : U \to \phi_\omega U$$

is a diffeomorphism and $\phi_t U \cap U = \emptyset$ for $\varepsilon < t < \omega - \varepsilon$. Restrict Σ to $\phi_\omega^{-1} U \cap U$.

Consider the map

$$\psi :]-\varepsilon + \omega, \omega + \varepsilon[\times \Sigma \to T_1 M; \qquad (t, X) \mapsto \phi_t X.$$

For $\varepsilon > 0$ sufficiently small, im $\psi \subset U$. Since $D\psi$ at (ω, X_0) is of maximal rank we can choose ε and Σ so small that

$$\psi^{-1} : \text{im}\, \psi \to]-\varepsilon + \omega, \omega + \varepsilon[\times \Sigma$$

is a diffeomorphism. $X_0 \in$ im ψ implies that some open neighborhood Σ_ω of X_0 on Σ belongs to im ψ.

Define

$$\chi : \Sigma_\omega \to \chi(\Sigma_\omega) =: \Sigma_0 \subset \Sigma \quad \text{and} \quad \eta : \Sigma_\omega \to \mathbb{R}$$

by

$$\psi(\omega + \eta(X'), \chi(X')) = \phi_{\omega + \eta(X')} \chi(X') = X'.$$

Then χ^{-1} is the desired Poincaré map $\mathscr{P}$ and $\delta(X) := \eta \circ \chi^{-1}(X)$.

It remains to show that $\mathscr{P}$ is symplectic. To see this we again consider the map ψ and observe that

$$\mathscr{P}(X) = \psi(\omega + \delta(X), X).$$

Put

$$D\psi | T(]-\varepsilon + \omega, \omega + \varepsilon[) = D_1 \psi,$$

$$D\psi | T\Sigma = D_2 \psi.$$

$D_2 \psi$ is symplectic.

Moreover, since $2\alpha\left(\frac{d}{dt} \phi_t X, \right) = dH(\phi_t X)$, and hence $2\alpha(D_1\psi \cdot \xi,) = D_1\psi \cdot \xi dH$ and $H|\Sigma = \text{const}$, and so $dH \circ D_2\psi = 0$, we find that

$$2\alpha(D\mathscr{P}.\xi, D\mathscr{P}.\eta)=2\alpha(D_1\psi.\xi+D_2\psi.\xi, D_1\psi.\eta+D_2\psi.\eta)$$
$$=0+D_1\psi.\xi dH\circ D_2\psi.\eta-D_1\psi.\eta dH\circ D_2\psi.\xi$$
$$+2\alpha(D_2\psi.\xi, D_2\psi.\eta)=2\alpha(\xi,\eta). \quad \square$$

Remark. The importance of the Poincaré map $\mathscr{P}$ is due to the fact that the periodic points X of $\mathscr{P}$ are in 1 : 1 correspondence with the periodic orbits of the flow near the given periodic orbit $\phi_t X_0$, $0 \leqslant t \leqslant \omega$. In fact, if $X \in \Sigma$ such that there exists an integer $N = N(X) > 0$ with

$$\mathscr{P}^N X = X,$$

then $\phi_t X$ is periodic. If we choose $N(X)$ to be the smallest such integer, then the period of $\phi_t X$ is approximately $N\omega$.

3.2 The Index Theorem for Closed Geodesics

In this section we are going to relate the index of a closed geodesic, defined in (2.4), with properties of the linearized geodesic flow along the corresponding periodic orbit. Our results will have some important implications. The material of this section is contained in [Kl 15]. The normal form (3.2.4) for symplectic transformations was first (in a somewhat different manner) derived by Williamson [Wi].

We begin by considering a *linear symplectic transformation* $P: V \to V$ of a *real symplectic vector space*, $\dim V = 2n$. That is to say, on V a non-degenerate skew-symmetric 2-form α will be given.

It will be convenient to complexify V, i.e. to consider the space $V_{\mathbb{C}} := V \otimes \mathbb{C} \cong V \oplus iV$. We extend α on $V \otimes \mathbb{C}$ to be a skew-Hermitian form

$$(X, Y) \in V_{\mathbb{C}} \times V_{\mathbb{C}} \mapsto \alpha(X, \bar{Y}) \in \mathbb{C}.$$

We extend $P: V \to V$ to $P: V_{\mathbb{C}} \to V_{\mathbb{C}}$ by $Pi = iP$. For the sake of simplicity we shall again write V instead of $V_{\mathbb{C}}$.

A subspace $U \subset V$ is called *real* if $\bar{U} = U$. It is called *non-degenerate* if $\alpha | U$ is non-degenerate, i.e. if for every $X \in U$, $X \neq 0$, there exists a $Y \in U$ with $\alpha(X, \bar{Y}) \neq 0$. U is called *isotropic* if $\alpha | U \equiv 0$.

3.2.1 Proposition. *Let $P: V \to V$ be a symplectic transformation.*

(i) *If ρ is an eigenvalue of P so are $\bar{\rho}, \rho^{-1}, \bar{\rho}^{-1}$.*

(ii) *Let $V(\rho)$ be the generalized eigenspace of the eigenvalue ρ of P, i.e. $V(\rho) = \{X \in V$, such that there exists an integer $k \geqslant 0$ with $(P\rho^{-1} - 1)^k X = 0\}$. Then V is the orthogonal sum of non-degenerate subspaces of the form*

$$V(\rho) \oplus V(\bar{\rho}^{-1}), \quad \text{if} \quad \rho\bar{\rho} \neq 1;$$

$$V(\rho), \quad \text{if} \quad \rho\bar{\rho} = 1.$$

Proof. (i) Since P is real, $\bar{\rho}$, as well as ρ, is an eigenvalue. Consider the isomorphism

$$\alpha_*: V \to V^*: X \mapsto \alpha(X, \ ).$$

Then $\alpha(PX, \bar{Y}) = \alpha(X, P^{-1}\bar{Y})$, i.e. $\alpha_* \circ P = {}^tP^{-1} \circ \alpha_*$; ${}^tP^{-1}$ is conjugate to P. Thus ρ^{-1}, as well as ρ, is an eigenvalue.

(ii) We show that, for $\rho\bar{\sigma} \neq 1$,

$$\alpha\big(V(\rho), \overline{V(\sigma)}\big) = 0.$$

To see this we denote by $V_j(\rho)$, j integer $\geqslant 0$, the subspace of $V(\rho)$ which is annihilated by $(P\rho^{-1} - 1)^j$. For $X \in V_1(\rho)$, $Y \in V_1(\sigma)$ we have

$$\alpha(X, \bar{Y}) = \alpha(PX, P\bar{Y}) = \rho\bar{\sigma}\alpha(X, \bar{Y}) = 0.$$

Assume that we already know that

$$\alpha\big(V_{r-1}(\rho), V_s(\bar{\sigma})\big) = \alpha\big(V_r(\rho), V_{s-1}(\bar{\sigma})\big) = 0.$$

For $(r, s) = (1, 2)$ or $= (2, 1)$ this is true. For $X \in V_r(\rho)$, $Y \in V_s(\sigma)$, we have

$$\alpha\big((P\bar{\rho}^{-1} - 1)X, P\bar{Y}\big) = \alpha\big(X, (P\bar{\sigma}^{-1} - 1)\bar{Y}\big) = 0,$$

thus

$$\alpha(X, \bar{Y}) = \alpha(PX, P\bar{Y}) = \rho\bar{\sigma}\alpha(X, \bar{Y}) = 0.$$

Finally, since for every $X \in V(\rho)$, $X \neq 0$, there exists a X^* satisfying $\alpha(X, \bar{X}^*) \neq 0$, $\alpha | V(\rho) \oplus V(\bar{\rho}^{-1})$ is non-degenerate. □

3.2.2 Proposition. *Let $\rho \neq 0$. Then*

$$(*) \qquad \alpha\big((P\rho^{-1} - 1)^j X, \bar{Y}\big) = (-1)^j \alpha\big((P\rho^{-1})^j X, (P\rho - 1)^j \bar{Y}\big).$$

Proof. $\alpha\big((P\rho^{-1} - 1)^j X, \bar{Y}\big) = \alpha\big(P\rho^{-1}(P\rho^{-1} - 1)^{j-1} X, \bar{Y}\big)$,
$-\alpha\big((P\rho^{-1} - 1)^{j-1} X, \bar{Y}\big) = -\alpha\big(P\rho^{-1}(P\rho^{-1} - 1)^{j-1} X, (P\rho - 1)\bar{Y}\big)$. □

3.2.3 Lemma. *Let ρ be an eigenvalue of $P, \rho\bar{\rho} = 1$. Let $k = k_\rho$ be the greatest integer $\geqslant 0$ such that $(P\bar{\rho} - 1)^k \neq 0$.*

Claim. (i) *If $k = 2l - 1$ then there exists $X \in V(\rho)$ such that the elements $(P\bar{\rho} - 1)^j X$, $0 \leqslant j \leqslant k$, form the basis of a non-degenerate subspace and*

$$\alpha\big((P\bar{\rho} - 1)^l X, (P\rho - 1)^{l-1} \bar{X}\big) = \pm 1$$

with the sign at the right hand side being well-determined.

(ii) *If $k = 2l$, then there exists $X \in V(\rho)$ such that the elements $(P\bar{\rho} - 1)^j X$, $0 \leqslant j \leqslant k$, form the basis of a non-degenerated subspace and*

$$\alpha\big((P\bar\rho-1)^l X,(P\rho-1)^l\bar X\big)=\pm i$$

with the sign on the right hand side being well-determined.

Proof. Define $V_1^k(\rho)=(P\bar\rho-1)^k V(\rho)$. From (3.2.2) we have, using $(P\bar\rho)^r=(1+P\bar\rho-1)^r=1+r(P\bar\rho-1)+\dots$, that

$$\alpha\big((P\bar\rho-1)^{k-r}X,(P\rho-1)^r\bar Y\big)=(-1)^{k-r}\alpha\big(X,(P\rho-1)^k\bar Y\big),$$

for $X, Y\in V(\rho)$.

Assume that for all $(P\bar\rho-1)^k X\in V_1^k, \neq 0$, $\alpha\big((P\bar\rho-1)^k X,\bar X\big)=0$. There exists $Y\in V(\rho)$ such that

$$\alpha\big((P\bar\rho-1)^k X,\bar Y\big)=1(k \text{ odd}),=i(k\text{ even})$$

$$\alpha\big((P\bar\rho-1)^k Y,\bar X\big)=(-1)^k\alpha\big(Y,(P\rho-1)^k\bar X\big)$$

$$=(-1)^{k+1}\overline{\alpha(P\bar\rho-1)^k X,\bar Y\big)}=1\,(k\text{ odd}),=i(k\text{ even}).$$

Thus $\alpha\big((P\bar\rho-1)^k(X+Y),\bar X+\bar Y\big)=\alpha\big((P\bar\rho-1)^k X,\bar Y\big)+\alpha\big((P\bar\rho-1)^k Y,\bar X\big)$
$=2\,(k\text{ odd}),=2i\,(k\text{ even}).$

Hence we can assume that there exists $X\in V_1^k(\rho)$ satisfying

$$\alpha\big((P\bar\rho-1)^l X,(P\rho-1)^{l-1}\bar X\big)\neq 0,\quad k=2l-1$$

$$\alpha\big((P\bar\rho-1)^l X,(P\rho-1)^l\bar X\big)\neq 0,\qquad k=2l.$$

The right-hand side is determined up to a positive real number. The first expression is real, whereas the second expression is purely imaginary. Hence (3.2.3). □

3.2.4 Theorem. *Let $P: V\to V$ be a real symplectic transformation. Then there exists an orthonormal decomposition*

$$V=V_{nc}\oplus V_{co}$$

into non-degenerate invariant subspaces such that $P|V_{nc}$ is non-compact and $P|V_{co}$ is compact, i.e. it belongs to a compact subgroup of $Sp(V)$.

V_{nc} has a decomposition

$$V_{nc}=V_{su}\oplus V_{od}\oplus V_{ev}$$

into non-degenerate invariant real subspaces. Here

V_{su} is the direct sum of subspaces $V(\rho)\oplus V(\bar\rho^{-1}), |\rho|\neq 1$;

V_{od} is the direct sum of subspaces $V_{od,\rho}$ having a basis of the type $(P\bar\rho-1)^j X$, $0\leqslant j\leqslant 2l-1$, $|\rho|=1$;

V_{ev} is the direct sum of subspaces $V_{ev,\rho}$ having a basis of the type $(P\bar\rho-1)^j X$, $0\leqslant j\leqslant 2l$, $l>1$, $|\rho|=1$.

3.2.5 Complement. *V_{su} has a maximal, isotropic, invariant real subspace V_s generated by the $V(\rho)$ with $|\rho|<1$;*

V_{od} has a maximal isotropic real invariant subspace $V_{od,in}$ generated by the $V_{od,in,\rho}\subset V_{od,\rho}$ having as basis the elements $(P\bar{\rho}-1)^j X$, $l\leqslant j\leqslant 2l-1$;
V_{ev} has an isotropic, real, invariant subspace $V_{ev,in}$ generated by the $V_{ev,in,\rho}\subset V_{ev,\rho}$ having as basis the elements $(P\bar{\rho}-1)^j X$, $l+1\leqslant j\leqslant 2l$.

Note. $V_{ev,in}$ is not maximal isotropic. For each of the k_{ev} real subspaces $V_{ev,\rho}\oplus\bar{V}_{ev,\rho}$ which constitute the space V_{ev}, the subspaces $V_{ev,in,\rho}\oplus\bar{V}_{ev,in,\rho}$ have dimension two less than the dimension of a maximal isotropic subspace.

Proof. The complement is clear once one has proved the theorem.

Obviously, V_{su} is well-defined. So consider an eigenvalue ρ with $|\rho|=1$.

Let $k=k_\rho$ be the maximum of the l with $(P\bar{\rho}-1)^l\neq 0$. Assume $k>0$. From (3.2.3), we have the existence of a subspace $V_{od,\rho}$ or $V_{ev,\rho}$ depending on whether k is odd or even. We apply the same argument to the orthogonal complement of this space. If $k=0$, then we obtain a subspace having a base X with $PX=\rho X$, $\alpha(X,\bar{X})=\pm i$. X, together with $\bar{X}$ is then added to V_{co}. This proves (3.2.4). □

3.2.6 Lemma. *Let*

$$V^{2n}=V_{in}^{2p}\oplus V_{un}^{2q}$$

be an orthogonal decomposition into non-degenerate real subspaces. Let $V_{in}^p\subset V_{in}^{2p}$ be an invariant real isotropic subspace and $V_v^n\subset V^{2n}$ an isotropic subspace.

Claim. (i) *The projection of $V_v^n\cap(V_{in}^p\oplus V_{un}^{2q})$ modulo V_{in}^p yields a maximal real isotropic subspace $V_{un}^q\subset V_{un}^{2q}$.*

(ii) $\quad V_e^n:=V_{in}^p\oplus V_{un}^q$ *is maximal isotropic in V^{2n}.*

(iii) $\quad$ *There exists a $V_v^q\subset V_v^n$ such that $V_e^n=V_{in}^p+V_v^q$.*

Proof. Put $V_v^n\cap V_{in}^p=V_{in}^k$, k-dimensional subspace. Let V_{in}^{*p} be an isotropic complement of V_{in}^p in V_{in}^{2p}. Let V^{*l} be the projection of V_v^n in V_{in}^{*p} mod $V_{in}^p\oplus V_{un}^{2q}$. Since any $X^*\in V^{*l}$ has a supplement $X'\in V_{in}^p\oplus V_{un}^{2q}$ so as to give $X^*+X'=X_v\in V_v^n$, for every $X\in V_{in}^k\subset V_v^n$ we have

$$\alpha(X^*,X)=\alpha(X^*+X',X)=0.$$

Since $\alpha|V_{in}^p\oplus V_{in}^{*p}$ is non-degenerate, $k+l\leqslant p$, i.e. $\dim(V_v^n\cap(V_{in}^p\oplus V_{un}^{2q}))=n-l\geqslant n+k-p$. Hence, the projection $V_{un}^{q'}$ of this space in V_{un}^{2q} has dimension $q'\geqslant n-p=q$.

However, since $V_{un}^{q'}$ is isotropic, elements X,Y of $V_{un}^{q'}$ can be supplemented by elements X',Y' of V_{in}^p so as to yield elements of V_v^n, i.e.

$$0=\alpha(X+X',\ Y+Y')=\alpha(X,Y)$$

– note that $\alpha(V_{in}^p,V_{un}^{2q})=0$. Therefore $q'=q$, and thus we have proved (i).

(ii) and (iii) follow immediately from our constructions. □

3.2.7 Proposition. *Let $V^q \subset V^{2q}$ be a maximal isotropic subspace. Let ρ not be an eigenvalue of the symplectic isomorphism $P: V^{2q} \to V^{2q}$.*

Claim. *$Q_\rho(X, Y) := -2\alpha\big(X, (P\rho - 1)^{-1}\bar{Y}\big)$ is a Hermitian form on V^q. The nullspace of Q_ρ consists of $V^q \cap (P\bar{\rho} - 1)V^q$.*

Proof. Put $(P\bar{\rho}-1)^{-1}X = S$, $(P\bar{\rho}-1)^{-1}Y = T$. Then

$$0 = 2\alpha\big((P\bar{\rho}-1)S, (P\rho-1)\bar{T}\big) = -2\alpha\big(X, (P\rho-1)^{-1}\bar{Y}\big)$$

$$+2\alpha\big((P\bar{\rho}-1)S, P\rho\bar{T}\big) = Q_\rho(X, Y) + 2\alpha(S, \bar{T}) - 2\alpha(S, P\rho\bar{T})$$

$$= Q_\rho(X, Y) + 2\alpha\big((P\rho-1)\bar{T}, S\big) = Q_\rho(X, Y) - \overline{Q_\rho(Y, X)}.$$

$Y \in$ nullspace $Q_\rho \Leftrightarrow (P\bar{\rho}-1)^{-1} Y = Y' \in V^q$ and $Y \in V^q$. □

We now come to an important generalization of the index of a closed geodesic c. Recall that index $c = I_c$ is defined as the dimension of the negative eigenspace of the index form $D^2E(c)$, on the space $T_c\Lambda = T_c\Lambda M$ of periodic H^1-vector fields along c, cf. (2.4.1).

Let $\rho \in \mathbb{C}$, $|\rho| = 1$. We denote by ${}_\rho T_c\Lambda = {}_\rho T_c\Lambda M$ the Hilbert space of complex valued H^1-vector fields $\zeta(t)$ along $c(t)$ satisfying $\bar{\rho}\zeta(1) = \zeta(0)$ with scalar product

$$\langle \zeta, \bar{\zeta}' \rangle_1 = \langle \zeta, \bar{\zeta}' \rangle_0 + \langle \nabla\zeta, \nabla\bar{\zeta}' \rangle_0,$$

cf. (1.2). Here,

$$\langle \zeta, \bar{\eta} \rangle_0 = \int_S \langle \zeta(t), \bar{\eta}(t) \rangle dt.$$

Moreover, we have on ${}_\rho T_c\Lambda$ the *ρ-index form*

$$D^2E_\rho(c)\,(\zeta, \bar{\zeta}) = \langle \zeta, \bar{\zeta}' \rangle_1 - \langle (K_c^{\sim} + \mathrm{id})\zeta, \bar{\zeta}' \rangle_0.$$

As in (2.4.2), one shows that the corresponding self-adjoint operator ${}_\rho A_c$ is of the form $\mathrm{id} + k_c$, with k_c a compact operator. In particular, the dimension of the negative eigenspace of ${}_\rho A_c$ and the dimension of the 0-eigenspace of ${}_\rho A_c$ are finite. These dimensions are called the *ρ-index of c*, $I_c(\rho)$, and the *ρ-nullity of c*, $N_c(\rho)$. Here we assume that $\rho \neq 1$. For $\rho = 1$, we define $N_c(1)$ to be the 0-eigenspace of A_c minus one, in accordance with our earlier definition, cf. (2.4).

See (4.1.5) for an interpretation of the ρ-index.

The importance of these concepts was recognized by Bott [Bo 3]. It is based on the following facts. Recall that we define the m-fold covering c^m of a closed geodesic by $c^m(t) = c(tm)$, cf. (2.2).

3.2.8 Proposition. *Let ρ be a primitive m^{th} root of unity. Let $c = c(t)$, $0 \leqslant t \leqslant 1$, be a closed geodesic.*

There exists a linear isomorphism

$$i: T_{c^m}\Lambda \to \bigoplus_{1\leq l\leq m} {}_{\rho^l}T_c\Lambda; \qquad \zeta \mapsto (\zeta_1, \ldots, \zeta_m)$$

with

$$\zeta_l(t) = \frac{1}{m}\sum_{j=1}^{m} \rho^{l(1-j)}\zeta\left(\frac{t+j-1}{m}\right).$$

The inverse is the composition $\oplus j_l$ *of the maps*

$$j_l: {}_{\rho^l}T_c\Lambda \to T_{c^m}\Lambda; \qquad \zeta_l(t) \mapsto \zeta_l(mt).$$

Proof. (cf. [Kl. 15]) One easily verifies that $\zeta_l \in {}_{\rho^l}T_c\Lambda$. The map i is clearly linear. The inverse is given by

$$(\zeta_1(t), \ldots, \zeta_l(t)) \to \sum_{l=1}^{m} \zeta_l(mt).$$

Indeed,

$$\sum_l \sum_j \frac{1}{m}\rho^{l-lj}\zeta(t+j-1) = \zeta(t)$$

since

$$\sum_l \rho^{l(1-j)} = 0 \quad \text{for} \quad j \neq 1, m > 1, \quad \text{and} \quad = m \quad \text{for} \quad j = 1. \quad \square$$

3.2.9 Theorem. *Let c be a closed geodesic and* c^m *its m-fold covering. Then index* $c^m = \sum_{\rho^m=1}$ *ρ-index c, and nullity* $c^m = \sum_{\rho^m=1}$ *ρ-nullity c.*

Proof. Write $\zeta(t) = \sum_k \zeta_k(mt)$, $\zeta'(t) = \sum_l \zeta'_l(mt)$, ρ a primitive m^{th} root of unity, c.f. (3.2.8). Then

$$D^2E(c^m)(\zeta, \bar{\zeta}') = \sum_{k,l} \int_0^{1/m} (\langle \nabla\zeta_k(mt), \nabla\bar{\zeta}'_l(mt)\rangle - \langle K_c\tilde{}(\zeta_k(mt)), \bar{\zeta}'_l(mt)\rangle)dt$$

$$\times [1 + \rho^{(k-l)} + \cdots + \rho^{(m-1)(k-l)}] = \sum_k D^2E(c^m)(j_k\zeta_k, j_k\bar{\zeta}'_k).$$

Thus $D^2E(c^m)$ is the direct sum of the restrictions $D^2E(c^m)|j_k({}_{\rho^k}T_c\Lambda)$ of $D^2E(c^m)$ to the direct summands of $T_{c^m}\Lambda$ described in (3.2.8). $\square$

Remark. The previous result is due to Bott [Bo 3]. It indicates the relevance of the function $\rho \in S^1 \subset \mathbb{C} \to I_c(\rho) \in \mathbb{N}$. The knowledge of this function gives complete information on the index of all the iterates (multiple coverings) of a closed geodesic. The main goal of this section is to show how this function can be determined by data given by the linearized geodesic flow along the corresponding periodic orbit in the geodesic flow.

Before we start this program we want to point out the connection with the Maslov index, cf. Arnold [Ar 3]. We start with a decomposition $V^{2q}=V_h^q\oplus V_v^q$ of V^{2q} into two complementary isotropic subspaces. On V_h^q we choose a scalar product $\langle\,,\rangle$. The linear isomorphism $\alpha_*: V^{2q}\to V^{*2q};\ X\mapsto\alpha(X,\ )$ carries V_h^q, V_v^q into the dual spaces V_v^{*q}, V_h^{*q}, respectively. Using $\langle\,,\rangle$ to identify V_h^{*q} with V_h^q, we get from α_* an isomorphism $J: V_v^q\to V_h^q$. We use J to define a scalar product on V_v^q and extend $\langle\,,\rangle$ to all of V^{2q} by letting $V_h^q\oplus V_v^q$ be an orthogonal decomposition.

Using the scalar product on V_v^q we use $\alpha_*: V_h^q\to V_v^{*q}$ to define $J: V_h^q\to V_v^q$. Since

$$\alpha_* X_v\cdot Y_h=\alpha(X_v, Y_h)=-\alpha(Y_h, X_v)=-\alpha_* Y_h\cdot X_v$$

we have $\langle JX_v, Y_h\rangle=-\langle JY_h, X_v\rangle$. Hence, since J is an isometry, $\langle J^2X_h, Y_h\rangle=-\langle JX_h, JY_h\rangle=-\langle X_h, Y_h\rangle$. Thus $J^2=-id$, ${}^tJ=-J$. The symplectic form α can therefore be written as

$$2\alpha(X,Y)=\langle X, JY\rangle=\langle X_h, Y_v\rangle-\langle Y_h, X_v\rangle.$$

Compare this with (3.1.3).

$P: V^{2q}\to V^{2q}$ symplectic now means that ${}^tPJP=J$. For $\rho, |\rho|=1$, not an eigenvalue of P, we define

$$S_\rho=(P\bar\rho+1)\,(P\bar\rho-1)^{-1},\ \text{i.e. } (P\bar\rho-1)^{-1}=(S_\rho-1)/2.$$

Then ${}^t\bar S_\rho=-JS_\rho J^{-1}$; i.e. if we write $S_\rho=K_\rho+iL_\rho$ with K_ρ, L_ρ real, and write K_ρ, L_ρ according to the decomposition $V_h^q\oplus V_v^q$ of V^{2q} we obtain

$$K_\rho=\begin{pmatrix} k_\rho & k'_\rho\\ k''_\rho & -{}^tk_\rho\end{pmatrix},\qquad L_\rho=\begin{pmatrix} l_\rho & l'_\rho\\ l''_\rho & {}^tl_\rho\end{pmatrix},$$

with ${}^tk'_\rho=k'_\rho$, ${}^tk''_\rho=k''_\rho$, ${}^tl'_\rho=-l'_\rho$, ${}^tl''_\rho=-l''_\rho$. It follows that the Hermitian form Q_ρ, (3.2.7), on V_v^q can be written as

$$Q_\rho(X_v, Y_v)=-\alpha(X_v, (\bar S_\rho-1)Y_v)=\langle k'_\rho Y_v, X_v\rangle.$$

A (real) isotropic subspace V^q of V^{2q} is also called a *Lagrangian subspace.* The set $\Lambda(q)$ of all Lagrangian subspaces can be identified with the symmetric space $U(q)/O(q)$. Denote by $\Lambda^0(q)$ the subspace of $\Lambda(q)$ formed by the V^q with $V^q\cap V_h^q=0$. Such a V^q can be written as

$$V^q=\{(kY_v, Y_v);\ Y_v\in V_v^q\}$$

with $k={}^tk$ symmetric. Thus, $\Lambda^0(q)$ can be parametrized by the symmetric (q,q)-matrices.

$\Lambda(q)-\Lambda^0(q)$ is a stratified algebraic variety of codimension 1 in $\Lambda(q)$. Its regular (=interior) points consist of the V^q with dim $(V^q \cap V_h^q)=1$. The singular points have codimension 3 in $\Lambda(q)$; $\Lambda(q)-\Lambda^0(q)$ represents an integer cohomology class of codimension 1.

Let $\{\rho_j\}$ be the set of eigenvalues having modulus 1.

$$(*) \qquad \rho \in S^1 - \{\rho_j\} \mapsto V_\rho^q = \{k_\rho' Y_v, Y_v); \; Y_v \in V_v^q\},$$

with k_ρ' determined by $S_\rho = K_\rho + iL_\rho$ as above, is a curve in $\Lambda^0(q)$. It is possible to extend the domain of definition to include the set $\{\rho_j\}$. The corresponding Lagrangian spaces belong to the cycle $\Lambda(q)-\Lambda^0(q)$. The Maslov index, associated to the symplectic transformation P, is the sum of the intersection numbers of the closed curve (*), extended to all of S^1, with the cycle $\Lambda(q)-\Lambda^0(q)$. For further details see [Ar 3], [Du], [CuD].

We now direct our attention again to a closed geodesic c of length ω and the corresponding periodic orbit $\phi_t X_0$, $X_0 = \dot{c}(0)/|\dot{c}(0)|$ of period ω. Here we do not assume that c or $\phi_t X_0$ is prime.

In the tangent bundle $\tau_{T_1 M}: TT_1 M \to T_1 M$ we have the subbundles

$$\tau_h^n: T_h^n T_1 M \to T_1 M, \; \tau_v^n: T_v^n T_1 M \to T_1 M.$$

The fibre in τ_h^n over $X_0 \in T_1 M$ consists of the $(X,0) \in T_{X_0 h} TM$ satisfying $\langle X, X_0 \rangle = 0$, and the fibre in τ_v^n over $X_0 \in T_1 M$ consists of the $(0, Y) \in T_{X_0 v} TM$ satisfying $\langle Y, X_0 \rangle = 0$.

The differentiable map $t \in \mathbb{R} \to \phi_t X_0 \in T_1 M$ induces the bundles $\tau_h^n: V_h^n \to \mathbb{R}$, $\tau_v^h: V_v^n \to \mathbb{R}$. We denote the direct sum $\tau_h^n \oplus \tau_v^h$ also by $\tau^{2n}: V^{2n} \to \mathbb{R}$.

Since $\phi_\omega X_0 = X_0$, we can form the quotient bundles over $S_\omega = \mathbb{R} \bmod \omega \mathbb{Z}$:

$$\tau_h^n: V_h^n \to S_\omega, \; \tau_v^n: V_v^n \to S_\omega, \; \tau^{2n}: V^{2n} \to S_\omega.$$

On the fibres of the bundle τ^{2n} we have the symplectic form α, cf. (3.1.4). Moreover, the differential $D\phi_t$ of the geodesic flow operates as 1-parameter group of sympletic linear fibre maps on τ^{2n}, cf. (3.1.7), (3.1.8). In particular,

$$D\phi_\omega: V^{2n}(0) \to V^{2n}(\omega) = V^{2n}(0)$$

is a linear sympletic transformation. It is the differential of the Poincaré map $\mathscr{P}$ at the point $X_0 \in \Sigma$, cf. (3.1.11). Therefore, we also write P instead of $D\phi_\omega$.

To the symplectic transformation $P: V^{2n}(0) \to V^{2n}(0)$ we now apply the results from the beginning of this section. According to (3.2.4), we can write $V^{2n}(0) = V_{su} \oplus V_{od} \oplus V_{ev} \oplus V_{co}$ and can find a real invariant isotropic subspace $V_{in}^p = V_s \oplus V_{od,in} \oplus V_{ev,in}$, cf. (3.2.5).

For every $V_{ev,\rho,\bar\rho} := V_{ev,\rho} \oplus \bar{V}_{ev,\rho}$ having a basis of the form $(P_{\bar\rho}-1)^j \tilde{X}, (P_\rho - 1)^j \bar{\tilde{X}}$, $0 \leq j \leq 2l$, $l > 0$, we consider the 2-dimensional subspace

$$V_{ev,un,\rho,\bar\rho} := \{(P_{\bar\rho}-1)^l \tilde{X}, (P_\rho - 1)^l \bar{\tilde{X}}\}.$$

Note: $PV_{ev,un,\rho,\bar\rho} = V_{ev,un,\rho,\bar\rho}$ mod $V_{ev,in,\rho} \oplus V_{ev,in,\bar\rho}$. We define $V_{ev,un}$ to be the direct sum of these $V_{ev,un,\rho,\bar\rho}$. dim $V_{ev,un} = 2k_{ev}$, k_{ev} = number of subspaces $V_{ev,\rho,\bar\rho}$. (Compare the note after (3.2.5)).

We put

$$V_{un}^{2q} = V_{co} \oplus V_{ev,un}$$

and define $V_{un}^q \subset V_{un}^{2q}$ as the projection of $V_v^n(0) \cap (V_{in}^p \oplus V_{un}^{2q})$, cf. (3.2.6). We put $V_e^n = V_{in}^p \oplus V_{un}^q$. This is a Lagrangian subspace of V^{2n}.

For $t \in [0, \omega]$, we put $W(t) := V_v^n(t) \cap D\phi_t V_e^n$.

3.2.10 Proposition. *Assume $i_0 = \dim W(t_0) > 0$. Choose a basis*

$$\tilde{Y}_j(t) = (Y_j(t), \nabla Y_j(t)), \ 0 \leqslant j \leqslant n,$$

of Jacobi fields for $D\phi_t V_e^n$ such that $\tilde{Y}_j(t_0)$, $1 \leqslant j \leqslant i_0$, is a basis for $W(t_0)$.

Claim. (i) *The elements $\nabla Y_j(t_0)$, $1 \leqslant j \leqslant i_0$; $Y_j(t_0), j > i_0$, are linearly independent.*

(ii) *For $|t - t_0| \neq 0$, sufficiently small, the $Y_j(t)$, $1 \leqslant j \leqslant n$, are linearly independent. In particular, there are only finitely many $t \in [0, \omega]$ with $W(t) \neq 0$.*

Proof. $\tilde{Y}(t_0) \in W(t_0)$ means that $Y(t_0) = 0$. Hence, the vectors $\{\nabla Y_j(t_0); 1 \leqslant j \leqslant i_0\}$ as well as the vectors $\{Y_j(t_0);\ i_0 + 1 \leqslant j \leqslant n\}$ are linearly independent. Since $0 = \alpha(\tilde{Y}_j, \tilde{Y}_k) = -\langle \nabla Y_j(t_0), Y_k(t_0)\rangle$ for $j \leqslant i_0$, $k > i_0$, we have (i). (ii) follows from the observation that for $\tilde{Y}(t) \neq 0$, $Y(t_0)$ implies $\nabla Y(t_0) \neq 0$. □

Let $\rho \in S^1$ not be an eigenvalue of $P = D\phi_\omega$. We wish to define a map ζ_ρ of the space $W := \bigoplus_{0 \leqslant t < \omega} W(t)$ into the space ${}_\rho T_c \Lambda = {}_\rho T_c \Lambda M$ where c is the closed geodesic corresponding to the periodic orbit $\phi_t X_0$. Note that $\dim W < \infty$.

For $\tilde{Y}(t_0) \in W(t_0)$ we define $\tilde{Z}(t) := (P\bar\rho - 1)^{-1} \tilde{Y}(t)$. This again is a Jacobi field satisfying $Z(t_0 + \omega) = Z(t_0)$. Now define $\zeta = \zeta_\rho(\tilde{Y}(t_0)) \in {}_\rho T_c \Lambda$ by

$$\zeta(t) = \bar\rho Z(\omega t + \omega), \quad 0 \leqslant t \leqslant t_0/\omega,$$

$$\zeta(t) = Z(\omega t), \quad t_0/\omega \leqslant t \leqslant 1.$$

Then $\zeta(t_0/\omega -) - \zeta(t_0/\omega +) = 0$, $\nabla\zeta(t_0/\omega -) - \nabla\zeta(t_0/\omega +) = \omega \nabla Y(t_0)$, $\bar\rho\zeta(1) - \zeta(0) = \bar\rho \nabla\zeta(1) - \nabla\zeta(0) = 0$, i.e. ζ is indeed an element of ${}_\rho T_c \Lambda$ with $\nabla\zeta$ smooth except possibly at t_0.

3.2.11 Proposition. *The map*

$$\zeta_\rho : W \to {}_\rho T_c \Lambda M$$

is linear injective. Moreover, if $\zeta = \zeta_\rho(\tilde{Y}(t_0))$, $\zeta' = \zeta_\rho(\tilde{Y}'(t_0'))$ then $D^2 E_\rho(\zeta, \bar\zeta') = Q_\rho(\tilde{Y}_{un}, \tilde{Y}'_{un})\omega$ with Q_ρ as in (3.2.7) and $\tilde{Y}_{un}$, $\tilde{Y}'_{un}$ the components of $\tilde{Y}(0)$, $\tilde{Y}'(0) \in V_e^n = V_{in}^p \oplus V_{un}^q$ in V_{un}^q.

Proof. Clearly, $\zeta_\rho|W(t_0)$ is injective. Moreover, $\zeta_\rho(\tilde{Y}(t_0))=\zeta_\rho(\tilde{Y}'(t_0'))$ and $\tilde{Y}(t_0)\neq 0$, $\tilde{Y}'(t_0')\neq 0$ implies $t_0=t_0'$, i.e. ζ_ρ is injective.
Now assume $t_0'\leqslant t_0$. Then, using partial integration,

$$D^2E_\rho(c)(\zeta,\bar{\zeta}')=\int_S \frac{d}{dt}\langle\nabla\zeta,\bar{\zeta}'\rangle\,dt-\int_S\langle\nabla^2\zeta+K_c^{\sim}(\zeta),\bar{\zeta}'\rangle dt$$

$$=\omega\langle\nabla Y(t_0),\bar{Z}'(t_0)\rangle=-2\alpha(\tilde{Y},(P\rho-1)^{-1}\tilde{Y}')\omega=Q_\rho(\tilde{Y}_{un},\tilde{Y}_{un}')\omega,$$

since $\alpha(V_{in}^P, V_e^n)=0$. □

We are now ready to prove the *ρ-index Theorem*, cf. [Kl 15], [Cu D].

3.2.12 Theorem. *Let c be a closed geodesic. Let P be the linear Poincaré map of the corresponding periodic orbit in the geodesic flow. Let* $\rho\in\mathbb{C}$, $|\rho|=1$.

Claim. (i) *The ρ-nullity of c is given by* $N_c(\rho)=$ *dimension of eigenspace of P belonging to the eigenvalue ρ.*
(ii) *Let ρ not be an eigenvalue of P. Then the ρ-index of c is given by*

$$I_c(\rho)=J_c+M_\rho,\quad \textit{with}$$
$$J_c=\sum_{0<t<\omega}\dim W(t)+\dim\,(V_v^n(0)\cap V_{in}^P),$$
$$M_\rho=(\textit{index plus nullity})\ Q_\rho.$$

Remark. Note that $I_c(\rho)$ consists of a term J_c which does not depend on ρ and a term M_ρ which depends only on ρ and P. Thus, if none of the eigenvalues of P is a root of unity, we obtain the following formula for the index of all the coverings c^m of c, cf. (3.2.5),

$$\text{index } c^m=mJ_c+\sum_{\rho^m=1}M_\rho.$$

The existence of such a formula was already proved by Bott [Bo 3]. However, he did not determine the constant J_c, nor was the description of the second term M_ρ explicit.

Finally, we observe that the theorem demonstrates the importance of the function $\rho\in S^1-\{\rho_j\}\mapsto M_\rho$. As we saw in the remark following (3.2.9), M_ρ is determined by an element in the space $\Lambda(q)$ and, being a continuous integer-valued function, it remains constant on the connected components of $S^1-\{\rho_j\}$. The "jumps" will occur at the intersection of this curve with the Maslov cycle $\Lambda(q)-\Lambda^0(q)$. For more details, see (3.2.15), [Kl 15].

Proof. The nullspace of the operator ${}_\rho A_c=\mathrm{id}+{}_\rho k_c$ consists of the solutions of

$$(\dagger)\qquad \nabla^2\zeta(t)+K_c^{\sim}(\zeta)(t)=0,\qquad\text{with}\qquad \bar{\rho}\zeta(1)=\zeta(0).$$

For the case $\rho=1$, this follows from (2.4.4). For arbitrary ρ, $|\rho|=1$, this can be proved in the same way as (2.4.4). In fact, $\zeta\in$ nullspace $D^2E_\rho(c)$, means that

$$D^2E_\rho(c)(\zeta,\bar{\eta})=0,\qquad\text{for all}\qquad \eta\in{}_\rho T_c\Lambda.$$

ζ being differentiable, we can write this in the following form, using partial integration,

$$(\dagger\dagger)\qquad \langle\nabla^2\zeta+K_c^{\sim}(\zeta),\eta\rangle_0=0,\qquad \text{all}\qquad \eta\in{}_\rho T_c\Lambda.$$

But (††) is equivalent to (†).

A solution ζ of (†) satisfying $\langle\zeta(t),\dot c(t)\rangle=0$ determines a Jacobi field $Y(t)=\zeta(t/\omega)$ with $\bar\rho\tilde Y(\omega)=\tilde Y(0)$, i.e. $(P\bar\rho-1)\ \tilde Y=0$. Conversely, such a $\tilde Y$ determines a solution $\zeta(t):=Y(t\omega)$ of (†). Finally, note that for $\rho\neq 1$, the Jacobi field $\dot c(t)$ does not satisfy (†). Thus we have proved (i).

To prove (ii) consider the map $\lambda_\rho\colon \zeta_\rho W\to V^q_{un}$ which associates to the element $\zeta_\rho(\tilde Y(t_0))$ the component $\tilde Y_{un}(0)$ of $\tilde Y(0)\in V^n_e$ in the decomposition $V^n_e=V^p_{in}\oplus V^q_{un}$. From (3.2.11), we know that it is well-defined. Moreover, λ_ρ is surjective since every $\tilde Y_{un}(0)\in V^q_{un}$ occurs as the component of an element in $V^q_v\subset W(0)=V^n_v(0)\cap V^n_e$, cf. (3.2.6) (iii).

$\zeta_\rho W\subset{}_\rho T_c\Lambda$ does not contain elements of the nullspace of $D^2E_\rho(c)$. From (3.2.11), we know that $D^2E_\rho(c)\ (\zeta,\bar\zeta)=Q_\rho(\lambda_\rho\zeta,\lambda_\rho\zeta')\ \omega$. Therefore

$$I_c(\rho)\geqslant(\text{ind}+\text{null})\ D^2E_\rho(c)|\zeta_\rho W=\dim\ker\lambda_\rho+(\text{ind}+\text{null})\ Q_\rho.$$

Since $\dim\ker\lambda_\rho=J_c$, it only remains to show that the extension of $\zeta_\rho W$ by a differentiable element $\xi\in{}_\rho T_c\Lambda$, satisfying $D^2E_\rho(c)\ (\zeta,\bar\xi)=0$ for all $\zeta\in\zeta_\rho W$, implies that $D^2E_\rho(c)\ (\xi,\bar\xi)\geqslant 0$, and is actually >0 if $\xi\neq 0$.

Let $\zeta=\zeta_\rho(\tilde Y(t_0))$. Then $0=D^2E_\rho(c)\ (\zeta,\bar\xi)=\langle\nabla Y(t_0),\bar\xi(t_0/\omega)\rangle$. If $\tilde Y_j(0)$, $1\leqslant j\leqslant n$, is a real basis for V^n_e, we therefore have from (3.2.10) that $\xi(t)$ can be written as $\xi(t)=\sum_j w^j(t)\,Y_j(t\omega)$ with $w^j(t)$ differentiable.

We may assume that $\tilde Y_j(0)\in V^p_{in}$, $1\leqslant j\leqslant p$, $\tilde Y_k(0)\in V^q_v$, $k>p$, cf. (3.2.6). Hence $w^k(0)=0$ for $k>p$. Since $PV^p_{in}=V^p_{in}$, $\tilde Y_l(\omega)=\sum_j a^j_l\tilde Y_j(0)$, $1\leqslant j,\ l\leqslant p$. $\bar\rho\xi(1)=\xi(0)$ $\in V^p_{in}$ implies that $w^k(1)=0$, $k>p$, and $w^j(0)=\sum_l\bar\rho a^j_l w^l(1)$, $1\leqslant j,\ l\leqslant p$. Then we obtain

$$\begin{aligned}D^2E_\rho(c)\ (\xi,\bar\xi)&=\textstyle\int_S|\sum_j\dot w^j(t)\,Y_j(t\omega)|^2dt\\&\quad+\int_0^1\frac{d}{dt}\langle\sum_k w^k(t)\nabla Y_k(t\omega),\ \sum_j\bar w^j(t)\,Y_j(t\omega)\rangle dt\,\omega\\&\geqslant\sum_{1\leqslant j,k,l\leqslant p}\{w^k(1)\bar w^l(1)a^j_l\langle\nabla Y_k(\omega),\ Y_j(0)\rangle\\&\quad-\bar w^k(1)w^l(1)a^j_l\langle Y_k(\omega),\ \nabla Y_j(0)\rangle\}\omega\\&=\sum_{1\leqslant j,k,l\leqslant p}w^k(1)\bar w^l(1)a^j_l 2\alpha(\tilde Y_j(0),\ \tilde Y_k(\omega))\omega\end{aligned}$$

$$+2i\ \text{imaginary}\ \Big(\sum_{1\leqslant j,k,l\leqslant p}w^k(1)\bar w^l(1)a^j_l\langle\nabla Y_j(0),\ Y_k(\omega)\rangle\Big)\omega=0.\qquad\square$$

Remark. In special cases we might have $V^{2q}_{un}=0$, i.e. $V^n_e=V^n_{in}$. This will be the case if in the decomposition (3.2.4) of V^{2n} with respect to P the subspaces V_{ev} and V_{co} are both zero. An important example is the case in which P is *hyperbolic*, i.e. P has no eigenvalue ρ with $|\rho|=1$. Then we also call the underlying closed geodesic *hyperbolic*.

3.2.13 Corollary 1. *Assume that the closed geodesic c is hyperbolic. Then*

$$\rho\text{-}index\ c=\sum_{0\leqslant t<\omega}\dim\left(V_v^n(t)\cap V_s^n(t)\right),\ index\ c^m=m\ index\ c.$$

Here, $V_s^n(t)=D\phi_t V_s^n(0)$ *is the so-called stable subspace formed by the Jacobi fields* $Y(t)$ *orthogonal to* $\dot c(t)$ *satisfying*

$$|\tilde Y(\omega)|^2=|Y(\omega)|^2+|\nabla Y(\omega)|^2\leqslant c|\tilde Y(0)|^2,$$

some c, $0<c<1$.

Proof. (cf. [Kl 12]) By our hypothesis, we have $V^{2n}(0)=V_s^n(0)+V_u^n(0)$, cf. (3.2.5). Here, $V_s^n(0)$ is generated by the eigenspaces $V(\rho)$ with $|\rho|<1$, whereas $V_u^n(0)$ is generated by the $V(\rho)$ with $|\rho|>1$. □

3.2.14 Corollary 2. *Let* $c=(c(t),\ 0\leqslant t\leqslant\omega)$ *be a non-degenerate elliptic closed geodesic on a surface M, i.e.* $\dim M=2$. *Denote by k the number of conjugate points of* $c(0)$ *along* $c(t)$ *in the interval* $]0,\omega[$. *If c is orientable {non-orientable} the index of c is* $k+1$ $\{k\}$ *or* k $\{k+1\}$, *depending on whether k is even or odd. In particular, on an orientable compact surface the index of an elliptic closed geodesic is always odd.*

Proof. (cf. [He 1], [Kl 19]). Consider the formula (3.2.12) for $\rho=1$, $n=1$, $p=0$, i.e. $V_e^n=V_v^1(0)$. Thus, $I_c=k$. It remains to determine the number $M_1=$(index plus nullity) Q_1 on $V_v^1(0)$.

We restrict ourselves to the case in which c is orientable, i.e., the normal bundle of the immersion $c:S_\omega\to M$ is trivial. We choose a unit normal vector E orthogonal to $\dot c(0)$. Using a parallel translation along c we have a canonical identification of the normal line over $t\in S_\omega$ with the normal line over $0\in S_\omega$. Moreover, the canonical identifications of the bundles τ_h^1,τ_v^1 over S_ω with the normal bundle yield a canonical identification of $\tau^2=\tau_h^1\oplus\tau_v^1: V^2\to S_\omega$ with $S_\omega\times\mathbb{R}^2$, where the fibre $\mathbb{R}^2$ possesses the orthonormal basis $\tilde E_h=(E,0)$, $\tilde E_v=(0,E)$.

Denote by

$$\tilde X(t)=x(t)\tilde E_h+\dot x(t)\tilde E_v;\ \tilde Y(t)=y(t)\tilde E_h+\dot y(t)\tilde E_v$$

the fundamental system of solutions of the Jacobi equations, i.e. $(x(0),\dot x(0))=(1,0)$; $(y(0),\dot y(0))=(0,1)$. The matrix representation of the Poincaré map P_c is given by

$$\begin{pmatrix} x(\omega) & y(\omega)\\ \dot x(\omega) & \dot y(\omega)\end{pmatrix},\ \text{with } 2-x(\omega)-\dot y(\omega)>0,$$

since P_c is elliptic.

Put $(P-1)^{-1}\tilde{Y}(t)=\tilde{Z}(t)$, $\tilde{Z}(t)=z(t)\tilde{E}_h+\dot{z}(t)\tilde{E}_v$. Then $Q_1(\tilde{Y}(0),\tilde{Y}(0))=\langle\nabla Y(0),Z(0)\rangle=z(0)$. We compute $z(0)$ from the linear equation $(P-1)\tilde{Z}(0)=\tilde{Y}(0)$ and find that

$$(*)\qquad z(0)=-y(\omega)/(2-x(\omega)-\dot{y}(\omega));\qquad 2-x(\omega)-\dot{y}(\omega)>0.$$

The number k of conjugate points of $c(0)$ in $]0,\omega[$ is the number of intersections of $\tilde{Y}(t)$ with the vertical axis $\mathbb{R}\tilde{E}_v$. Each such intersection is transversal and clockwise.

Let k be even. This is equivalent to $y(\omega)\geq 0$ where $y(\omega)=0$ occurs precisely if $\tilde{Y}(0)$ is an eigenvector of P_c for the eigenvalue -1. From (*) we have $Q_1(\tilde{Y}(0),\tilde{Y}(0))\leq 0$, i.e. $M_1=1$. On the other hand, if k is odd, this is equivalent to $y(\omega)<0$, i.e. $z(0)>0$. Thus, $M_1=0$. This completes the proof of (3.2.14). □

Let $P=P_c$ be the Poincaré map associated to the closed geodesic c. We write the eigenvalues of P of modulus 1 in the form $(\rho_j,\bar{\rho}_j)=(e^{2\pi i a_j},e^{-2\pi i a_j})$, $1\leq j\leq l-1$, with $a_0=0\leq a_1<\ldots<a_{l-1}\leq a_l=\frac{1}{2}$. Here we do not exclude the possibility that $l-1=0$, i.e. that P_c has no eigenvalue of modulus 1.

From (3.2.12), we see that the ρ-index $I_c(\rho)$ of c, considered as function on S, is constant on the connected components of $S-\{\rho_j,\bar{\rho}_j;\ 1\leq j\leq l-1\}$. Denote by $I_{c,j}$ the value of $I_c(\rho)$ for $\rho=e^{2\pi i a}$, $a_{j-1}<a<a_j$, $1\leq j\leq l-1$. Define

$$\alpha_c=2\sum_{j=1}^{l} I_{c,j}(a_j-a_{j-1});\qquad \beta_c=2\sum_{j=1}^{l} I_{c,j}.$$

With this we obtain the following estimates for the growth of the index of the iterates c^m, $m=1,2,\ldots$, of c.

3.2.15 Lemma. *Let c be a closed geodesic. Then the index I_{c^m} of the m-fold covering of c satisfies the relations*

$$(*)\qquad m\alpha_c-\beta_c\leq I_{c^m}\leq m\alpha_c+\beta_c.$$

Remark. A weak form of (*) is already contained in Bott's formula for the "average index" $\bar{I}_c=\lim_{m\to\infty} I_{c^m}/m$; see [Bo 3]. Gromoll and Meyer [GM 2] observed that a formula similar to (*) can be obtained from Bott's result. Also in [Kl 16] we derived a weak form of (*). The present formulation is due to Ziller [Zi]. One can show that this is the best possible general estimate.

Proof. We shall need that $I_c(\rho_j)\leq I_c(\rho)$, for ρ sufficiently near ρ_j. This follows from the continuity in ρ of the spectrum of the operator ${}_\rho A_c$ determined by $D^2E_\rho(c)$

$${}_\rho A_c=-(\mathrm{id}-\nabla^2)^{-1}\cdot(K_c^{\sim}+\mathrm{id}):\ {}_\rho T_c\Lambda\to{}_\rho T_c\Lambda;$$

see [Bo 3]. The subsequent formulas (3.2.16) also imply this. If we want to avoid this result we can still get a relation of the type (*) by replacing the $I_{c,j}$ by the number $\max(I_{c,j},I_c(\rho_j))$ in the definition of β_c.

We know from (3.2.9) that $I_{c^m}=\Sigma I_c(\rho)$, $\rho^m=1$. For the number $k_{m,j}$ of mth roots of unity $\rho=e^{2\pi il/m}$ satisfying $a_{j-1}<l/m<a_j$ the following estimates hold:

$$[m(a_j-a_{j-1})-1]\leqslant k_{m,j}\leqslant[m(a_j-a_{j-1})+1].$$

Thus,

$$m\alpha_c-\beta_c\leqslant 2\sum_{j=1}^{l} I_{c,j}[m(a_j-a_{j-1})-1]\leqslant \sum_{\rho^m=1} I_c(\rho)=I_{c^m}$$

$$\leqslant 2\sum_{j=1}^{l} I_{c,j}[m(a_j-a_{j-1})+1]\leqslant m\alpha_c+\beta_c.$$

Here we use that, if the term $I_c(\rho_j)\leqslant I_{c,j}$ occurs in $\Sigma I_c(\rho), \rho^m=1$, i.e. if $\rho_j^m=1$, then

$$k_{m,j}=[m(a_j-a_{j-1})]. \quad \square$$

Remark. It remains to investigate $I_c(\rho_0)$ for ρ_0, $|\rho_0|=1$, being an eigenvalue of P. We already know that $I_c(\rho)$ remains constant if ρ varies in a connected component of $S^1-\{\rho_j\}$ where $\{\rho_j\}$ is the set of eigenvalues of P in S^1. Following Bott [Bo 3], we introduce the *splitting numbers at* $\rho_0\in S^1$:

$$S_c^{\pm}(\rho_0)=\lim_{\rho\to\rho_0}(I_c(\rho)-I_c(\rho_0)),\quad \arg\rho\bar{\rho}_0\begin{cases}>0\\<0\end{cases}.$$

From (3.2.12) one can presume that these splitting numbers depend only on ρ_0 and P. That this is actually the case was proved by Bott [l.c.] using topological intersection theory. An entirely different proof of a more elementary algebraic nature was given in [Kl 15]; at the same time, in this paper explicit expressions are established for the $S_c^{\pm}(\rho_0)$ in terms of the "normal form" of P derived in (3.2.4).

For an eigenvalue ρ_0 of T, $|\rho_0|=1$, we introduce the following non-negative integers:

Consider first the case $\rho_0^2\neq 1$. Define $k_{co}^{\pm}(\rho_0,\bar{\rho}_0)$ to be the number of 2-dimensional subspaces of the type (3.2.3) (ii), $l=0$, $\rho=\rho_0$, with $\pm i$ on the right-hand side, respectively. Define $k_{ev}^{\pm}(\rho_0,\bar{\rho}_0)$ to be the number of subspaces of type (3.2.3) (ii), $l>0$, $\rho=\rho_0$, with $\pm i$ on the right-hand side, respectively. Finally, define $k_{od}^{-}(\rho_0,\bar{\rho}_0)$ to be the number of subspaces of type (3.2.3) (i) with $\rho=\rho_0$ and -1 on the right-hand side.

In the case $\rho_0^2=1$, we denote by $2k_{co}(\rho_0,\bar{\rho}_0)$ the number of all the subspaces of type (3.2.3) (ii) with $\rho=\rho_0$. We define $2k_{ev}(\rho_0,\bar{\rho}_0)$ similarly. Finally, $k_{od}^{-}(\rho_0,\bar{\rho}_0)$ denotes the number of subspaces of type (3.2.3) (i) with -1 on the right-hand side.

Using these invariants of a symplectic transformation P we have, cf. [Kl 15]:

3.2.16 Theorem. *The splitting numbers* $S_c^{\pm}(\rho_0)$ *of the* ρ*-index at an eigenvalue* ρ_0 *of the Poincaré map* P *associated to* c *are given by*

$$S_c^{\pm}(\rho_0)=k_{co}^{\pm}(\rho_0,\bar{\rho}_0)+k_{ev}^{\pm}(\rho_0,\bar{\rho}_0)+k_{od}^{-}(\rho_0,\bar{\rho}_0),\qquad \text{if}\quad \rho_0^2\neq 1;$$

$$S_c^{\pm}(\rho_0)=k_{co}(\rho_0,\bar{\rho}_0)+k_{ev}(\rho_0,\bar{\rho}_0)+k_{od}^{-}(\rho_0,\bar{\rho}_0),\qquad \text{if}\quad \rho_0^2=1.\quad \square$$

3.3 Properties of the Poincaré Map

In this section we shall investigate in more detail the Poincaré map associated to a periodic orbit of the geodesic flow. In particular, we shall investigate under what conditions the Poincaré map possesses periodic points. As observed earlier, such periodic points are in 1:1 correspondence with periodic orbits of the geodesic flow near the given periodic orbit.

The linear part P of a Poincaré map $\mathscr{P}:(\Sigma_0, X_0)\to(\Sigma_\omega, \Sigma_0)$ at the fixed point X_0 already supplies some information about our problem. As we shall see, however, only approximations of $\mathscr{P}$ at X_0 of third order (the so-called 3-jet of $\mathscr{P}$) will provide sufficient information to guarantee the existence of periodic points; this is the content of the fixed point theorem of Birkhoff and Lewis, cf. (3.3.3) and 3.3-Appendix.

Let $P: V^{2n}\to V^{2n}$ be a symplectic transformation. From (3.2.4), we can see that the eigenvalues $+1$ and -1 each occur with even multiplicity. Indeed, taking $+1$, dim $V_{od,1}$ and dim $(V_{ev,1}\oplus\overline{V_{ev,1}})$ are both even.

In the light of (3.2.1), we pick from the $2n$ eigenvalues of P (counted with their multiplicity) a set of n so-called *principal eigenvalues* as follows. ρ is a principal eigenvalue if $|\rho|<1$ or if $\rho=e^{2\pi i a}$ with $0\leqslant a\leqslant 1/2$; and here we take only half of the ρ in case $a=0$ or $a=1/2$.

We call a linear symplectic transformation P *N-elementary*, N some integer >0, if a set $\{\rho_1,\dots,\rho_n\}$ of principal eigenvalues of P satisfies the following condition. Whenever $k_1,\dots,k_n$ are integers with $1\leqslant\Sigma|k_j|\leqslant N$, then $\prod_j \rho_j^{k_j}\neq 1$.

1-elementary means that 1 is not an eigenvalue of P. We will be interested mainly in the 4-elementary case; this means in particular that there are no eigenvalues which are roots of unity of order $\leqslant 4$ and that there are no multiple eigenvalues.

Before we can explain our interest in 4-elementary transformations we formulate a useful method for describing "formal" symplectic transformations by a so-called generating function. Note that any "formal" symplectic transformation agrees up to order k with a symplectic transformation of class C^l, $l\geqslant k$, or C^ω. We follow the exposition in [Mos 5].

Let (x^k, y^k), $1\leqslant k\leqslant n$, or briefly (x, y), be canonical coordinates of $\mathbb{R}^{2n}$ with the canonical 2-form $\Sigma_k dx^k\wedge dy^k$. Denote by

$$(*)\qquad \begin{aligned}\xi^k &= x^k + a^k_{(r)}(x,y) + -\\ \eta^k &= y^k + b^k_{(r)}(x,y) + \dots\end{aligned}$$

a formal local symplectic transformation $\psi:(\mathbb{R}^{2n},0)\to(\mathbb{R}^{2n},0)$ where the $a^k_{(r)}$, $b^k_{(r)}$ are the homogeneous parts of order $r\geqslant 2$.

3.3.1 Proposition. *Let* (*) *be a formal symplectic transformation* ψ. *Then there exists a homogeneous polynomial* $H_{(r+1)}(x,y)$ *of degree* $r+1$ *such that*

$$(**)\qquad a^k_{(r)}(x,y)=\partial H_{(r+1)}(x,y)/\partial y^k;\qquad b^k_{(r)}(x,y)=-\partial H_{(r+1)}(x,y)/\partial x^k.$$

Conversely, given such a homogeneous polynomial $H_{(r+1)}(x,y)$, *there is a formal symplectic transformation* ψ *defined by* (*) *and* (**).

By introducing complex-conjugate coordinates $(\zeta,\bar{\zeta})=(\xi+i\eta,\xi-i\eta)$ *and* $(z,\bar{z})=(x+iy,x-iy)$, *we can write* (*), (**) *as*

$$(*)_{\mathbb{C}} \qquad \zeta^k = z^k + c^k_{(r)}(z,\bar{z}) + \dots$$

$$(**)_{\mathbb{C}} \qquad \begin{aligned} &c^k_{(r)}(z,\bar{z}) = \partial K_{(r+1)}(z,\bar{z})/\partial \bar{z}^k \\ &\overline{K_{(r+1)}(z,\bar{z})} = -K_{(r+1)}(z,\bar{z}). \end{aligned}$$

Proof. Since ψ, (*), is symplectic, the 1-form

$$\Sigma_k(\xi^k - x^k)d\eta^k + \Sigma_k(y^k-\eta^k)dx^k$$

is closed. Using (x,η) as independent variables (which is possible since $D\psi(0)$ is the identity), we can write this form as $dH(x,\eta)$ with $H(0,0)=0$, i.e.

$$(***) \qquad \xi^k = x^k + \partial H(x,\eta)/\partial\eta^k; \qquad y^k = \eta^k + \partial H(x,\eta)/\partial x^k.$$

$H(x,\eta)$ is called a generating function for ψ.

A comparison of (***) with (*) yields that $H(x,\eta)$ starts with a homogeneous polynomial $H_{(r+1)}(x,\eta)$ of degree $(r+1)$ such that (**) is satisfied.

Conversely, given a homogeneous polynomial $H_{(r+1)}(x,\eta)$, we can define a formal local symplectic transformation ψ implicitly by (***). This ψ is clearly of the form (*), (**).

The expression $(*)_{\mathbb{C}}$, $(**)_{\mathbb{C}}$ follows by simple computation.

$$K_{(r+1)}(z,\bar{z}) = -2iH_{(r+1)}((z+\bar{z})/2,(z-\bar{z})/2i),$$

$$c^k_{(r)}(z,\bar{z}) = a^k_{(r)}((z+\bar{z})/2,\ (z-\bar{z})/2i) + ib^k_{(r)}((z+\bar{z})/2,\ (z-\bar{z})/2i),$$

and

$$\partial/\partial\bar{z}^k = (\partial/\partial x^k + i\partial/\partial y^k)/2. \quad \square$$

We now can derive the *Birkhoff Normal Form* of a 4-elementary symplectic transformation. See [Bi 2], [Bi 3] for the basic ideas and [Mos 5] for the present proof.

3.3.2 Lemma. *Let*

$$\mathscr{P}:(\mathbb{R}^{2n},0)\to(\mathbb{R}^{2n},0)$$

be a local symplectic transformation of class C^1 *which is of class* C^3 *at the origin* 0. *Assume that* $P=D\mathscr{P}(0)$ *has only eigenvalues of modulus* 1 *and is* 4*-elementary.*

Then there exist real analytic symplectic complex-conjugate coordinates $(z,\bar{z})$ *near* $0\in\mathbb{R}^{2n}$ *such that this transformation is given in the so-called Birkhoff Normal Form*

$$z^{*k}=z^k \exp 2\pi i\big(a^k+\sum_l b_l^k z^l \bar{z}^l\big)+w^k(z,\bar{z}).$$

Here, $w^k(z,\bar{z})$ *is of class* C^1 *with vanishing derivatives up to order 3 at 0;* a^k *and* b_l^k *are real with* $\rho_k=\exp(2\pi i a^k)$ *a principal eigenvalue. If we choose the* a^k *in strictly increasing order, the matrix* (b_k^l) *is uniquely determined. In particular, the property* $\det(b_k^l)\neq 0$ *is independent of the particular choice of the normal form.*

Proof. In choosing appropriate linear complex-conjugate coordinates $(z^*,\bar{z}^*)$ in $\mathbb{R}^{2n}$, we can write $\mathscr{P}$ in the form

$$(\dagger)\qquad \zeta^{*k}=\rho_k\big(z^{*k}+p^k_{(2)}(z^*,\bar{z}^*)\big)+\dots$$

with $p^k_{(2)}$ homogeneous of degree 2. According to (3.3.1), there exists a third degree homogeneous polynomial $S_{(3)}(z^*,\bar{z}^*)=-\overline{S_{(3)}(z^*,\bar{z}^*)}$ such that $p^k_{(2)}(z^*,\bar{z}^*)=\partial S_{(3)}(z^*,\bar{z}^*)/\partial\bar{z}^{*k}$.

We subject ($\dagger$) to a real analytic symplectic coordinate transformation $\psi:(z,\bar{z})\to(z^*,\bar{z}^*)$ of the form $(*)_{\mathbb{C}}$, $(**)_{\mathbb{C}}$. The resulting mapping $\mathscr{Q}=\psi^{-1}\circ\mathscr{P}\circ\psi$ will be of the form

$$\zeta^k=\rho_k\big(z^k+q^k_{(2)}(z,\bar{z})\big)+\dots,$$

$$q_2^{(k)}(z,\bar{z})=\partial T_{(3)}(z,\bar{z})/\partial\bar{z}^k,$$

where $T_{(3)}(z,\bar{z})=-\overline{T_{(3)}(z,\bar{z})}$ is homogeneous of degree 3. We claim that

$$(\S)\qquad T_{(3)}(z,\bar{z})=S_{(3)}(z,\bar{z})-K_{(3)}(\rho z,\bar{\rho}\bar{z})+K_{(3)}(z,\bar{z}).$$

To see this, we compare the quadratic terms in the Taylor expansions of $\psi\circ\mathscr{Q}=\mathscr{P}\circ\psi$

$$\rho_k\partial T_{(3)}(z,\bar{z})/\partial\bar{z}^k+\rho_k\partial K_{(3)}(\rho z,\bar{\rho}\bar{z})/\partial\bar{z}^k=$$

$$\rho_k\partial S_{(3)}(z,\bar{z})/\partial\bar{z}^k+\rho_k\partial K_{(3)}(z,\bar{z})/\partial\bar{z}^k.$$

Our claim follows since all terms are homogeneous and, under conjugation, only the signs change.

We write $K_{(3)}(z,\bar{z})=\Sigma K_{\alpha\bar{\beta}}z^\alpha\bar{z}^\beta$ where $z^\alpha=\Pi z_k^{a_k}$, $\bar{z}^\beta=\Pi\bar{z}_k^{b_k}$, α, β being multi-indices, $|\alpha|=\Sigma a_k$, $|\beta|=\Sigma b_k$. Using similar notation for $S_{(3)}$, $T_{(3)}$, (§) becomes

$$(\S\S)\qquad T_{\alpha\bar{\beta}}=S_{\alpha\bar{\beta}}+K_{\alpha\bar{\beta}}(1-\rho^{\alpha-\beta}).$$

Since P is 4-elementary and $|\alpha|+|\beta|=3$ $1-\rho^{\alpha-\beta}=0$ if and only if $\alpha=\beta$. Hence, this never occurs; i.e. we can choose $K_{\alpha\bar{\beta}}$ in ψ such that $T_{\alpha\bar{\beta}}=0$.

We may therefore assume that $\mathscr{P}$ is given by (†), with $p_{(2)}^k$ being replaced by $p_{(3)}^k(z^*,\bar{z}^*)=\partial S_{(4)}(z^*,\bar{z}^*)/\partial\bar{z}^{*k}$, where $S_{(4)}$ is homogeneous of degree 4. Subjecting $\mathscr{P}$ to a real analytic coordinate transformation ψ with $(*)_{\mathbb{C}}$, $(**)_{\mathbb{C}}$ and $k=3$ we obtain, just as before, a relation (§), this time with homogeneous polynomials of degree 4. In the corresponding relation (§§), $1-\rho^{\alpha-\beta}=0$, $|\alpha|+|\beta|=4$, can occur only for $\alpha=\beta$. That is to say, by an appropriate choice of $K_{\alpha\bar{\beta}}$ we can make all $T_{\alpha\bar{\beta}}$ equal to zero for $\alpha\neq\beta$. To fix ψ, we put $K_{\alpha\bar{\alpha}}=0$. Thus, $\mathscr{Q}=\psi^{-1}\circ\mathscr{P}\circ\psi$ is given by

$$\zeta^k=\rho_k\big(z^k+\partial\big(\sum_{l,m} T_{lm\bar{l}\bar{m}}z^l z^m \bar{z}^l \bar{z}^m\big)/\partial\bar{z}^k\big)+\dots$$

If we put $T_{lk\bar{l}\bar{k}}+T_{kl\bar{k}\bar{l}}=2\pi i b_l^k$ we obtain the desired normal form — note that $T_{lk\bar{l}\bar{k}}$ is purely imaginary. This concludes the proof of (3.3.2). □

Note. If one assumes that P is $(2L+1)$-elementary, then one can write $\mathscr{P}$ in the form

$$z^{*k}=z^k\exp 2\pi i\,(\text{polynomial of degree } L \text{ in } z\bar{z})+O\big(|z|^{2L+1}\big).$$

If P is N-elementary for all N, then one obtains in this way formal power series; however, in general this series will not be convergent, cf. [SM], [Mos 4].

The importance of the normal form (3.3.2) lies in the fact that we can now formulate the following property. A local symplectic diffeomorphism $\mathscr{P}:(\mathbb{R}^{2n},0)\to(\mathbb{R}^{2n},0)$ is said to be of *twist type* if all eigenvalues of the linear part P have modulus 1 and P is 4-elementary and, in the normal form (3.3.2), $\det(b_l^k)\neq 0$.

To see the geometric meaning of this property more clearly, we introduce outside $0\in\mathbb{R}^{2n}$ new symplectic coordinates $\tau^k>0$, $l\leqslant k\leqslant n$, and $\theta^k \bmod 2\pi$, $l\leqslant k\leqslant n$, as follows:

$$2\tau^k=z^k\bar{z}^k;\qquad 2i\theta^k=\log z^k-\log\bar{z}^k.$$

One verifies at once that

$$\sum_k dx^k\wedge dy^k=\frac{i}{2}\sum_k dz^k\wedge d\bar{z}^k=\sum_k d\tau^k\wedge d\theta^k,$$

i.e. (τ,θ) are indeed symplectic coordinates.

The normal form (3.3.2) is then

$$(\S)\qquad \tau^{*k}=\tau^k+O(|\tau|^3,\theta);\quad \theta^{*k}=\theta^k+2\pi\big(a^k+\sum_l 2b_l^k\tau^l\big)+O\big(|\tau^2,|\theta\big).$$

That is to say, we may view $\mathscr{P}$ as a perturbation of the map $\mathscr{P}_0$ given by

$$(\S_0)\qquad \tau^{*k}=\tau^k;\quad \theta^{*k}=\theta^k+2\pi\big(a^k+\sum_l 2b_l^k\tau^l\big).$$

$\mathscr{P}_0$ leaves the tori $\{\tau^k = \tau_0^k\}$ globally invariant and operates on each of these tori by a product of rotations $\theta^k \mapsto \theta^k + 2\pi(a^k + \sum_l 2b_l^k \tau_0^l)$. That $\mathscr{P}$ is of twist type simply means that these rotations as functions of the τ^k are transversal to the tori, i.e., det $(\partial\theta^{*k}/\partial\tau^l) \neq 0$.

The importance of maps of twist type lies in the fact that such maps have infinitely many periodic points in every neighborhood of the origin. Here we call a point $z \neq 0$ in some preassigned neighborhood U of $0 \in \mathbb{R}^{2n}$ *periodic* of *period* N if the points $\{\mathscr{P}^k z;\ 0 \leqslant k \leqslant N-1\}$ all belong to U, if $\mathscr{P}^N z = z$ and if $\mathscr{P}^k z \neq z$ for $0 < k < N$. $\{\mathscr{P}^k z;\ 0 \leqslant k \leqslant N-1\}$ is also called a *periodic orbit* of $\mathscr{P}$.

Since det $(b_l^k) \neq 0$, there exist in every neighborhood of 0 elements $\tau_0^k > 0$, $1 \leqslant k \leqslant n$, such that $\mathscr{P}_0$ operates on the torus $\{\tau^k = \tau_0^k\}$ by a product of rational rotations

$$N\left(a^k + \sum_l 2b_l^k \tau_0^l\right) = m^k \in \mathbb{Z}.$$

That is to say, all points of the torus $\{\tau^k = \tau_0^k\}$ are periodic points of period N. Here we assume that N and $(m^1, \ldots, m^n)$ have no common divisor >1.

Using certain *a priori* estimates one can show that $\mathscr{P}$ has at least a finite number $\geqslant 1$ of periodic points of period N near the torus $\{\tau^k = \tau_0^k\}$. This is the content of the *Birkhoff-Lewis Fixed Point Theorem:*

3.3.3 Theorem. *Let* $\mathscr{P}: (\mathbb{R}^{2n}, 0) \to (\mathbb{R}^{2n}, 0)$ *be a local symplectic transformation of twist type. Then there exist in every neighborhood of* $0 \in \mathbb{R}^{2n}$ *infinitely many periodic orbits. The number of orbits of bounded period is finite.* □

Notes. For the case $n = 1$ and $\mathscr{P}$ real analytic, this was proved by Birkhoff [Bi 3] who used an idea of Poincaré [Po 2]; cf. also the proof in [Mos 1] and [SM]. For the case $n > 1$, the basic ideas of a proof are indicated in [Bi 4], [Ar 2], [AA]. However, no complete proof has appeared in print. This is true in particular under the weaker hypothesis that $\mathscr{P}$ is not analytic but only differentiable. Moser has treated this case in his lectures at New York University; in an appendix we present his proof.

Closely related to our case are Hamiltonian systems with an isolated equilibrium point of "general stable type" which correspond to our "twist type". It was this case, for arbitrary n and real analytic systems, which was studied by Birkhoff and Lewis [BL]. A gap in this proof was closed by Harris [Ha]. The differentiable version was treated by Marzouk [Ma] and Zehnder [Ze].

Local symplectic transformations $\mathscr{P}$ of twist type have received considerable attention in recent years, the reason being that Kolmogorov, Arnold and Moser were able to prove the existence of large sets of invariant tori near $0 \in \mathbb{R}^{2n}$ for such that maps $\mathscr{P}$ leaves these tori globally invariant and operates on each of them as a quasi-periodic transformation, cf. [Ar 1], [AA], [Mos 2,3,4], [SM].

If we look at an "elementary twist map" $\mathscr{P}_0$, (§$_0$), these invariant tori correspond in a rather precise sense to those $\tau^k = \tau_0^k$ where $a^k + \sum_l 2b_l^k \tau_0^l$ is (strongly) irrational, whereas the tori $\tau^k = \tau_0^k$ with $a^k + \sum_l 2b_l^k \tau_0^l$ rational correspond to periodic points, as we saw above. For more details see [AA], [Mos 4], [Ze] and (3.3.A).

An interesting consequence of these theorems on invariant tori is that for $n=1$ the Birkhoff-Lewis Fixed Point Theorem can be strengthened as follows:

3.3.4 Complement. *Let $\mathscr{P}:(\mathbb{R}^2,0)\to(\mathbb{R}^2,0)$ be a local area preserving diffeomorphism of twist type, $\mathscr{P}$ of class C^4. Then the closure of the periodic points has positive measure.*

For the proof we refer to [Mos 4], p. 54 and Zehnder [Ze] who quotes a similar result for twist maps of arbitrary dimension. □

If we wanted to apply the Birkhoff-Lewis fixed point theorem (3.3.3) to the geodesic flow, we would have to start with a periodic orbit $\phi_t X_0$ for which the Poincaré map $\mathscr{P}$ is of twist type. In particular, all eigenvalues of the associated linear Poincaré map P would have to be on the unit circle.

This situation need not exist in general, however; and even small perturbations of the geodesic flow might not be sufficient to achieve this. In fact, consider in the linear symplectic group $Sp(n)$ for any q satisfying $0\leqslant q\leqslant n$, the set $Sp_{(q)}(n)$ of elements $P\in Sp(n)$ which have exactly $2q$ of the $2n$ eigenvalues on the unit circle and which are N-elementary, $N>1$. One easily checks that these $Sp_{(q)}(n)$ are non-empty open sets in $Sp(n)$.

In the case $P\in Sp_{(0)}(n)$, i.e. P having no eigenvalues on the unit circle or P hyperbolic (cf. (3.2.13)), one cannot expect periodic points near 0 for $\mathscr{P}$. However, for $P\in Sp_{(q)}(n)$, $q>0$, such periodic points may exist. The idea is to restrict $\mathscr{P}$ to an invariant local submanifold of dimension $2q>0$ for which the corresponding linear map has all its eigenvalues on the unit circle and then apply the Birkhoff-Lewis Fixed Point Theorem.

The existence of such a local submanifold is contained in the following *Theorem on Invariant Manifolds.*

3.3.5 Theorem. *Let $\mathscr{P}:(\mathbb{R}^{2n},0)\to(\mathbb{R}^{2n},0)$ be a local symplectic diffeomorphism. Let $P=D\mathscr{P}(0)$ be its linear part and let*

$$(V_s^p+V_u^p)\oplus V_{ce}^{2q}$$

be the direct decomposition of $V^{2n}=\mathbb{R}^{2n}$ into the stable, unstable and center subspace with respect to P. That is to say, these spaces are invariant under P and $P|V_s^p$ has only eigenvalues ρ with $|\rho|<1$, $P|\ V_u^p$ has only eigenvalues ρ with $|\rho|>1$, and $P|V_{ce}^{2q}$ has only eigenvalues ρ with $|\rho|=1$, cf. (3.2.4).

Claim. *There exist local embeddings W_s^p, $W_u^p:(\mathbb{R}^p,0)\to(\mathbb{R}^{2n},0)$, $W_{ce}^{2q}:(\mathbb{R}^{2q},0)\to(\mathbb{R}^{2n},0)$ such that $T_0W_s^p=V_s^p$, $T_0W_u^p=V_u^p$, $T_0W_{ce}^{2q}=V_{ce}^{2q}$, which are invariant under $\mathscr{P}$, i.e. $\mathscr{P}W_s^p$, $\mathscr{P}W_u^p$, $\mathscr{P}W_{ce}^{2q}$ are locally equal to W_s^p, W_u^p, W_{ce}^{2q}, respectively. They are called stable, unstable and center manifolds, respectively.*

3.3.6 Complement. *If $\mathscr{P}$ is of class C^k or C^∞, $\mathscr{P}_s=\mathscr{P}|W_s^p$ and $\mathscr{P}_u=\mathscr{P}|W_u^p$ are of class C^k and C^∞, respectively. $\mathscr{P}_{ce}=\mathscr{P}|W_{ce}^{2q}$ is of class C^k and of class C^l, respectively, with $l<\infty$ but arbitrarily large, if only W_{ce}^{2q} is chosen sufficiently small.*

Whereas W_s^p and W_u^p are locally unique, W_{ce}^{2q} generally is not.

Proof. For the proof we refer to Moser [Mos 5]; see also Kelley [Ke] and Hirsch-Pugh-Shub [HPS]. Using Perron's method [Pe], Duistermaat has given a proof [unpublished manuscript, August 1972]. Special cases have been considered by Hartman [Hart]; see also Irwin [Ir] and Lanford [La III]. In [AM] one also finds a formulation of this theorem. □

Note. Actually, in our application we obtained $\mathscr{P}$ as the Poincaré map associated with a periodic orbit $\phi_t X_0$, $0 \leqslant t \leqslant \omega$, of the geodesic flow $\phi_t: T_1 M \to T_1 M$. There is a version of the invariant manifold theorem for periodic orbits which runs as follows. Assume that the linear part P of $\mathscr{P}$ has the properties stated in (3.3.5). Then there exist globally defined immersions

$$\tilde{W}_s^{p+1},\ \tilde{W}_u^{p+1}: (\mathbb{R} \times \mathbb{R}^p, (0,0)) \to (T_1 M, X_0)$$

$$(t, X) \mapsto \phi_t X$$

with

$$T_{X_0} \tilde{W}_s^{p+1} | \{0\} \times \mathbb{R}^p = V_s^p,$$

$$T_{X_0} \tilde{W}_u^{p+1} | \{0\} \times \mathbb{R}^p = V_u^p.$$

$\tilde{W}_s^{p+1}$ and $\tilde{W}_u^{p+1}$ are called *stable* and *unstable manifolds*, respectively. Moreover,

$$W_{ss}^p := \tilde{W}_s^{p+1} | \{0\} \times \mathbb{R}^p,$$

$$W_{uu}^p := \tilde{W}_n^{p+1} | \{0\} \times \mathbb{R}^p$$

are carried into themselves by ϕ_ω; they are called *strong stable* and *strong unstable* manifolds, respectively.

ϕ_ω operates on W_{ss}^p and W_{uu}^p as a contraction and an expansion, respectively. These immersions are uniquely determined by the property

$$W_{ss}^p = \{X \in T_1 M,\ \lim_{t \to \infty} \phi_t X = X_0\},$$

$$W_{uu}^p = \{X \in T_1 M,\ \lim_{t \to \infty} \phi_t X = X_0\}.$$

One can define a counterpart of the center manifold as a locally defined immersion:

$$\tilde{W}_{ce}^{2q+1}: (\mathbb{R} \times \mathbb{R}^{2q}, (0,0)) \to (T_1 M, X_0)$$

$$(t,\ X) \mapsto \phi_t X,$$

where (t, X) belongs to some neighborhood of $[0, \omega] \times \{0\} \in \mathbb{R} \times \mathbb{R}^{2q}$, $T_{X_0} \tilde{W}_{ce}^{2q+1} | \{0\} \times \mathbb{R}^{2q} = V_{ce}^{2q}$. Moreover, $W_{ce}^{2q} := \tilde{W}_{ce}^{2q+1} | \{0\} \times \mathbb{R}^{2q}$ is invariant under the flow, in the sense that there exists a locally defined differentiable map $\delta: W_{ce}^{2q} \to \mathbb{R}$, $\delta(X_0)=0$, such that $\phi_{\omega+\delta(X)}(X) \in W_{ce}^{2q}$.

It is important to observe, however, that W_{ce}^{2q} is generally not locally unique and that generally W_{ce}^{2q} is not transformed into itself simply by ϕ_ω as is the case for W_{ss}^p and W_{uu}^p.

In either case, if $2q>0$, we define a *center Poincaré map*

$$\mathscr{P}_{ce}:(\Sigma_{ce,0}, X_0) \to (\Sigma_{ce,\omega}, X_0)$$

to be a diffeomorphism of a neighborhood $\Sigma_{ce,0}$ of X_0 on W_{ce}^{2q} onto a neighborhood $\Sigma_{ce,\omega}$ of X_0 on W_{ce}^{2q} as the restriction of $\mathscr{P}$ in the case (3.3.6), or as the map $X \in W_{ce}^{2q} \mapsto \phi_{\omega+\delta(x)} X \in W_{ce}^{2q}$ in the case described above.

In the light of the existence of an invariant center manifold, we generalize the concept of a map of twist type as follows.

A local symplectic map $\mathscr{P}:(\mathbb{R}^{2n},0) \to (\mathbb{R}^{2n},0)$ is said to be of *twist type* if

(i) the corresponding linear map P is not hyperbolic, i.e. it has $2q>0$ eigenvalues on the unit circle;

(ii) P is 4-elementary;

(iii) in the normal form (3.3.2) of $\mathscr{P}_{ce} := \mathscr{P}|W_{ce}^{2q}$, $\det(b_l^k) \neq 0$.

Our next goal is to investigate under what conditions the Poincaré map associated to a periodic orbit of the geodesic flow is of twist type in the generalized sense just defined.

Consider as an example the geodesic flow on the unit tangent bundle $T_1 S^{n+1}$ of the sphere S^{n+1} of constant curvature. Here, the Poincaré map is the identity, since all orbits are periodic of the same period 2π.

Thus we are lead to the question whether it would be possible to produce a Poincaré map of twist type, associated to a periodic orbit $\phi_t X_0$, if we change the Riemannian metric arbitrarily little and thereby preserve $\phi_t X_0$ as a periodic orbit.

The first thing we have to specify is the topology on the set $\mathscr{G} = \mathscr{G}M$ of all Riemannian metrics on a compact differentiable manifold M. Recall that a Riemannian metric on M is a differentiable section

$$g: M \to L_s^2(TM)$$

in the vector bundle $L_s^2(\tau)$ of symmetric bilinear forms associated to the tangent bundle τ of M such that $g(p)$ belongs to the open cone of positive definite forms on T_pM, cf. (1.1).

If we fix a Riemannian metric on M, this metric defines a Riemannian metric on every vector bundle $\alpha: A \to M$ associated to the tangent bundle $\tau: TM \to M$ and, in particular, it defines a metric on $L_s^2(\tau)$. Using covariant derivatives, we also obtain Riemannian metrics for the r-jet bundle $J^r(\alpha): J^r(A) \to M$ of α, cf. [AR], [El 3].

Now let $\Gamma^r(\alpha)$ be the vector space of C^r-sections in $\alpha: A \to M$. Then the norm derived from the Riemannian metric on $J^r(\alpha)$ makes $\Gamma^r(\alpha)$ into a Banach space.

The set $\mathcal{G}^r M$ of Riemannian metrics of class C^r in this way becomes an open subset in the Banach space $\Gamma^r(L_s^2(\tau))$ of C^r-sections of $L_s^2(\tau)$.

We also let $\mathcal{G}^\infty M$ be the set of C^∞ Riemannian metrics. $\mathcal{G}^\infty M$ can be made into an open subspace of a Fréchet space.

In either case, the spaces $\mathcal{G}M$ are metric spaces and the underlying topology is independent of the Riemannian metric which we used in the construction of the metric, since M is compact. In particular, $\mathcal{G}M$ is a *Baire space* in the sense that every residual set $\mathcal{G}^* \subset \mathcal{G}M$ is dense. Here $\mathcal{G}^*$ is called *residual* if it can be represented as the intersection of a countable number of open dense subsets.

A residual subset might be viewed as an analog for Baire spaces (or at least for metric spaces) of a set of measure 1 in a measure space with total measure 1. An example of a residual set in the space of real numbers is given by the set of non-rational numbers.

We call a property of a Riemannian metric g on M *generic* if the set $\mathcal{G}^*$ satisfying this property contains a residual set.

The remainder of this section is devoted to the proof of the fact that the following property is generic: For every periodic orbit of the geodesic flow, the associated Poincaré map is either hyperbolic or of twist type.

As a first step towards the proof of this theorem we show that:

3.3.7 Proposition. *Let $\phi_t X_0$, $0 \leqslant t \leqslant \omega$, be a periodic orbit of the geodesic flow. Put $\tau \circ \phi_t X_0 = c(t)$, i.e. $c(t)$ is a prime closed geodesic of length ω, $|\dot{c}| = 1$.*

Claim. *In an arbitrarily small tubular neighborhood of the immersed circle $t \in S_\omega \mapsto c(t) \in M$ there exist arbitrarily small perturbations of the Riemannian metric g on M (small in the C^r-topology, any $r \geqslant 0$, of $\mathcal{G}M$) such that in the perturbed metric $c(t)$ is still a geodesic and the associated linear Poincaré map P is N-elementary, N any given integer > 0.*

Proof. (cf. [KT]) We begin by introducing *Fermi coordinates* along c, cf. [GKM]. That is to say, we choose an orthonormal frame E_i, $0 \leqslant i \leqslant n$, in $T_{c(0)}M$ with $E_0 = \dot{c}(0)$. By $E_i(t_0)$ we denote the orthonormal frame in $T_{c(t_0)}M$ obtained from $E_i(0)$ by parallel translation along $c(t)$, $0 \leqslant t \leqslant t_0$. Consider the differentiable map

$$\Phi : [0, \omega] \times \mathbb{R}^n \to M$$

$$(t; x^1, \ldots, x^n) \mapsto \exp_{c(t)}\Big(\sum_{i>0} x^i E_i(t)\Big).$$

This map is of maximal rank at $(t_0; 0)$; Φ defines coordinates in the usual sense when restricted to a sufficiently small neighborhood of $(t_0; 0) \in]0, \omega[\times \mathbb{R}^n$.

From the definition, it follows that

$$g_{ik}(t; 0) = \delta_{ik}; \qquad \partial g_{ik}(t; 0)/\partial x^l = 0;$$

$$\partial^2 g_{00}(t; 0)/\partial x^i \partial x^k = -2 R_{0i0k}(t; 0).$$

If we want to preserve $c(t)$ as geodesic under small perturbations of the given metric g, the metric g^* must be of the form

$$g^*_{00}(t;x)=g_{00}(t;x)+\sum_{i,k>0}\alpha_{ik}(t;x)x^i x^k;$$

$$g^*_{0i}(t;x)=g_{0i}(t;x)+\sum_{k>0}\beta_{ik}(t;x)x^k,\quad i>0;$$

$$g^*_{ik}(t;x)=g_{ik}(t;x)+\gamma_{ik}(t;x),\quad i,k>0.$$

Indeed, one easily checks that the curve $x^0(t)=t$, $x^i(t)=0$, for $i>0$, satisfies the equations of a geodesic for the g^*-metric

$$\ddot{x}^i(t)+\sum_{k,l}\Gamma^{*i}_{kl}(x(t))\dot{x}^k\dot{x}^l=0$$

because $\Gamma^{*i}_{00}(t;0)=0$, $i\geqslant 0$.

To make sure that the $g^*(t;x)$ defined above can be viewed as a small perturbation of the given Riemannian metric g, we assume that the $\alpha_{ik}(t;x)$, $\beta_{ik}(t;x)$, and $\gamma_{ik}(t;x)$ have their carriers in a small neighborhood of $(t_0;0)\in]0,\omega[\times\mathbb{R}^n$.

In the light of a later application we denote by $\{(x^0;x),(y^0;y)\}$ with $y=(y^1,\ldots,y^n)$ the canonical coordinates in T^*M associated to the Fermi coordinates $(x^0;x)\equiv(t;x)$. The local transversal hypersurface $\Sigma=\Sigma(X_0)$ at $X_0=\dot{c}(0)=E_0(0)\in T_1M$ has coordinates $x^0=t=0$, x^i for $i>0$ in a neighborhood of $0\in\mathbb{R}^n$, y^i for $i>0$ in a neighborhood of $0\in\mathbb{R}^n$, and y^0 near 1 determined by $\sum_{i,k}g^{ik}(0;x)y^iy^k=1$, cf. the section preceding (3.1.9).

We define coordinates z^α, $1\leqslant\alpha\leqslant 2n$, by $z^i=x^i$, $z^{i+n}=y^i$. Let $J^{\alpha\alpha'}$ denote the elements of the matrix

$$(J^{\alpha\alpha'})=\begin{pmatrix}0 & E_n\\ -E_n & 0\end{pmatrix}.$$

We can then write the Hamilton equations as

$$(H^0)\quad \dot{x}^0=H_{y^0},\quad \dot{y}^0=-H_{x^0},$$

$$(H^\alpha)\quad \dot{z}^\alpha=\sum_{\alpha'}J^{\alpha\alpha'}H_{\alpha'},$$

where H_α stands for $\partial H/\partial z^\alpha$. We shall denote higher partial derivatives of H similarly, e.g. $H_{\alpha\beta}\equiv\partial^2H/\partial z^\alpha\partial z^\beta$.

Let ζ^α, $1\leqslant\alpha\leqslant 2n$, be the coordinates x^i, y^i, $i>0$, of $\Sigma=\Sigma(X_0)$ described above; in particular, the origin X_0 of Σ is given by $\zeta^\alpha=0$. Denote by

$$(*)\qquad x^0(t;\zeta);\quad y^0(t;\zeta);\quad z^\alpha(t;\zeta)$$

the solution of (H^0), (H^α) which for $t=0$ has the value $\zeta\in\Sigma$. Then the Poincaré map $\mathscr{P}$ is given by

$$(\zeta^\alpha)\mapsto(z^\alpha(\omega(\zeta),\zeta))$$

where $\omega(\zeta)=\omega+\delta(\zeta)$, cf. (3.1.10).

We denote the partial derivatives at $t=0$ of the solutions (*) with respect to the ζ^α by the subscript α. In particular,

$$z_\beta^\alpha(t):=\partial z^\alpha(t;0)/\partial\zeta^\beta$$

are the coordinates of the Jacobi field $\tilde{Y}_\beta(t)=(Y_\beta(t),\nabla Y_\beta(t))$ along $c(t)$ with the initial condition $Y_\beta(0)=\sum_{i>0} z_\beta^i(0)E_i(0)=\delta_{i\beta}E_i(0)$, $\nabla Y_\beta(0)=\sum_{i>0} z_\beta^{i+n}(0)E_i(0)=\delta_{i+n,\beta}E_i(0)$. That is to say, the matrix $(z_\beta^\alpha(t))$ is the fundamental solution matrix of the Jacobi equations

$$\text{(J)}\qquad \dot{x}^i(t)=y^i(t);\qquad \dot{y}^i(t)=-\sum_k R_{0i0k}(t)x^k(t).$$

This follows from (3.1.6). We may also see this directly by differentiating (H^0), (H^α) with respect to ζ^β and putting $t=0$.

To do this, we first observe that in our Fermi coordinates $H_{x^0x^i}=H_{x^0y^i}=H_{y^0x^i}=0$. Hence we obtain from (H^0)

$$(H_\beta^0)\qquad \begin{aligned}\dot{x}_\beta^0&=H_{y^0y^0}y_\beta^0+\sum_{i>0}H_{y^0y^i}y_\beta^i,\\ \dot{y}_\beta^0&=0.\end{aligned}$$

Since $y_\beta^0(0)=0$, $y_\beta^0(t)=x_\beta^0(t)=0$. Thus, we get from (H^α) by differentiating with respect to ζ^β

$$(H_\beta^\alpha)\qquad \dot{z}_\beta^\alpha(t)=\sum_{\alpha',\rho}J^{\alpha\alpha'}H_{\alpha'\rho}(t)z_\beta^\rho(t).$$

Now,

$$\begin{aligned}H_{ij}(t)&=(1/2)\partial^2 g_{00}(t;0)/\partial x^i\partial x^j=-R_{0i0j}(t),\\ H_{ij+n}(t)&=\partial g_{0j}(t;0)/\partial x^i=0,\\ H_{i+n,j+n}(t)&=g_{ij}(t;0)=\delta_{ij}.\end{aligned}$$

That is to say, the matrix $z_\beta^\alpha(t)$ satisfying (H_β^α) and $z_\beta^\alpha(0)=\delta_\alpha^\beta$ is the matrix of fundamental solutions of (J).

Consider now the fundamental solution matrix $(z_\beta^{*\alpha}(t))$ of the system

$$(H_\beta^{*\alpha})\qquad \dot{z}_\beta^{*\alpha}(t)=\sum_{\alpha',\rho}J^{\alpha\alpha'}H_{\alpha',\rho}^*(t)z_\beta^{*\rho}(t)$$

where the g_{ik} are replaced by the g^*_{ik} described above. Then the same computations show that the matrix $(\Delta z^\alpha_\beta(t)) := (z^{*\alpha}_\beta(t))\,(z^\alpha_\beta(t))^{-1}$ satisfies

$$(\Delta z^\alpha_\beta(t))^{\cdot} = \begin{pmatrix} \beta_{ij}(t;0) & \gamma_{ij}(t;0) \\ -\alpha_{ij}(t;0) & -\beta_{ij}(t;0) \end{pmatrix} (\Delta z^\alpha_\beta(t)).$$

Now, observe that the matrix in this equation represents the general element of the Lie algebra of the symplectic group $Sp(n)$.

To complete the proof of (3.3.7) we note that the condition N-elementary is an open, dense and invariant condition for P, i.e. the set of elements $P \in Sp(n)$ which are N-elementary is open, dense and invariant under inner automorphisms. Therefore, an arbitrarily small perturbation of the metric g into a metric g^* of the type described above can be found such that the corresponding linear Poincaré map P^* (given by the matrix $(z^{*\alpha}_\beta(\omega))$ is N-elementary. Thus we have proved (3.3.7). □

Note that this should be viewed as a statement that, by small perturbations of the Riemannian metric, the 1-jet of the Poincaré map can be made to satisfy an open, dense and invariant condition. The next lemma will prove a similar result for the 3-jet of the Poincaré map. For a general result of this type we refer to [KT].

3.3.8 Lemma. *Let $\phi_t X_0$, $0 \leqslant t \leqslant \omega$, be a periodic orbit and $c(t) := \tau \circ \phi_t X_0$ the corresponding prime closed geodesic of length ω, $|\dot{c}| = 1$. Let $\mathscr{P}$ be the associated Poincaré map. Assume that the linear Poincaré map P has $2q$ eigenvalues on the unit circle, $2q > 0$.*

Claim. *In an arbitrarily small tubular neighborhood of the immersed circle $t \in S_\omega \mapsto c(t) \in M$ there exist arbitrarily small perturbations of the Riemannian metric g on M such that, for the perturbed metric, $c(t)$ is still a geodesic and the associated Poincaré map is of twist type.*

Proof. We can assume that P is 4-elementary, cf. (3.3.7). As in the proof of (3.3.7), we introduce Fermi coordinates along c. Let us consider modifications of the fundamental tensor $g_{ik}(t;x)$ of the following type:

$$g^*_{00}(t;x) = g_{00}(t;x) + \sum_{i,j,k,l>0} \alpha_{ijkl}(t;x)\,x^i x^j x^k x^l/12;$$

$$g^*_{0i}(t;x) = g_{0i}(t;x) + \sum_{j,k,l>0} \beta_{i+njkl}(t;x)\,x^j x^k x^l/6, \quad i>0;$$

$$g^*_{ij}(t;x) = g_{ij}(t;x) + \sum_{k,l>0} \gamma_{i+n\;j+n\;kl}(t;x)\,x^k x^l, \quad i,j>0.$$

Modifications of this type will not change the 1-jet or the 2-jet of $\mathscr{P}$.

We obtain differential equations for the coefficients $z^\alpha_\beta(t)$, $z^\alpha_{\beta\gamma}(t)$, and $z^\alpha_{\beta\gamma\delta}(t)$ of the 1-jet, 2-jet and 3-jet of $\mathscr{P}$ by differentiating the equations (H^0) and (H^α) above.

First of all, we claim that $x^0_\beta(t)$, $x^0_{\beta\gamma}(t)$, $x^0_{\beta\gamma\delta}(t)$, $y^0_\beta(t)$, $y^0_{\beta\gamma}(t)$, and $y^0_{\beta\gamma\delta}(t)$ all vanish identically. Indeed, we have already shown this for $x^0_\beta(t)$ and $y^0_\beta(t)$. Differentiating (H^0_β) once more yields

$$(H^0_{\beta\gamma}) \qquad \dot{x}^0_{\beta\gamma}=H_{y^0y^0}y^0_{\beta\gamma}+\cdots=y^0_{\beta\gamma},$$
$$\dot{y}^0_{\beta\gamma}=0$$

Since $x^0_{\beta\gamma}(0)=0$, $x^0_{\beta\gamma}(t)=y^0_{\beta\gamma}(t)=0$. Similarly we obtain $x^0_{\beta\gamma\delta}(t)=y^0_{\beta\gamma\delta}(t)=0$.

Differentiating (H^α_β) we obtain

$$(H^\alpha_{\beta\gamma}) \qquad \dot{z}^\alpha_{\beta\gamma}=\sum_{\alpha',\rho} J^{\alpha\alpha'}H_{\alpha'\rho}z^\rho_{\beta\gamma}+\sum_{\alpha',\rho,\sigma} J^{\alpha\alpha'}H_{\alpha'\rho\sigma}z^\rho_\beta z^\sigma_\gamma;$$

$$(H^\alpha_{\beta\gamma\delta}) \qquad \dot{z}^\alpha_{\beta\gamma\delta}=\sum_{\alpha',\rho} J^{\alpha\alpha'}H_{\alpha'\rho}z^\rho_{\beta\gamma\delta}+\cdots+\sum_{\alpha',\rho,\sigma,\tau} J^{\alpha\alpha'}H_{\alpha'\rho\sigma\tau}z^\rho_\beta z^\sigma_\gamma z^\tau_\delta.$$

The same equations hold for the coefficients $z^*{}_{..}^{.}$ of the jets of the Poincaré map $\mathscr{P}^*$ associated to the modified metric g^*. Since we only consider modifications of the special type described above, the 1-jet and 2-jet of $\mathscr{P}$ and $\mathscr{P}^*$ are the same. For the difference $w^\alpha_{\beta\gamma\delta}(t):=z^{*\alpha}_{\beta\gamma\delta}(t)-z^\alpha_{\beta\gamma\delta}(t)$ between the 3-jet coordinates we obtain the equations

$$(\mathrm{W}) \qquad \dot{w}^\alpha_{\beta\gamma\delta}(t)=\sum_{\alpha',\rho} J^{\alpha\alpha'}H_{\alpha'\rho}(t)w^\rho_{\beta\gamma\delta}(t)+\sum_{\alpha',\rho,\sigma,\tau} J^{\alpha\alpha'}E_{\alpha'\rho\sigma\tau}(t)z^\rho_\beta(t)z^\sigma_\gamma(t)z^\tau_\delta(t),$$

with $w^\alpha_{\beta\gamma\delta}(0)=0,$

$$E_{\alpha'\rho\sigma\tau}(t)=H^*_{\alpha'\rho\sigma\tau}(t)-H_{\alpha'\rho\sigma\tau}(t).$$

The identity $x^0(\omega(\zeta),\zeta)=\omega$ yields

$$\dot{x}^0\omega_\alpha+x^0_\alpha=0, \quad \text{i.e. } \omega_\alpha=0;$$
$$\dot{x}^0\omega_{\alpha\beta}+\cdots+x^0_{\alpha\beta}=0, \quad \text{i.e. } \omega_{\alpha\beta}=0;$$

and also

$$\omega_{\alpha\beta\gamma}=0.$$

Hence, the coordinates of the 3-jet of $\mathscr{P}$ in the representation

$$(\zeta^\alpha)\mapsto(z^\alpha(\omega(\zeta),\zeta))$$

are indeed given by $z^\alpha_{\beta\gamma\delta}(\omega)$.

The equations (W) therefore define the coordinates of the modification of the 3-jet of the Poincaré map associated to the periodic orbit in a general Hamiltonian system. In our case, however, we have the geodesic flow, i.e. our Hamiltonian system is restricted due to the fact that in

$$H^*(x,y)-H(x,y)=\tfrac{1}{2}\sum_{i,k}\left(g^{*ik}(x)-g^{ik}(x)\right)y^iy^k$$

the y occur in only quadratic terms. This means that the $E_{\alpha'\rho\sigma\tau}(t)$ are subject to the following restrictions, cf. the definition of the g^* given above,

$$E_{ijkl}(t)=\alpha_{ijkl}(t): \quad \text{no restriction};$$

$$E_{i+n\ jkl}(t)=\beta_{i+n\ jkl}(t): \quad \text{no restriction};$$

$$E_{i+n\ j+n\ kl}(t)=\gamma_{i+n\ j+n\ kl}(t): \quad \text{no restriction};$$

$$E_{i+n\ j+n\ k+n\ l}(t)=E_{i+n\ j+n\ k+n\ l+n}(t)\equiv 0.$$

Note, however, that the factors $z^\rho_\beta(t)z^\sigma_\gamma(t)z^\tau_\delta(t)$ of $E_{\alpha'\rho\sigma\tau}(t)$ in (W) are the elements of the matrix of fundamental solutions of (H^α_β), i.e. $z^\rho_\beta(0)=\delta^\rho_\beta$,

$$(\dot z^\rho_\beta(t))=\begin{pmatrix} 0 & \delta_{ij} \\ -R_{0i0j}(t) & 0 \end{pmatrix}(z^\rho_\beta(t)).$$

Since $\dot z^{j+n}_i(0)=\delta^{j+n}_i$, $\dot z^j_i(0)=\dot z^{j+n}_{i+n}(0)=0$, we see that multiplication of $E_{\alpha'\rho\sigma\tau}(0)$ by $\dot z^\rho_\beta(0)$ will move the index ρ out of the "bad" range $[n+1, 2n]$, where $E_{\alpha'\rho\sigma\tau}(t)$ may be restricted, into the "good" range $[1,n]$.

Thus, once we have moved away from 0, the solutions $w^\alpha_{\beta\gamma\delta}(t)$ of (W) need not be restricted any longer by the restrictions imposed upon the $E_{\alpha'\rho\sigma\rho}(t)$. One can convince oneself of this by computing higher derivatives of $w^\alpha_{\beta\gamma\delta}$ at $t=0$. We only mention the results.

$$\dot w^{i+n}_{jkl}(0)=-\alpha_{ijkl}(0);$$

$$\dot w^i_{jkl}(0)=\beta_{i+n\ jkl}(0); \qquad \dot w^{i+n}_{j+n\ kl}(0)=-\beta_{j+n\ ikl}(0);$$

$$\dot w^i_{j+n\ kl}(0)=\gamma_{i+n\ j+n\ kl}(0); \qquad \dot w^{i+n}_{j+n\ k+n\ l}(0)=-\gamma_{j+n\ k+n\ il}(0).$$

All other $\dot w^\alpha_{\beta\gamma\delta}(0)=0$. But

$$\ddot w^i_{j+n\ k+n\ l}(0)=-\gamma_{j+n\ k+n\ il}(0)+\gamma_{i+n\ j+n\ kl}(0)+\gamma_{i+n\ k+n\ jk}(0);$$

$$\ddot w^{i+n}_{j+n\ k+n\ l+n}(0)=\gamma_{k+n\ l+n\ ij}(0)+\gamma_{j+n\ l+n\ ik}(0)+\gamma_{j+n\ k+n\ il}(0);$$

$$\ddot w^i_{j+n\ k+n\ l+n}(0)=0;$$

$$\dddot w^i_{j+n\ k+n\ l+n}(0)=-\ddot w^{i+n}_{j+n\ k+n\ l+n}(0) +2\left(\gamma_{i+n\ l+n\ jk}(0)+\gamma_{i+n\ k+n\ jl}(0)+\gamma_{i+n\ j+n\ kl}(0)\right). \qquad \square$$

We are now ready to prove the generic nature of the properties stated in (3.3.7) and (3.3.8).

More precisely, let $\kappa>0$, N an integer $\geqslant 1$. Then $\mathscr{G}^r(N,\kappa)$ will denote the set of Riemannian metrics on M which are of class C^r and have the property that, for every closed geodesic of E-value $\leqslant\kappa$, the associated linear Poincaré map is N-elementary.

By $\mathscr{G}^r(N) \subset \mathscr{G}^r M$ we denote the set of Riemannian metrics such that, for all closed geodesics, the associated linear Poincaré map is N-elementary.

3.3.9 Theorem. *Let M be a compact differentiable manifold.*

Claims. (i) *The set $\mathscr{G}^r(N,\kappa)$ is dense in $\mathscr{G}^r M$ and contained in the interior of $\mathscr{G}^r(N,\kappa/2)$.*

(ii) *$\mathscr{G}^r(N)$ contains a residual set in $\mathscr{G}^r M$, i.e. the property that the linear Poincaré maps of all closed geodesics is N-elementary, is generic.*

Remark. For $N=1$, (ii) is the *bumpy metric theorem* of Abraham [Ab]. Our proof is somewhat more elementary than Abraham's proof.

Proof. We first show that $\mathscr{G}^r(N,\kappa)$ is dense in $\mathscr{G}^r M$. Let g be an arbitrary metric. For every closed geodesic c on (M,g) with $E(c) \leqslant 2\kappa$ we can find a tubular neighborhood $T(c)$ of the immersion $t \in S \mapsto c(t) \in M$ and a small modification g_c^* of the metric g inside $T(c)$ such that c remains a geodesic and such that all geodesics of E-value $\leqslant 2\kappa$ in $T(c)$ have an N-elementary linear Poincaré map. This follows from (3.3.7) and the fact that N-elementary geodesics are non-degenerate critical points (the linear Poincaré map does not have 1 as eigenvalue) and such points are isolated ΛM. Moreover, the property "N-elementary" is open.

Observe also that it suffices to consider prime geodesics c only. Indeed, the multiplicity of the closed geodesics having E-value $\leqslant 2\kappa$ is bounded by, say, m. Hence, we only have to make all prime closed geodesics in $\Lambda^{2\kappa}$ Nm-elementary.

The set of closed geodesics in $\Lambda^{2\kappa}$ is compact, cf. (1.4.9). Hence, a finite number of the tubular neighborhoods $T(c)$, say $T(c_i)$, $1 \leqslant i \leqslant k$, will cover all closed geodesics of (M,g) of E-value $\leqslant 2\kappa$.

We now define a modification g^* of the metric g as follows. Start with the modification $g_{c_1}^*$. We may even choose $g_{c_1}^*$ closer to g in the $\mathscr{G}^r M$-topology than in the beginning and still preserve the desired properties, since "N-elementary" is open and dense. Next, superimpose the modification $g_{c_2}^*$ (recall that $\mathscr{G}^r M$ is a cone in a Banach space), or a smaller version of $g_{c_2}^*$. If there should appear in $T(c_1) \cup T(c_2)$ closed geodesics of E-value $\leqslant 2\kappa$ for which the linear Poincaré map is not N-elementary, an additional small perturbation will suffice to eliminate such closed geodesics. We proceed in this way until we obtain the deformation $g_{c_k}^*$. Since $\bigcup_i T(c_i)$ covers all the original geodesics of E-value $\leqslant 2\kappa$ in the old metric g, it will cover all geodesics of E-value $\leqslant \kappa$ in the deformed metric.

To complete the proof of (i) we observe that, for $g \in \mathscr{G}^r(N,\kappa)$, a sufficiently small perturbation of g will not destroy the N-elementary nature of the geodesics of E-value $\leqslant \kappa$ in (M,g). However, the E-value of these geodesics might be affected slightly. But at least geodesics of E-value $\leqslant \kappa/2$ will keep their E-value $< \kappa$, i.e. $\mathscr{G}^r(N,\kappa)$ is in the interior of $\mathscr{G}^r(N,\kappa/2)$.

To prove (ii) we need only observe that (i) implies that $\mathscr{G}^r(N,\kappa)$ contains an open, dense subset of $\mathscr{G}^r M$, i.e. the interior of $\mathscr{G}^r(N,2\kappa)$ which contains the dense subset $\mathscr{G}^r(N,4\kappa)$. Since $\mathscr{G}^r(N)$ can be represented as the intersection of the countable many $\mathscr{G}^r(N,\kappa)\, \kappa=1,2,\ldots$, we have proved (ii). □

We now can prove the final result of this section.

3.3.10 Theorem. *Let M be a compact differentiable manifold. Then the following property of a Riemannian metric g on M is generic.*

For every closed geodesic on M, endowed with the metric g, the associated Poincaré map is either hyperbolic or of twist type.

Proof. (cf. [KT] for a slightly different proof). We proceed as in the proof of (3.3.9). Actually, it suffices to prove the analog of statement (i) in (3.3.9), since the property (ii) then follows in exactly the same manner.

For $\kappa > 0$, we denote by $\mathscr{G}^r(\tau, \kappa)$ the set of $g \in \mathscr{G}^r M$ with the property that every closed geodesic on (M, g) with E-value $\leqslant \kappa$ has a Poincaré map which is either hyperbolic or of twist type.

From (3.3.9) we know that $\mathscr{G}^r(4)$ is dense in $\mathscr{G}^r M$. Thus, a given $g \in \mathscr{G}^r M$ can first be approximated by a g' which has only finitely many closed geodesics of E-value $\leqslant \kappa$. To each of these geodesics we can apply (3.3.8). Thus, $\mathscr{G}^r(\tau, \kappa)$ is dense in M. Since the property of the Poincaré map in question is also open, a small perturbation of $g \in \mathscr{G}^r(\tau, \kappa)$ will leave g in $\mathscr{G}^r(\tau, \kappa/2)$, i.e. $\mathscr{G}^r(\tau, \kappa)$ is in the interior of $\mathscr{G}^r(\tau, \kappa/2)$. □

3.3 Appendix. The Birkhoff-Lewis Fixed Point Theorem

By Jürgen Moser

We present a complete proof of the so-called Birkhoff-Lewis Fixed Point Theorem (3.3.3). This theorem is closely related to a theorem by Birkhoff and Lewis [BL] concerning periodic solutions of a real analytic Hamiltonian system near an equilibrium point of general stable type. Note that we prove this theorem under very weak differentiability assumptions. For further comments, see the note following (3.3.3).

Let $x \in \mathbb{R}^n$. Thus, (x_k), (y_k), $1 \leqslant k \leqslant n$, are coordinates on $\mathbb{R}^{2n}$ endowed with the canonical or symplectic 2-form $\sum_k dx_k \wedge dy_k$. We consider a canonical transformation

$$\phi : (\mathbb{R}^{2n}, 0) \to (\mathbb{R}^{2n}, 0)$$

of the form

$$
\begin{aligned}
& x_k = x_k \cos \Phi_k(x,y) - y_k \sin \Phi_k(x,y) + f_k(x,y); \\
(1) \qquad & y_k = x_k \sin \Phi_k(x,y) + y_k \cos \Phi_k(x,y) + f_{k+m}(x,y); \\
& \Phi_k(x,y) = \alpha_k + \sum_{l=1}^{n} \beta_{kl}(x_l^2 + y_l^2).
\end{aligned}
$$

The error terms $f_k(x,y)$, $f_{k+n}(x,y)$ are assumed to have vanishing derivatives up to order 3 at the origin, but are otherwise only assumed to be in C^1. One could simply assume the f to be in C^3, but the following proof does not require this and this fine point will be useful later on.

Thus with $\rho^2 = \sum_k (x_k^2 + y_k^2)$ we have that

$$\text{(2)} \qquad \rho^{-3}\{|f(x,y)| + \rho(|D_x f| + |D_y f|)\}$$

tends to zero as $\rho \to 0$.

If we assume the matrix (β_{kl}) to be non-singular, we obtain a local symplectic tranformation of twist type, cf. (3.3.2) and (3.3.3) where we used complex-conjugate coordinates

$$z^k = x_k + iy_k, \; \bar{z}^k = x_k - iy_k, \; 2\pi a^k = \alpha_k, \; 2\pi b_l^k = \beta_{kl}.$$

3.3.A.1 Theorem. *If the map $\phi: (\mathbb{R}^{2n}, 0) \to (\mathbb{R}^{2n}, 0)$ is a canonical transformation with non-singular matrix (β_{kl}), then every punctured neighborhood $0 < \Sigma_k (x_k^2 + y_k^2) < \varepsilon$ contains a periodic orbit. The period tends to infinity as ε tends to zero.*

Proof. We introduce new coordinates (τ_k, θ_k) by

$$\text{(3)} \qquad x_k = \sqrt{2\tau_k \varepsilon} \cos \theta_k; \qquad y_k = \sqrt{2\tau_k \varepsilon} \sin \theta_k$$

with θ_k determined modulo 2π. Observe that this transformation is singular if one or more of the τ_k vanish. Therefore we restrict our attention to a ball

$$B_\rho : \sum_k \left(2\tau_k - \frac{1}{2n}\right)^2 < \rho^2, \quad \text{with} \quad 0 < \rho < \frac{1}{2n},$$

on which $2\tau_k > \frac{1}{2n} - \rho > 0$. Moreover, on this ball we have

$$0 < \sum_k (x_k^2 + y_k^2) = \varepsilon \sum_k 2\tau_k < \varepsilon(\tfrac{1}{2} + n\rho) < \varepsilon$$

so that it lies in the considered punctured neighborhood. The mapping ϕ is defined in the domain $B_\rho \times T^n$ where T^n denotes the torus $\{\theta_k \bmod 2\pi\}$.

The transformation (3) takes the 2-form $\alpha = \sum_k dx_k \wedge dy_k$ into $\varepsilon \sum_k d\tau_k \wedge d\theta_k$, i.e. the transformation ϕ is again canonical with respect to the symplectic 2-form $\sum_k d\tau_k \wedge d\theta_k$ in the new variables. Moreover, ϕ is what is called "exact cannonical", i.e. for any closed curve Γ in $B_\rho \times T^n$ we have

$$\int_\Gamma \sum_k \tau_k d\theta_k = \int_{\phi(\Gamma)} \sum_k \tau_k d\theta_k.$$

Indeed, we have $2\varepsilon\tau_k d\theta_k = y_k dx_k - x_k dy_k$, by Stokes' theorem

$$2\varepsilon \int_\Gamma \sum_k \tau_k d\theta_k = \int_{\partial\Omega} \sum_k (y_k dx_k - x_k dy_k) = -2 \int_\Omega \alpha,$$

where Ω is a two-dimensional manifold in $|x|^2+|y|^2<\varepsilon$ with boundary $\partial\Omega=\Gamma$. Since our original mapping preserves α, hence $\int_\Omega \alpha$, the new mapping preserves $\int\sum_k \tau_k d\theta_k$ for closed curves Γ in $B_\rho\times T^m$. Note that we used in this argument the fact that $|x|^2+|y|^2<\varepsilon$ is simply connected, which is, of course, not the case for $B_\rho\times T^m$. The concept "exact canonical" is indeed stronger than "canonical", as the example $\tau_k\mapsto\tau_k+c_k,\theta_k\mapsto\theta_k$ illustrates.

The new mapping which we again call ϕ is actually defined on the covering surface of $B_\rho\times T^m$ which is parametrized by τ_k,θ_k. The mapping takes the form

$$(4)\qquad \begin{aligned} \tau_k^{(1)}&=\tau_k+o_1(\varepsilon);\\ \theta_k^{(1)}&=\theta_k+\psi_k(\tau,\varepsilon)+o_1(\varepsilon);\\ \psi_k&=\alpha_k+\varepsilon\sum_{l=1}^n 2\beta_{kl}\tau_l. \end{aligned}$$

Here $o_1(\varepsilon)$ stands for a function $f=f(\tau,\theta,\varepsilon)$ of period 2π in the θ_k such that $\varepsilon^{-1}f$ and its first derivatives with respect to r,θ tend uniformly to zero as $\varepsilon\to 0$. If we use the notation $|f|_s$ for the maximum of all derivatives of order $\leqslant s$ in $B_\rho\times T^n$, we can express this by

$$\varepsilon^{-1}|f|_1\to 0 \quad \text{as} \quad \varepsilon\to 0.$$

Functions for which $\varepsilon^{-1}|f|_0\to 0$, are symbolically denoted by $o(\varepsilon)$.

3.3.A.2 Theorem. *Let ϕ be an exact canonical mapping in $B_\rho\times T^n$ of the form* (4) *with non-singular (β_{kl}). Then, for sufficiently small $\varepsilon>0$, the domain $B_\rho\times T^n$ contains at least one periodic point.*

Clearly, this theorem implies Theorem 3.3.A.1. We give the proof of 3.3.A.2 in several steps (see also Zehnder [Ze]).

3.3.A.3 Lemma. *Given a ρ' in $0<\rho'<\rho$ and a positive constant c_1, we show that the iterates ϕ^j for $j=1,2,\dots,N$ satisfy*

$$(5)\qquad \phi^j: B_{\rho'}\times T^n\to B_\rho\times T^n \quad \text{for} \quad j=1,2,\dots,N$$

if $\varepsilon N<c_1$ and $0<\varepsilon<\varepsilon_0$ where ε_0 depends on ρ,ρ',c_1. Moreover,

$$(6)\qquad \phi^j-\psi^j=\varepsilon^{-1}o_1(\varepsilon)\to 0 \quad \text{for} \quad j=1,2,\dots,N$$

where

$$\psi:(\tau,\theta)\mapsto(\tau,\theta+\psi(\tau,\varepsilon)).$$

The purpose of this lemma is to show that iterates $\phi^j(\tau,\theta)$ do not "escape" from the domain of definition, if $j=O(\varepsilon^{-1})$. It will be important to have the estimates for the first derivatives of ϕ^j as they are given by (6).

Proof. Using the usual Euclidian norm in the (τ, θ)-space we have

$$|\tau^{(j)} - \tau| < |\tau^{(1)} - \tau| + \ldots + |\tau^{(j)} - \tau^{(j-1)}| < j o(\varepsilon)$$

and since $j < N < \frac{c_1}{\varepsilon}$ we can obtain $|\tau^{(j)} - \tau| < c_1 \varepsilon^{-1} o(\varepsilon) < \rho - \rho'$ if ε is small enough. This proves (5).
Also

$$\begin{aligned}|\theta^{(j)} - \theta - j\psi(\tau, \varepsilon)| &\leqslant |\theta^{(1)} - \theta - \psi(\tau, \varepsilon)| + \ldots + |\theta^{(j)} - \theta^{(j-1)} - \psi(\tau, \varepsilon)| \\ &\leqslant o(\varepsilon) + \big(o(\varepsilon) + |\psi(\tau^{(1)}, \varepsilon) - \psi(\tau, \varepsilon)|\big) + \ldots \\ &\qquad + \big(o(\varepsilon) + |\psi(\tau^{(j-1)}, \varepsilon) - \psi(\tau, \varepsilon)|\big) \\ &\leqslant N o(\varepsilon) + \|\beta\| \varepsilon \sum_{l=1}^{j-1} |\tau^{(l)} - \tau| \\ &\leqslant N o(\varepsilon) + \varepsilon \sum_{l=1}^{j-1} l o(\varepsilon) \\ &= N o(\varepsilon) + N^2 \varepsilon o(\varepsilon) = (1 + c_1) N o(\varepsilon) = \varepsilon^{-1} o(\varepsilon).\end{aligned}$$

Here, $\|\beta\|$ is the norm of the matrix $\beta = (\beta_{kl})$. This proves $|\phi^j - \psi^j|_0 = o(\varepsilon)$, and to prove (6) it remains to estimate the Jacobians $d\phi^j, d\psi^j$.

We denote the Jacobian $d\phi$ at $\tau^{(j)}, \theta^{(j)}$ by F_j, and that of $d\psi$ at $\tau^{(j)}, \theta^{(j)}$ by G_j. But since $\tau^{(j)} = \tau$ and the Jacobian $d\psi$ is independent of θ, we have

$$G_j = G = \begin{pmatrix} I & \varepsilon\beta \\ 0 & I \end{pmatrix}.$$

We note that $|G - I|_0 = O(\varepsilon)$ and

$$|F_j - I|_0 = |F_j - G|_0 + |G - I| = o(\varepsilon) + O(\varepsilon) = O(\varepsilon).$$

Therefore the product of N such mappings F_j or G can be estimated by

$$(1 + O(\varepsilon))^N = e^{N O(\varepsilon)} = O(1),$$

i.e. it is bounded if $N\varepsilon$ is bounded. Therefore the Jacobians

$$|d\phi^j - d\psi^j|_0 = \Big| \sum_{k=0}^{k=j-1} F_j F_{j-1} \ldots F_{k+1} (F_k - G) G^{k-1} \Big|_0 \leqslant N c_2 o(\varepsilon) = \varepsilon^{-1} o(\varepsilon).$$

This proves (3.3.A.3). □

Next we search for a fixed point of ψ^N in $B_{\rho'} \times T^n$. This requires us to find a $\tau \in B_{\rho'}$ such that $(2\pi)^{-1} N\psi_k(\tau, \varepsilon)$ are integers for $k=1,\ldots,n$, i.e.

(7) $$(2\pi)^{-1} N\psi(\tau, \varepsilon) \in \mathbb{Z}^n.$$

3.3.A.4 Lemma. *If* $0 < \rho'' < \rho' < \rho < \frac{1}{2n}$ *and* $\varepsilon N \rho'' > 2\pi\sqrt{n}\,\|\beta^{-1}\|$ *and* ε *is sufficiently small, then there exists a* $\tau^* \in B_{\rho'}$ *satisfying* (7), *i.e. the entire torus* $\tau = \tau^*$ *consists of fixed points of* ψ^N.

Proof. The image of the ball $B_{\rho''}$ under the mapping $\tau \mapsto \psi(\tau, \varepsilon) = \alpha + \varepsilon\beta\tau$ contains a ball of radius

$$\varepsilon\|\beta^{-1}\|^{-1}\rho''.$$

Indeed, if $\sigma = \psi(\tau, \varepsilon)$, $\sigma' = \psi(\tau', \varepsilon)$ then

$$|\tau - \tau'| < \varepsilon^{-1}\|\beta^{-1}\|\ |\sigma - \sigma'|$$

and, taking τ' as the center of $B_{\rho'}$ and σ' as its image, then $|\sigma - \sigma'| < \rho''\varepsilon\|\beta^{-1}\|^{-1}$ implies that $|\tau - \tau'| < \rho''$, which proves the statement.

Thus the image of $B_{\rho''}$ under the map $\tau \mapsto (2\pi)^{-1} N\psi(\tau, \varepsilon)$ contains a ball of radius

$$R = (2\pi)^{-1}\|\beta^{-1}\|^{-1} N\varepsilon\rho'$$

and hence a cube of side length $n^{-\frac{1}{2}}R$. Thus, if $n^{-\frac{1}{2}}R > 1$ this cube certainly contains a lattice point of $\mathbb{Z}^n$, which proves the lemma. □

3.3.A.5 Lemma. *Setting*

$$c_0 = 2\pi\sqrt{n}\,\|\beta^{-1}\|/\rho'', c_1 = c_0 + 1,$$

taking an integer N in

$$c_0 < \varepsilon N < c_1$$

and ε *small enough, there exists a torus* $\tau = \tau(\theta, \varepsilon) = \tau^* + \varepsilon^{-1} o_1(\varepsilon)$ *in* $B_{\rho'} \times T^n$ *which is mapped by* ϕ^N *"radially", that is, in such a way that*

$$\theta^{(N)} - \theta \in 2\pi\,\mathbb{Z}^n.$$

Proof. By Lemma (3.3.A.3),

$$\theta^{(N)} - \theta = N\psi(\tau, \varepsilon) + \varepsilon^{-1} o_1(\varepsilon),$$

and, by the preceding lemma, there exists a $\tau^* \in B_{\rho'}$ and a lattice point $h \in \mathbb{Z}^m$ such that

$$(2\pi)^{-1} N\psi(\tau^*, \varepsilon) = h \in \mathbb{Z}^n.$$

To prove our lemma we have to solve the equations

$$\theta^{(N)} - \theta = N\psi(\tau, \varepsilon) + \varepsilon^{-1} o_1(\varepsilon) = 2\pi h$$

for τ. We can apply the implicit function theorem. If we drop the error term $\varepsilon^{-1} o_1(\varepsilon) = o_1(1)$ we have the solution $\tau = \tau^*$, and, since the Jacobian of $N\psi$ has an inverse

$$\|(\varepsilon N\beta)^{-1}\| \leqslant c_0^{-1} \|\beta^{-1}\|$$

which is bounded, there exists a C'-solution $\tau = \tau^* + \varepsilon^{-1} o_1(\varepsilon) \in B_{\rho'}$ which is what we wanted to show. At this point it was important to have an estimate for the first derivative of ϕ^j for $j < N$.

To find the desired fixed point of ϕ^N we have to solve the equation

$$\theta^{(N)} - \theta = 2\pi h, \quad h \in \mathbb{Z}^n;$$
$$\tau^{(N)} - \tau = 0.$$

So far we have solved only the first half of the equation, which holds on an m-dimensional torus $\mathscr{T}: \tau = \tau(\theta, \varepsilon)$. Geometrically speaking, we have to find a point of intersection of $\mathscr{T} \cap \phi^N(\mathscr{T})$, and any point of this intersection is a fixed point of ϕ^N.

Generally speaking, two nearby n-dimensional tori in a $2n$-dimensional space need not intersect, and it is here that the exact canonical character of ϕ comes into play.

In fact, on $\tau = \tau(\theta, \varepsilon)$ there exists a function $W(\theta, \varepsilon) \in C''$, of period 2π in $\theta_1, \ldots, \theta_n$, such that

$$\tau_k^{(N)} - \tau_k = \frac{\partial W}{\partial \theta_k}. \tag{8}$$

Indeed, since ϕ is an exact canonical mapping we have

$$\int_\Gamma \sum_k \tau_k^{(N)} d\theta_k^{(N)} - \sum_k \tau_k d\theta_k = 0$$

for any closed curve Γ on $\tau = \tau(\theta, \varepsilon)$. Since $\theta^{(N)} - \theta = 2\pi h$, and hence $d\theta^{(N)} = d\theta$, we have

$$\int_\Gamma \sum_k (\tau_k^{(N)} - \tau_k) d\theta_k = 0$$

for any closed curve Γ on the torus. Thus the integral

$$W(\theta) = \int_0^\theta \sum_k (\tau_k^{(N)} - \tau_k) d\theta_k$$

is independent of the path and defines the desired function W.

To find a fixed point of ϕ^N we have to pick a point $\tau=\tau(\theta,\varepsilon)$ with $\tau^{(N)}=\tau$, i.e. by (8) a critical point of W. For example, the maximum or minimum of W provides such a point in $B_{\rho'} \times T^n$, and the corresponding orbit lies in $B_\rho \times T^n$. This proves the Theorem (3.3.A.1). □

Chapter 4. On the Existence of Many Closed Geodesics

In this chapter we wish to attack the problem of the existence of many, even infinitely many, closed geodesics on a compact Riemannian manifold. Note that, up to now, we have proved the existence of only one closed geodesic, see (2.1.3) and (2.1.6). This result of Lusternik and Fet [LF] in 1951 was not extended until 1965 by Fet [Fe 3] to the existence of at least two closed geodesics, provided that all closed geodesics are non-degenerate.

The first section contains a complete proof of Fet's theorem. In preparation for the proof we first give a detailed exposition of the local topological structure of the space of unparameterized closed curves near a non-degenerate critical submanifold due to Švarc [Šv].

The second section is devoted to a simplified proof of a slightly strengthened version of the celebrated theorem of Gromoll and Meyer [GM] which asserts the existence of infinitely many closed geodesics on a compact Riemannian manifold M, provided the Betti numbers of ΛM are unbounded.

In Section 3 we prove that, on any compact Riemannian manifold with finite fundamental group, there exist infinitely many prime closed geodesics. The proof uses, besides the results of Gromoll and Meyer presented in (4.2), Sullivan's theory of the rational homotopy type and, in an essential manner, the properties of the Morse complex. An appendix by J. Sacks gives a detailed description of the minimal model of ΛM.

Section 4 contains some generic existence theorems, using properties of the geodesic flow.

4.1 Critical Points in ΠM and the Theorem of Fet

We begin with an exposition of the local topological structure in the space of unparameterized curves near a non-degenerate ciritical manifold. For a long time the fact was overlooked that the presence in general of a non-trivial isotropy group causes this structure to be much more complicated than in the case of a non-degenerate critical manifold in the space of parameterized curves. It was Švarc [Šv] who first pointed this out. Our presentation goes back to [Kl 4, 5, 6] where some special cases were considered in full detail.

As a first application of these results we shall give a complete proof of Fet's Theorem, which asserts the existence of at least two prime closed geodesics; here one has to assume that, in general, all closed geodesics are non-degenerate.

As in (2.2), we consider a non-degenerate critical submanifold B. We assume that B is connected and that all elements of B have the same multiplicity, say m. We therefore also write B_m instead of B. Let $k=k_m$ be the index of $B=B_m$ and put $E|B_m=\kappa=\kappa_m$. Clearly, $\kappa_m>0$.

Denote by B_1 the submanifold of ΛM formed by the underlying prime closed geodesics of B_m. Assume also that B_1 is non-degenerate of index k_1. Put $E|B_1=\kappa_1$.

Let

$$\mu=\mu_m: N=N_m\to B=B_m$$

be the *normal bundle of the submanifold* B. The fibre $\mu^{-1}(c)$ over $c\in B_m$ is the subspace of $T_c\Lambda M$ which is orthogonal to the tangent space T_cB_m of B_m at c.

μ_m splits into the positive and negative subbundles

$$\mu_m^{\pm}: N_m^{\pm}\to B_m.$$

Here, $(\mu^{\pm})^{-1}(c)$ is given by $T_c^{\pm}\Lambda M$.

Let $\tilde{\pi}: \Lambda M\to\tilde{\Pi}M=\Lambda M/S$ be the quotient with respect to the S-action $\tilde{\chi}$. We want to give a description of the map $\tilde{\pi}|N_m^-$. A similar description will hold for $\tilde{\pi}|N_m^+$.

For this purpose we go back to (3.2.8). Let $c_1\in B_1$ be the prime closed geodesic underlying $c=c_m\in B$, i.e. $c_m=c_1^m$. Let $\rho=e^{2\pi il/m}$, $0\leq l<m$, be a m - th root of unity. We define the bundle

$$_\rho\mu_1: {}_\rho N_1\to B_1$$

by taking as a fibre over c_1 the space $_\rho T_{c_1}\Lambda M$ of complex-valued H^1-vector fields ζ along c_1 satisfying $\bar{\rho}\zeta_1(1)=\zeta(0)$.

That this is indeed a bundle follows from the standard properties of such vector fields.

We associate with $_\rho\mu_1$ a bundle

$$_\rho\lambda_m: {}_\rho L_m\to B_m$$

by taking as a fibre over $c_m=c_1^m\in B_m$ the H^1-vector fields

$$({}_\rho\eta_m(t), c_m(t))=(\bar{\rho}^{mt}{}_\rho\zeta_1(mt), c_1(mt)), t\in S.$$

Here, $({}_\rho\zeta_1(t), c_1(t))$, $t\in S$, is an element in the fibre over c_1 of the bundle $_\rho\mu_1$.

Note that

$$\left({}_\rho\eta_m\left(t+\frac{1}{m}\right), c_m\left(t+\frac{1}{m}\right)\right)=(\bar{\rho}^{mt}\bar{\rho}\,{}_\rho\zeta_1(mt+1), c_1(mt+1))=({}_\rho\eta_m(t), c_m(t)).$$

That is to say, if we restrict the canonical S-action "change of parameter" to the subgroup $\mathbb{Z}_m$, then $\mathbb{Z}_m$ operates as the identity.

This allows us to define a free $S/\mathbb{Z}_m$-action on ${}_\rho\lambda_m$ by associating with $z=e^{2\pi i r}$, $0\leqslant r<\frac{1}{m}$, the map

$$({}_\rho\eta_m(t), c_m(t))\in{}_\rho\lambda_m^{-1}(c_m)\mapsto({}_\rho\eta_m(t+r), c_m(t+r))\in{}_\rho\lambda_m^{-1}(z\,.\,c_m).$$

We use this to define the bundle ${}_\rho\nu_m$ over $\tilde{\Delta}_m=B_m/S$ as quotient space of this action:

$$\begin{array}{ccc} {}_\rho L_m & \xrightarrow{/(S/\mathbb{Z}_m)} & {}_\rho O_m \\ \big\downarrow {}_\rho\lambda_m & & \big\downarrow {}_\rho\nu_m \\ B_m & \xrightarrow{/(S/\mathbb{Z}_m)} & \tilde{\Delta}_m \end{array}$$

Here, the horizontal maps are circle bundles.

Recall that $\mathbb{Z}_m\subset S$ operates on B_m as the identity. Thus, $\tilde{\Delta}_m=B_m/S$ can indeed be viewed as quotient of B_m by the action of $S/\mathbb{Z}_m$.

We also define the m-fold covering ${}_\rho\mu_1^m={}_\rho\mu_m$ of the bundle ${}_\rho\mu_1$ by

$${}_\rho\mu_m : {}_\rho N_m\to B_m .$$

Here, the fibre over $c_m=c_1^m$ is given by the vector fields

$$({}_\rho\zeta_m(t), c_m(t))=({}_\rho\zeta_1(mt), c_1(mt)),\ t\in S,$$

with

$$({}_\rho\zeta_1(t), c_1(t))\in({}_\rho\mu_1)^{-1}(c_1).$$

From (3.2.8) we know that the normal bundle $\mu_m : N_m\to B_m$ is the direct sum of the ${}_\rho\mu_m$, $\rho^m=1$.

The restriction of the canonical S-action on ${}_\rho\mu_m$ to $\mathbb{Z}_m$ gives a $\mathbb{Z}_m$-action where the generator $z_0=e^{2\pi i/m}$ of $\mathbb{Z}_m$ operates by multiplication with ρ

$$z_0\,.\,({}_\rho\zeta_m(t), c_m(t))=\left({}_\rho\zeta_m\left(t+\frac{1}{m}\right), c_m\left(t+\frac{1}{m}\right)\right)=(\rho\,{}_\rho\zeta_m(t), c_m(t)).$$

Denote by

$${}_\rho\mu_m/\mathbb{Z}_m : {}_\rho N_m/\mathbb{Z}_m\to B_m/\mathbb{Z}_m\cong B_1$$

the quotient bundle with respect to this $\mathbb{Z}_m$-action. The direct sum of the ${}_\rho\mu_m/\mathbb{Z}_m$, $\rho^m=1$, defines the bundle

$$\mu_m/\mathbb{Z}_m : N_m/\mathbb{Z}_m\to B_1 .$$

We define a $\mathbb{Z}_m$-action on ${}_\rho\lambda_m$ by using the isomorphism between the fibres ($c_m = c_1^m$)

$$ {}_\rho\lambda_m^{-1}(c_m) \leftrightarrow {}_\rho\mu_1^{-1}(c_1) \leftrightarrow {}_\rho\mu_m^{-1}(c_m) $$

and carrying the $\mathbb{Z}_m$-action just described over from ${}_\rho\mu_m^{-1}(c_m)$ to ${}_\rho\lambda_m^{-1}(c_m)$

$$ z_0 \cdot ({}_\rho\eta_m(t), c_m(t)) = (\rho\,{}_\rho\eta_m(t), c_m(t)). $$

This $\mathbb{Z}_m$-action on ${}_\rho\lambda_m$ commutes with the $(S/\mathbb{Z}_m)$-action on ${}_\rho\lambda_m$. We thus get an induced $\mathbb{Z}_m$-action on ${}_\rho\nu_m = {}_\rho\lambda_m/(S/\mathbb{Z}_m)$. The orbit $\left({}_\rho\eta_m(t+r), c_m(t+r);\ 0 \leqslant r < \frac{1}{m}\right)$ is carried into the orbit $\left(\rho\,{}_\rho\eta_m(t+r), c_m(t+r);\ 0 \leqslant r < \frac{1}{m}\right)$.

Denote by

$$ {}_\rho\nu_m/\mathbb{Z}_m : {}_\rho O_m/\mathbb{Z}_m \to \tilde{\Delta}_m/\mathbb{Z}_m \cong \tilde{\Delta}_1 $$

the quotient bundle with respect to this $\mathbb{Z}_m$-action. The direct sum of the ${}_\rho\nu_m/\mathbb{Z}_m$, $\rho^m = 1$, yields the bundle

$$ \nu_m/\mathbb{Z}_m : O_m/\mathbb{Z}_m \to \tilde{\Delta}_1 . $$

The S-orbits of the induced S-action on ${}_\rho\mu_m/\mathbb{Z}_m$ are in 1:1 correspondence with the elements of $\nu_m/\mathbb{Z}_m$: Indeed, the elements of $({}_\rho\mu_m/\mathbb{Z}_m)/S \cong {}_\rho\mu_m/S$ can be represented by

$$ \left(\rho^l\,{}_\rho\zeta_m(t+r), c_m(t+r);\ 0 \leqslant r < \frac{1}{m};\ 0 \leqslant l < m\right) $$

whereas the corresponding element of $\nu_m/\mathbb{Z}_m$ is given by:

$$ \left(\rho^l \bar{\rho}^{mt}\,{}_\rho\zeta_m(t+r), c_m(t+r);\ 0 \leqslant r < \frac{1}{m};\ 0 \leqslant l < m\right). $$

We therefore get the following result.

4.1.1 Lemma. *Assume that the set of critical points of ΛM at E-level $\kappa = \kappa_m > 0$ contains among its connected components a non-degenerate critical submanifold $B = B_m$ of index $k = k_m$, where all elements have the same multiplicity m.*

Let

$$ \mu^- = \mu_m^- : N^- = N_m^- \to B = B_m $$

be the negative bundle over B_m, where the fibre over $c_m \in B_m$ consists of the negative space $T_{c_m}^- \Lambda M$ as defined in (2.2).

The canonical S-action ΛM induces a S-action on μ_m^-. Let $\tilde{\Delta}=\tilde{\Delta}_m=B_m/S$ be the quotient manifold of B_m.

There exists a bundle

$$\nu^- = \nu_m^- : O^- = O_m^- \to \tilde{\Delta} = \tilde{\Delta}_m$$

over $\tilde{\Delta}_m$ such that the quotient of μ_m^- by the S-action can be factored into the commutative diagram:

$$\begin{array}{ccccc}
N_m^- & \xrightarrow[/\mathbb{Z}_m]{} & N_m^-/\mathbb{Z}_m & \xrightarrow[/(S/\mathbb{Z}_m)]{} & O_m^-/\mathbb{Z}_m \cong N_m^-/S \\
\downarrow{\scriptstyle \mu_m^-} & & \downarrow{\scriptstyle \mu_m^-/\mathbb{Z}_m} & & \downarrow{\scriptstyle \nu_m^-/\mathbb{Z}_m} \\
B_m & \xrightarrow[/\mathbb{Z}_m]{} & B_m/\mathbb{Z}_m & \xrightarrow[/(S/\mathbb{Z}_m)]{} & \tilde{\Delta}_1
\end{array}$$

Here, the first horizontal maps come from the action of the isotropy group $\mathbb{Z}_m$ of B_m on the fibres; hence, in particular, $B_m \to B_m/\mathbb{Z}_m$ is an isomorphism.
The second horizontal maps are circle bundles.

A corresponding diagram exists for the positive eigenbundle μ_m^+. □

The lemma allows us to give a complete description of the relative homology of $\tilde{\Pi}^\kappa$ mod $\tilde{\Pi}^{\kappa-}$ at a non-degenerate critical level.

4.1.2 Theorem. *Assume that the set of critical points of $\Lambda=\Lambda M$ at the E-level $\kappa>0$ is a non-degenerate critical submanifold B of index k and that all elements of B have the same multiplicity, say m. Let $\tilde{\pi}B=\tilde{\Delta}$; $\tilde{\Delta}$ is a manifold. Let $D\nu^-$ be the disc bundle of the bundle ν^- over $\tilde{\Delta}$, cf.* (4.1.1). *Let $\tilde{\Pi}=\tilde{\Pi}M$.*

Claim. $H^*(\tilde{\Pi}^\kappa, \tilde{\Pi}^{\kappa-}) = H^*(D\nu^-/\mathbb{Z}_m, \partial D\nu^-/\mathbb{Z}_m) = H^*(\tilde{\Delta}) \otimes H^*(D^k/\mathbb{Z}_m, \partial D^k/\mathbb{Z}_m)$ *where one has to take $\mathbb{Z}_2$-homology in the second equation, unless ν^- is orientable.*

Proof. The first equality follows from (2.4.10), (2.4.11) and (4.1.1) since all attachments are made equivariantly. The second equality is proved in the same way in which one proves the Thom isomorphism of the homology of a disc bundle modulo its boundary, where the homology of the base is shifted by the dimension k = dimension of the fibre (c.f. [Kl 4]). □

Of particular interest is the case in which B is the $O(2)$-orbit of a single non-degenerate closed geodesic c.

4.1.3 Corollary. *Assume that the set of critical points in $\Lambda=\Lambda M$ at E-level $\kappa>0$ consists of the $O(2)$-orbit $S.c \cup S.\theta c$ of a non-degenerate closed geodesic of index k.*

Claim. *In arbitrary coefficients*

$$H_*(\Pi^\kappa, \Pi^{\kappa-}) = H_*(D^k/\mathbb{Z}_m, \partial D^k/\mathbb{Z}_m).$$

Proof. Note that $(\Pi^\kappa, \Pi^{\kappa-}) = (\tilde{\Pi}^\kappa, \tilde{\Pi}^{\kappa-})/_\theta \mathbb{Z}_2$, cf. (2.2). Under θ, the two parts of $\Delta = O(2) \,.\, c/_\chi\!\sim\! S$ are identified. Thus, (4.1.3) follows from (4.1.2). The negative bundle is obviously orientable in this case. □

Remark. The previous results show the importance of computing the homology of the space $(D^k/\mathbb{Z}_m, \partial D^k/\mathbb{Z}_m)$, when the quotient is being taken with respect to an orthogonal action of the cyclic group $\mathbb{Z}_m$ on $\mathbb{R}^k$. That the homology of this space can be much richer than the homology of $(D^k, \partial D^k)$ is shown in the following lemma.

4.1.4 Lemma. *Let there be given an orthogonal representation of the cyclic group $\mathbb{Z}_m$ of order m on the Euclidean vector space $\mathbb{R}^k$.*

Claim. (i) *Consider an homology with rational coefficients. Then*

$$H_i = H_i(D^k/\mathbb{Z}_m, \partial D^k/\mathbb{Z}_m) = 0, \quad \text{if} \quad i \neq k \quad \text{or if} \quad i = k$$

and the subspace on which a generator T of $\mathbb{Z}_m$ is represented as $-\mathrm{id}$ has odd dimension. This is equivalent to saying that $(D^k/\mathbb{Z}_m, \partial D^k/\mathbb{Z}_m)$ is not orientable. In all other cases, $H_k = \mathbb{Q}$.

(ii) *Consider an homology with $\mathbb{Z}_2$-coefficients. If $\mathbb{Z}_m$ has representations of odd order only, then $H_i(D^k/\mathbb{Z}_m, \partial D^k/\mathbb{Z}_m) = \mathbb{Z}_2$ for $i = k$ and $= 0$ otherwise.*

If the subspace of $\mathbb{R}^k$ on which $\mathbb{Z}_m$ is represented as group of odd order has dimension $a < k$, then

$$H_i(D^k/\mathbb{Z}_m, \partial D^k/\mathbb{Z}_m) = \begin{cases} \mathbb{Z}_2, & a+2 \leqslant i \leqslant k, \\ 0 & \text{otherwise} \end{cases}$$

Proof. We first consider rational coefficients. Then it is well known that $D^k/\mathbb{Z}_m \bmod \partial D^k/\mathbb{Z}_m$ is homologically the same as $D^k \bmod \partial D^k$, except when T is operating orientation reversing, cf. [Bor 2]. This happens precisely when the dimension of the subspace on which T operates as $-\mathrm{id}$ is odd.

For $\mathbb{Z}_2$-coefficients we observe that the homology of a space does not change if we form the quotient with respect to a cyclic group which operates with odd order. Hence, if we write $m = r2^s$, r odd, it suffices to compute the homology with respect to the subgroup-operation $\mathbb{Z}_{2^s} \subset \mathbb{Z}_m$, since the subsequent quotient with respect to the group $\mathbb{Z}_r$ will not change the homology any more.

So we may assume that $m = 2^s$, $s \geqslant 0$. If a generator T operates with odd order, this means it operates as the identity: $H_i = H_i(D^k/\mathbb{Z}_m, \partial D^k/\mathbb{Z}_m) = \mathbb{Z}_2$ for $i = k$ and $= 0$ otherwise.

It remains to consider the case where we have an orthogonal decomposition

$$\mathbb{R}^k = \mathbb{R}^a \oplus \mathbb{R}^b \oplus \mathbb{C}^c$$

into subspaces invariant under a generator T of $\mathbb{Z}_m$ such that $T|\mathbb{R}^a = \mathrm{id}$, $T|\mathbb{R}^b = -\mathrm{id}$, and $\mathbb{C}^c = \mathbb{C}_1 \oplus \ldots \oplus \mathbb{C}_c$ such that $T|\mathbb{C}_j$ operates as multiplication with $e^{i\alpha_j}$, $0 < \alpha_j < \pi$, $a + b + 2c = k$, $b + 2c > 0$.

Denote the coordinates of $\mathbb{R}^a$ by x_k, $1 \leqslant k \leqslant a$, those of $\mathbb{R}^b$ by y_i, $1 \leqslant i \leqslant b$, and those of $\mathbb{C}_j$ by z_j, $1 \leqslant j \leqslant c$. Define a sequence of spaces in which each space contains the preceding one, by

$$u^a = \mathbb{R}^a \cap D^k;$$

$$u^{a+i} = \{y_i \geqslant 0, y_{i+1} = \cdots = y_b = z_1 = \cdots = z_c = 0\} \cap D^k, 1 \leqslant i \leqslant b;$$

$$u^{a+b+2j-1} = \{\arg z^j = 0, z_{j+1} = \cdots = z_c = 0\} \cap D^k;$$

$$u^{a+b+2j} = \{0 \leqslant \arg z_j \leqslant \alpha_j;\ z_{j+1} = \cdots = z_c = 0\} \cap D^k, 1 \leqslant j \leqslant c.$$

One checks that these u mod ∂u give a cell decomposition of $D^k/\mathbb{Z}_m$ mod $\partial D^k/\mathbb{Z}_m$. For the boundary operator one finds that.

$$\partial u^{a+1} = u^a;\ \partial u^{a+v} = 0,\ 2 \leqslant v \leqslant b + 2c.$$

Hence, the homology groups of $D^k/\mathbb{Z}_m$ mod $\partial D^k/\mathbb{Z}_m$ are $=\mathbb{Z}_2$ in dimensions i, $a+2 \leqslant i \leqslant k$, and $=0$ otherwise. □

Note. Lemma (4.1.4) is due originally to Švarc [Šv]. He made a slight error in the computation, which was corrected in [Al 2], [Kl 6]. Actually, Švarc computed the homology with coefficients in an arbitrary prime field $\mathbb{Z}_p$; the result depends on the decomposition $m = rp^s$, $(r, p) = 1$.

If we wish to apply the lemma to the computation of the relative homology at an isolated non-degenerate critical $O(2)$-orbit as in (4.1.3), we can turn to the formula (3.2.9) for the index I_{c^m} of the m-fold covering of c in terms of the ρ-index $I_c(\rho)$:

$$I_{c^m} = \sum_{\rho^m = 1} I_c(\rho).$$

The negative eigenspace $T_c^- \Lambda$ is the fibre over c of the negative bundle μ^- over $O(2) \cdot c$. Actually, we may speak of the negative bundle μ^- over $O(2) \cdot c$ also in the case that c is degenerate, the fibre over $z \cdot c$ being $T_{z.c}^- \Lambda$.

In any case, we obtain from (3.2.5) and the definition of the ρ-index $I_c(\rho)$ of c:

4.1.5 Proposition. *Let $c = c_0^m$ be a closed geodesic of multiplicity m. Then the ρ-index $I_c(\rho)$, $\rho = m^{th}$-root of unity, describes the representation of the isotropy group $\mathbb{Z}_m$ on the fibre $T_c^- \Lambda$ of the negative bundle over $O(2) \cdot c$ as follows.*

$I_c(1)$ gives the dimension of the subspace on which $\mathbb{Z}_m$ operates as identity.

Let T be a generator of $\mathbb{Z}_m$. Let $0 < p < [(m+1)/2]$. Then $2I_c(e^{2\pi i p/m})$ gives the dimension of the subspace on which T operates by multiplication with $e^{\pm 2\pi i p/m}$. If m is even, then $I_c(-1)$ gives the dimension of the subspace on which T operates as $-\mathrm{id}$. □

4.1.6 Corollary. *Let $c = c_1$ be a prime closed geodesic and assume that all iterates $c_m := c^m$, $m \geqslant 1$, are non-degenerate. For each integer $m \geqslant 1$, denote by $D\mu_m^-$ the negative disc bundle over the non-degenerate critical submanifold $O(2) \cdot c_m$. Let $D_{c_m}\mu_m^-$ denote the typical fibre of $D\mu_m^-$. Consider homology with rational coefficients.*

Claim. (i) *Assume that* $I_{c_1}(-1) = \text{index } c_2 - \text{index } c_1$ *is odd. Then, for all even* m,

$$H_*(D_{c_m}\mu_m^- / \mathbb{Z}_m, \partial D_{c_m}\mu_m^- / \mathbb{Z}_m) = 0.$$

Thus, if the critical set at E-level $\kappa_m = E(c_m)$ *consists precisely of* $O(2).\, c_m$ *and* m *is even, then*

$$H_*(\Pi^{\kappa_m}, \Pi^{\kappa_m -}) = 0.$$

(ii) *Let index* $c = k$. *Then, for all* $m \geq 1$,

$$H_{k+\text{odd}}(D_{c_m}\mu_m^- / \mathbb{Z}_m, \partial D_{c_m}\mu_m^- / \mathbb{Z}_m) = 0.$$

Proof. (i) follows from (4.1.4), (4.1.3) and (4.1.2).

To prove (ii) we first observe that $H_{k+\text{odd}} \neq 0$ implies $\dim D_{c_m}\mu_m^- = \text{index } c_m = \text{index } c_1 + \text{odd}$. Now,

$$\text{index } c_m = \text{index } c_1 + \sum_{\rho^m = 1} I_c(\rho),\ \rho \neq 1,$$

and since $I_c(\rho) = I_c(\bar{\rho})$, the right-hand side is of the form: $k + \text{odd}$ if an only if $I_c(-1)$ is odd and m is even. We now apply (i). □

Note. (4.1.6) (ii) will play a crucial role in the proof of Fet's theorem (4.1.8).

As an application of (4.1.2) we want to compute the relative homology at the critical level $\kappa_q = 2\pi^2 q^2 > 0$ of ΠS^n, cf. [Kl 4]. First of all, we know from (2.5.3) that the critical manifold B_q in ΛS^n at κ_q is the Stiefel manifold $V(2, n-1) = O(n+1)/O(n-1)$. Hence $\Delta_q := \pi B_q = O(n+1)/O(n-1) \times O(2)$ is the Grassmann manifold $G(2, n-1)$ of 2-planes (or $(n-1)$-planes) in $\mathbb{R}^{n+1}$.

The negative bundle μ_q^- over B_q is given by

$$\mu_q^- = \eta^{n-1} \oplus (\sigma_1^{2n-2} \oplus \cdots \oplus \sigma_{q-1}^{2n-2}).$$

Here, η^{n-1} is the bundle induced by π from the canonical $(n-1)$ bundle ζ^{n-1} over $G(2, n-1)$, cf. [Mi 1],

$$\begin{array}{ccc}
E(\eta^{n-1}) & \longrightarrow & E(\zeta^{n-1}) \\
\Big\downarrow \ \eta^{n-1} = \pi^* \zeta^{n-1} & & \Big\downarrow \ \zeta^{n-1} \\
V(2, n-1) \cong B_q & \xrightarrow{\ \pi\ } & G(2, n-1) \cong \Delta_q
\end{array}$$

The isotropy group $\mathbb{Z}_q$ of B_q operates as the identity on η^{n-1}.

The $(q-1)$ bundles σ_p^{2n-2}, $1 \leq p \leq q-1$, are all isomorphic to the bundle σ^{2n-2}, induced by π from the tangent bundle $\tau = \tau^{2m-2}$ of $G(2, n-1)$

$$\begin{array}{ccc}
\pi^* TG(2,n-1) & \xrightarrow{\tau^*\pi} & TG(2,n-1) \\
\Big\downarrow \sigma^{2n-2} := \pi^*\tau^{2n-2} & & \Big\downarrow \tau = \tau^{2n-2} \\
V(2,n-1) \cong B_q & \xrightarrow{\pi} & G(2,n-1) \cong \Delta_q
\end{array}$$

To describe the action of $\mathbb{Z}_q$ on σ_p^{2n-2} we note that the 2-fold covering $\tilde{G}(2,n-1)$ of $G(2,n-1)$, consisting of the oriented 2-planes in $\mathbb{R}^{n+1}$, has a natural complex structure; thus, we can define on $T\tilde{G}(2,n-1)$ the operation of $\mathbb{Z}_q$ on a generator as being multiplication by $e^{2\pi i p/q}$. The induced operation on $TG(2,n-1)$ will then be the operation of $\mathbb{Z}_q$ on σ_p^{2n-2}.

4.1.7 Theorem. *The negative bundle $\pi\mu_q^-$ over $\Delta_q = \pi B_q$ is given by*

$$\nu_q^-/\mathbb{Z}_q = \zeta^{n-1} \oplus (\tau_1^{2n-2} \oplus \cdots \oplus \tau_{q-1}^{2n-2})/\mathbb{Z}_q .$$

Here, ζ^{n-1} is the canonical $(n-1)$-bundle over $\Delta_q \cong G(2,n-1)$. All the τ_p^{2n-2}, $1 \leqslant p \leqslant q-1$, are isomorphic to the tangent bundle τ^{2n-2} of $G(2,n-1)$ with a generator of $\mathbb{Z}_q$ acting on τ_p^{2n-2} by multiplication with $e^{2\pi i p/q}$.

The homology in $\mathbb{Z}_2$-coefficients of $\Pi^{\kappa_q} S^n$ mod *$\Pi^{\kappa_q -} S^n$ is therefore given as follows. Set $q = 2^{q_0} q_1$ with q_1 odd.*

If $q_0 = 0$, i.e. q odd,

$$H_*(D\nu_q^-/\mathbb{Z}_q, \partial D\nu_q^-/\mathbb{Z}_q) = \tilde{H}_{*-(2q-1)(n-1)}(G(2,n-1));$$

if $q_0 > 0$,

$$H_*(D\nu_q^-/\mathbb{Z}_q, \partial D\nu_q^-/\mathbb{Z}_q) = \bigoplus_{i=(2q_1-1)(n-1)+2}^{(2q-1)(n-1)} \tilde{H}_{*-i}(G(2,n-1)).$$

Proof. Apply (4.1.2) and (4.1.4). □

As a further consequence of (4.1.4) we can now prove Fet's Theorem [Fe 3]. In (4.3) we shall prove a generalization of Fet's Theorem.

4.1.8 Theorem. *Let M be a compact Riemannian manifold.*

(i) *If $\pi_1 M$ is not finite then there exist at least two prime closed geodesics on M.*

(ii) *If $\pi_1 M$ is finite, assume that all closed geodesics are non-degenerate. Then there exist at least two prime closed geodesics on the compact universal covering $\tilde{M}$, having index k_0 and k_0+1, where $k_0+1 \leqslant k+1$, $k+1 \geqslant 2$, equal to the dimension of the first non-vanishing homotopy group of M. These two geodesics project onto closed geodesics on M which have different underlying prime closed geodesics.*

Proof. We first assume that the universal covering $\tilde{M}$ of M is contractible. The existence of two closed geodesics in this case is claimed by Fet [Fe 2]. At the moment, we are not able to present a proof. We can only mention a few facts which make it very plausible that the assumption that only one closed geodesic on M exists leads to a contradiction. In fact, this assumption implies that $\pi_1 M$ has no element of finite order, that the first Betti number is zero and that there is an element $\alpha \in \pi_1 M$ such that every element in $\pi_1 M$ is conjugate to some power of α.

Next assume that $\pi_1 M$ is infinite and the universal covering $\tilde{M}$ is not contractible. Then there must be a first $k+1$, $2 \leqslant k+1 \leqslant n = \dim M$, with $\pi_{k+1} M \neq 0$. With the help of a homotopically non-trivial map $f: S^{k+1} \to M$ we construct, according to (2.1.8), a closed nullhomotopic geodesic on M. A non-trivial element of $\pi_1 M$ of infinite order gives rise to a second closed geodesic. The underlying prime closed geodesics of these two geodesics must differ from one another, since no iterate of the second geodesic will be nullhomotopic.

Now assume that $\pi_1 M = 0$. The first non-vanishing homotopy group appears in dimension $k+1$, $2 \leqslant k+1 \leqslant n = \dim M$. From (2.1.6) or (2.1.8) we obtain the existence of a non-degenerate closed geodesic c of index k.

Among the closed geodesics choose one, say c_0, which has the smallest index, say k_0, and the smallest E-value, say κ_0. Then c_0 is prime and $k_0 \leqslant k$. The negative disc $D^-(\tilde{\pi} c_0)$ at $\tilde{\pi} c_0 \in \tilde{\Pi} = \tilde{\Pi} M$ is in 1:1 correspondence with the negative disc $D^-(c_0)$ of an element $c_0 \in \Lambda M$ covering $\tilde{\pi} c_0$.

Via the strong unstable manifold, $D^-(\tilde{\pi} c_0)$ represents a non-trivial k_0-cycle of $\tilde{\Pi}^{\kappa_0}$ mod $\tilde{\Pi}^{\kappa_0 -}$, cf. (2.5.1). We consider the inclusion

$$(\tilde{\Pi}^{\kappa_0}, \tilde{\Pi}^{\kappa_0 -}) \to (\tilde{\Pi}, \tilde{\Pi}^{\kappa_0 -}).$$

If the cycle $D^-(\tilde{\pi} c_0)$ is hereby killed, there must be another critical point $\tilde{\pi} c_1 \in \tilde{\Pi}$ of E-value $> \kappa_0$ and index $k_0 + 1$.

Otherwise, we show that there exists a homotopy h^* from the cycle $D^-(\tilde{\pi} c_0)$ to $D^-(\theta \tilde{\pi} c_0)$. To see this we first observe that the negative cycle $D^-(c_0)$ of $c_0 \in \Lambda M$ can be induced in the manner of (2.1), cf. in particular (2.1.7), by a map

$$f: S^{k_0+1} \to M$$

which determines a map

$$f^* = \Lambda f \circ \delta^{k_0} : (D^{k_0}, \partial D^{k_0}) \to (\Lambda M, \Lambda^0 M)$$

such that $f^* D^{k_0} = D^-(c_0)$. The midpoint m_0 of D^{k_0} corresponds to c_0 and nearby points go into elements of the strong unstable manifold.

If we consider D^{k_0} as half-equator of S^{k_0+1}, we can fix two orthonormal vectors (e_1, e_2) tangent to S^{k_0+1} at m_0, one orthogonal to D^{k_0}. Consider the rotation which carries (e_1, e_2) into $(-e_1, -e_2)$. This gives a homotopy h from δ^{k_0} into $-\theta \delta^{k_0}$, which we consider as a chain in ΛS^{k_0+1}. Put $h^* = \Lambda f \circ h$. The ϕ-family $\{\phi_s h^*\}$ generated by h^* then has a critical point = closed geodesic c_1 of index $k_0 + 1$, since $-\partial \phi_s h^* = -\phi_s \partial h^* = \phi_s D^-(c_0) + \phi_s \theta D^-(c_0) = D^-(c_0) + D^-(\theta c_0)$.

Thus we always have, besides a closed geodesic c_0 of index k_0, a closed geodesic c_1 of index $k_0 + 1$. But then, according to (4.1.6), c_1 cannot be a multiple covering of a closed geodesic of index k_0. Hence, c_1 is prime.

Finally, consider the case $\pi_1 M < \infty$. Then the universal covering $\tilde{M}$ of M is compact and there exist two prime closed geodesics $\tilde{c}_0$ and $\tilde{c}_1$ on $\tilde{M}$ of index k_0 and k_0+1, respectively. Their images in M must have different underlying prime closed geodesics. □

Note. The homotopy which we used in the proof will play an essential role in the generalization (5.1.1) of Fet's Theorem.

4.2 The Theorem of Gromoll-Meyer

In this section we will give a somewhat simplified proof of a slightly strengthened version of the Gromoll-Meyer Theorem: If there exists a prime field for which the sequence of Betti numbers of ΛM is unbounded, then there are infinitely many prime closed geodesics on M.

The crucial step in the proof of this theorem is a semi-local result concerning the sequence c^m, $m=1,2,\ldots$, of iterates of a closed geodesic c: Assume that index $c^{m_0}>0$ for some $m_0 \geqslant 1$ and that the $S\,.\,c^m$ are all isolated critical sets. Then the sum $\sum_m b_i(c^m)$ of the i-th type numbers is bounded from above by a constant independent of i, for all i sufficiently large.

Let $c=(c(t))$ be a closed geodesic of multiplicity $m \geqslant 1$. Denote by $T_c'=T_c'\Lambda M$ the subspace of $T_c\Lambda M$ of codimension 1 which is orthogonal to $\dot{c} \in T_c\Lambda M$. From the decomposition (2.4.3) we get the orthogonal decomposition

$$(*)\qquad T_c'\Lambda = T_c^+\Lambda \oplus T_c^-\Lambda \oplus T_c'^0\Lambda$$

where $T_c'^0\Lambda = T_c^0\Lambda \cap T_c'\Lambda$ consists of the periodic Jacobi fields along c which are orthogonal to $\dot{c}$.

Recall from (2.4) that index $c=\dim T_c^-\Lambda$, nullity $c=\dim T_c'^0\Lambda$. We do not assume that nullity $c=0$, i.e. that c is non-degenerate.

Let $\mu=\mu(S\,.\,c): N \mapsto S$ be *the normal bundle* of c over S, induced from the embedding $z \in S \to z^{\frac{1}{m}}\,.\,c \in \Lambda$, where $m=$multiplicity of c. This is the same as the normal bundle of the submanifold $S\,.\,c$ formed by the S-orbit of c, cf. (4.1). Note, however, that here we do not assume that $S\,.\,c$ is a non-degenerate critical submanifold.

Let $\mu=\mu^+ \oplus \mu^- \oplus \mu^0$ be the splitting of the normal bundle, determined by the splitting (*) of the fibre. Using the exponential map we can identify the total space $D=D(S\,.\,c)$ of a sufficiently small disc bundle $D\mu$ of μ with an open neighborhood of $S \subset N$. We use the exponential map to pull the Riemannian metric and the energy integral E back onto D. The action of S on D respects these quantities.

For $z \in S$, we denote by D_z the fibre over z in $D\mu$. The restriction of E to D_z will be denoted by E_z and the gradient field of E_z with respect to the induced metric will be denoted by grad E_z.

Actually, for our purposes a different metric on D is useful. Let m be the multiplicity of c. Define on D the following modification of the Riemannian metric

$$\langle \xi, \xi' \rangle_{m,1} := \langle \xi, \xi' \rangle_0 + \frac{1}{m^2} \langle \nabla\xi, \nabla\xi' \rangle_0 .$$

On the tangent space of each $\xi \in D$, the metric $\langle\ ,\ \rangle_{m,1}$ is clearly equivalent to the metric $\langle\ ,\ \rangle_1$. The index and nullity of c are not affected by this change of the metric.

The gradient field of E_z with respect to $\langle\ ,\ \rangle_{m,1}$ will be denoted by $\operatorname{grad}_m E_z$. For $\xi \in N$ we denote by $T'_\xi N \subset T_\xi N$ the subspace of the tangent space which is tangent to the fibre.

4.2.1 Proposition. *Let c be a closed geodesic of nullity $l \geq 0$ and multiplicity $m \geq 1$. Using the previously defined concepts, we put $\mu^+ \oplus \mu^- = \mu^*$,* i.e. $\mu = \mu^* \oplus \mu^0$, *fibre dimension of $\mu^0 = l$. Denote by $0_\sim$ the zero section of the bundle $\mu^{\sim}$, and by $D_\sim$ the total space of a disc bundle associated to $\mu^{\sim}$, $\sim \in \{-, 0, +, *\}$.*

Claim. *There exists a local equivariant embedding*

$$W_{ca} : (D_0, 0_{D_0}) \to (D, 0_D)$$

characterized by

$$T'_{0_z} W_{ca} = T'_{0_z} D_0 = (\mu^0)^{-1}(z),$$

and by the property that $\operatorname{grad}_m E_z(W_{ca}(\xi_0)) \in T'_{W_{ca}(\xi_0)} D$ *is parallel to the fibre* $(\mu^0)^{-1}(z)$.

Moreover, there exist a local equivariant diffeomorphism ψ of D and a section $z \in S \mapsto P_z \in L(\mu^*; \mu^*)$ with P_z being an orthogonal bundle projection such that, for $(\xi, \xi_0) \in D_* \oplus D_0$,

$$E_z \circ \psi(\xi, \xi_0) = \| P_z \xi \|^2_{m,1} - \| (\mathrm{id} - P_z) \xi \|^2_{m,1} + E_z \circ W_{ca}(\xi_0)$$

with $z = \mu^*(\xi) = \mu^0(\xi_0)$, $\dim \ker P_z =$ *index c = fibre dimension μ^-.*

Note. (4.2.1) was proved by Gromoll and Meyer [GM 1]. It is the Hilbert space version (based on Palais's proof of the generalized Morse Lemma (2.4.8)) of a more precise result of Thom [Th] who also proved the uniqueness up to conjugation of the degenerate part $E_z \circ W_{ca} : D_0 \to \mathbb{R}$ of E_z. We call W_{ca} the *characteristic manifold at c* and also write $W_{ca}(c)$.

Proof. Define a local fibre map $\lambda : D_* \oplus D_0 \to D_* \oplus D_0$ by

$$\lambda(\eta, \xi_0) = \mathrm{pr}_1 \operatorname{grad}_m E_z(\eta, \xi_0) + \xi_0$$

where pr_1 is the bundle projection $\mu = \mu^* \oplus \mu^0 \to \mu^*$.

We claim that in

$$D\lambda(0_z) = D(\mathrm{pr}_1 \operatorname{grad}_m E_z)(0_z) + \mathrm{id}_{\mu_0}(0_z)$$

the first map is invertible. Indeed, for $\|\eta\|_{m,1}$, $\|\xi_0\|_{m,1}$ sufficiently small, the product $\langle\,,\,\rangle_{m,1}$ in $T'_{(\eta,\xi_0)}\mu$ can be approximated by the product in $T'_{0_z}D=\mu^{-1}(z)$. Hence, since the bilinear form

$$\langle D\,\mathrm{grad}_m E_z(0)_z\,.\,\eta',\eta\rangle_{m,1}=D^2E_z(0_z)\,.\,(\eta,\eta'),$$

for $\eta,\eta'\in(\mu^*)^{-1}(0)$, is non-degenerate, the claim follows.

Thus, λ is a locally invertible fibre map. Put $\lambda^{-1}(\xi_0)=W_{ca}(\xi_0)$. Then $DW_{ca}(0_z)=\mathrm{id}_{\mu^0}(0_z)$. Moreover, $\mathrm{pr}_1\,\mathrm{grad}_m E_z(W_{ca}(\xi_0))=\mathrm{pr}_1\,\mathrm{grad}_m E_z(\lambda^{-1}(\xi_0))=\lambda\circ\lambda^{-1}(\xi_0)-\xi_0=0$.

Now define, for $(\eta,\xi_0)\in D_{*,z}\oplus D_{0,z}$,

$$F_z(\eta,\xi_0):=E_z(\eta,\xi_0)-E_z\circ W_{ca}(\xi_0).$$

Denote by D^*F_z the restriction of the differential DF_z to the subspace parallel to $\mu^*\subset\mu^*\oplus\mu^0=\mu$. We may write

$$F_z(\eta,\xi_0)=\int_0^1(1-t)D^{*2}F_z(t\eta,\xi_0)\,.\,(\eta,\eta)dt.$$

Define bilinear forms $h_z(\eta,\xi_0)$ on T'_zD_* by

$$h_z(\eta,\xi_0)\,.\,(\xi_1,\xi_2):=\int_0^1(1-t)D^{*2}F_z(t\eta,\xi_0)\,.\,(\xi_1,\xi_2)dt.$$

We define self-adjoint operators $k_z(\eta,\xi_0)$ on T'_zD_* by

$$\langle k_z(\eta,\xi_0)\,.\,\xi_1,\xi_2\rangle_{m,1}=h_z(\eta,\xi_0)\,.\,(\xi_1,\xi_2).$$

$k_z(0,0)$ is invertible. Hence, as in the proof of (2.4.7), we may define $m_z(\eta,\xi_0):=\sqrt{k_z(\eta,\xi_0)k_z(0,0)^{-1}}$ and thus obtain

$$F_z(\eta,\xi_0)=\langle k_z(\eta,\xi_0)\,.\,\eta,\eta\rangle_{m,1}=\langle k_z(0,0)m_z(\eta,\xi_0)\,.\,\eta,m_z(\eta,\xi_0)\,.\,\eta\rangle_{m,1}.$$

Therefore, with

$$\phi'^{-1}:(\eta,\xi_0)\in D_{*,z}\oplus D_{0,z}\mapsto(\xi',\xi_0):=(m_z(\eta,\xi)\,.\,\eta,\xi_0)\in D_*\oplus D_0,$$
$$E_z\circ\phi'(\xi',\xi_0)=\langle k_z(0,0)\xi',\xi_0\rangle_{m,1}+E_z\circ W_{ca}(\xi_0).$$

Since $k_z(0,0)$ is the operator representing $D^{*2}E_z(0_z)$, we can conclude the proof of (4.2.1) as in (2.4.8). □

Note that all our constructions were made equivariantly.

4.2.2 Corollary. *Assume, in addition to the hypotheses of* (4.2.1), *that* $S\,.\,c$ *is an isolated set of critical points. Then* $E_z:D_z\to\mathbb{R}$ *has* $0_z\in D_z$ *as isolated critical point.*

Proof. Assume that $\xi\in D_z$ is critical point for E_z, i.e. $\mathrm{grad}_m E(\xi)\in T_\xi D$ is orthogonal to D_z. From (1.4.13) we have a vector field u with $u(z\,.\,c)=\partial(z\,.\,c)$ i.e.

$u(z\,.\,c)$ is orthogonal to $T_{0_z}D_z$, and $\langle u, \operatorname{grad} E\rangle_1 = 0$. We can construct a vector field u_m having the same properties with respect to the metric $\langle\ ,\ \rangle_{m,1}$. Thus, for $\|\xi\|_{m,1}$ sufficiently small, $\operatorname{grad}_m E(\xi) = 0$, i.e. $\xi = 0_z$. □

The result (4.2.1) leads us to the consideration of the following situation. Let $D_+ \oplus D_- \oplus D_0$ be the sum of three Hilbert discs, $\dim D_+ = \infty$, $\dim D_- = k < \infty$, $\dim D_0 = l < \infty$. Denote this sum by B. Let a Riemannian metric $\langle\ ,\ \rangle_1$ on B be given such that at the origin $0_B = (0_{D_+}, 0_{D_-}, 0_{D_0})$ of B the splitting $D_+ \oplus D_- \oplus D_0 \subset T_{0_B}B$ is orthogonal.

Moreover, let a differentiable function

$$E(\xi) = \|\xi_+\|_1^2 - \|\xi_-\|_1^2 + E_0(\xi_0)$$

on B be given, with $\xi = (\xi_+, \xi_-, \xi_0)$ and $\|\ \|_1$ the norm derived from $\langle\ ,\ \rangle_1$ at 0_B. 0_B shall be an isolated critical point of E, and the origin 0_{D_0} of D_0 shall be totally degenerate with $E_0(0_{D_0}) = 0$, i.e.

$$E_0(0_{D_0}) = DE_0(0_{D_0}) = D^2E_0(0_{D_0}) = 0.$$

Finally, the gradient field of E, derived from the Riemannian metric, shall satisfy the condition (C), cf. (1.4).

Denote by $\phi_s\xi$ the integral curve of the vector field $-\operatorname{grad} E$, starting at ξ.

For every subset $A \subset B$ we define

$$A^0 := A \cap E^{-1}(]-\infty, 0]), \quad A^{0-} := A \cap E^{-1}(]-\infty, 0[).$$

B_ε denotes the ε-ball around $0_B \in B$.

4.2.3 Lemma. *There exists a fundamental system of neighborhoods Z_ε of the origin $0_B \in B$ with $B_\varepsilon^0 \subset Z_\varepsilon$ having the property that $(Z_\varepsilon, B_\varepsilon{}^{0-})$ can be retracted into $(0_B \cup B_\varepsilon^{0-}, B_\varepsilon^{0-})$, leaving 0_B invariant. Hence,*

$$H_*(Z_\varepsilon, B_\varepsilon^{0-}) = H_*(0_B \cup B_\varepsilon^{0-}, B_\varepsilon^{0-})$$

is a local invariant.

Here we take as coefficients an arbitrary prime field $\mathbb{Z}_p$, with $\mathbb{Z}_\infty = \mathbb{Q}$ = field of rational numbers.

There exists a retraction of $(0_B \cup B_\varepsilon^{0-}, B_\varepsilon^{0-})$ into $(0_B \cup (D_0 \oplus D_-)^{0-}, (D_0 \oplus D_-)^{0-})$ and this pair is homologically equivalent to the product of $(0_B \cup D_0^{0-}, D_0^{0-})$ with the relative disc $(D_-, D_-^{0-}) = (D_-^0, D_-^{0-})$. Thus, with $k = \dim D_-$,

$$(\dagger) \qquad H_*(0_B \cup B_\varepsilon^{0-}, B_\varepsilon^{0-}) = H_{*-k}(0_{D_0} \cup D_0^{0-}, D_0^{0-}).$$

Finally, all homology groups are finitely generated; they vanish for all sufficiently high dimensions.

Remark. The rank of $H_i(0_B \cup B_\varepsilon^{0-}, B_\varepsilon^{0-})$ in $\mathbb{Z}_p$-coefficients is called the ith *type number* (mod $\mathbb{Z}_p$) of the function $E: (B, 0_B) \to (\mathbb{R}, 0)$ at 0_B, denoted by $b_i(0_B)$. Similarly, the ith *singular type number* (mod $\mathbb{Z}_p$) of E at $0_B, b_i^0(0_{D_0})$, is defined as the rank of $H_i(0_{D_0} \cup D_0^{0-}, D_0^{0-})$. With these concepts we have the following corollary.

4.2.4 Corollary.

$$b_{i+k}(0_B) = b_i^0(0_{D_0})$$

where k = *index of the critical point* 0_B *for the function* $E:(B,0_B) \to (\mathbb{R},0)$.

Note. Our proof differs from the one given for a very similar result by Gromoll-Meyer [GM 1], inasmuch as we closely follow Seifert-Threlfall [ST]. The formula (†) is the so-called shifting theorem of [GM 1]. We shall prove this by referring to a shift by $+k$ in the dimension of the homology of a pair if one forms the product with the pair (k-disc, boundary of the k-disc).

Proof. We define, for sufficiently small $\varepsilon > 0$, the set Z_ε as follows. First of all, $B_\varepsilon^0 \subset Z_\varepsilon$. Moreover, those $\xi \in B_\varepsilon$ belong to Z_ε which satisfy $0 \leqslant E(\xi) \leqslant \varepsilon$ and one of the following two properties:

(i) $\lim \phi_s \xi = 0_D$;
(ii) there exists a finite $s \geqslant 0$ such that $\phi_s \xi \in B_\varepsilon \cap E^{-1}(0)$.

We first show that Z_ε is actually a neighborhood of $0_B \in B$. In fact, otherwise there would exist a sequence $\{\xi_m\}$ with $\xi_m \notin Z_\varepsilon$, $\lim \xi_m = 0_B$ and $E(\xi_m) > 0$.

We claim that the trajectory $\phi_s \xi_m$, $s \geqslant 0$, must meet the boundary of B_ε at a point $\eta_m, E(\eta_m) \geqslant 0$, for all m sufficiently large. Indeed, $\xi_m \notin Z_\varepsilon$ implies that otherwise we would have $E(\phi_s \xi) \geqslant 0$ for all $s \geqslant 0$, $\phi_s \xi_m \in B_\varepsilon$. Then, $\|\text{grad } E(\phi_s \xi_m)\|_1 \to 0$ for $s \to \infty$. Since 0_B is the only critical point in B_ε it would follow that $\lim \phi_s \xi_m = 0_B$, which contradicts $\xi_m \notin Z_\varepsilon$.

$E(\xi_m) > E(\eta_m) \geqslant 0$ and $\lim \xi_m = 0_B$ imply $\lim E(\eta_m) = 0$. On the other hand, $\|\text{grad } E(\eta)\|_1$ is bounded away from zero for all $\eta \in B_\varepsilon - B_{\varepsilon/2}$. Hence, for sufficiently large m, when $E(\xi_m)$ is close to 0, the E-value will decrease along the trajectory $\phi_s \xi_m$ from ξ_m to η_m which passes through $B_\varepsilon - B_{\varepsilon/2}$ by so much as to push it below the 0-level of E.

Thus, Z_ε is indeed a neighborhood of 0_B. Since $Z_\varepsilon \subset B_\varepsilon$ and since the B_ε form a fundamental system of neighborhoods of $0_B \in B$, the same is true for the Z_ε.

We now define a retraction

$$\Phi_\tau : (Z_\varepsilon, Z_\varepsilon^0, 0_B) \to (0_B \cup Z_\varepsilon^0, Z_\varepsilon^0, 0_B), \quad 0 \leqslant \tau \leqslant 1,$$

as follows. The $\xi \in Z_\varepsilon^0 = B_\varepsilon^0$ remain fixed. For every $\xi \in Z_\varepsilon$ with $E(\xi) > 0$ there exists a first $s \geqslant 0$, possibly $s = \infty$, such that $E(\phi_s \xi) = 0$. We define $\Phi_\tau \xi$ to be $\phi_{s(\tau)} \xi$ where $s(\tau) \geqslant 0$ is determined by the condition that $E(\phi_{s(\tau)} \xi) = (1-\tau) E(\xi)$. Thus, $\Phi_1 \xi \in Z_\varepsilon^0$.

To see that Φ_τ is indeed a deformation we consider a sequence $\{\xi_m\}$ in Z_ε, $E(\xi_m) > 0$, with $\lim \xi_m = \xi_0$ and a sequence $\{\tau_m\}$ in $[0,1]$ with $\lim \tau_m = 1$. We must show that $\lim \Phi_{\tau_m} \xi_m = \Phi_1 \xi_0$.

Clearly, it suffices to show that $\lim \Phi_1 \xi_m = \Phi_1 \xi_0$. This is certainly true if $\Phi_1 \xi_0 \neq 0_B$, due to the continuous dependence of the integral curves $\phi_s \xi_m$ from their initial values ξ_m. In the case $\Phi_1 \xi_0 = 0_B$, we note that otherwise there would exist a subsequence of $\{\xi_m\}$, which we denote again by $\{\xi_m\}$, and $\varepsilon' > 0$, $\varepsilon' \ll \varepsilon$, such that $\Phi_1\xi_m \notin Z_{2\varepsilon'} \cap E^{-1}(0)$. But then also $\Phi_{1-\varepsilon'}\xi_m \notin Z_{2\varepsilon'}$, whereas $\lim \Phi_{1-\varepsilon'}\xi_m = \Phi_{1-\varepsilon'}\xi_0 \in Z^0_{2\varepsilon'}$.

It is now clear how the deformation Φ_τ, $0 \leqslant \tau \leqslant 1$, may be followed by a deformation Φ_τ, $1 \leqslant \tau \leqslant 2$, of $(0_B \cup B^0_\varepsilon, B^0_\varepsilon)$ into $(0_B \cup B^{0-}_\varepsilon, B^{0-}_\varepsilon)$, leaving 0_B invariant.

Finally, the deformation Φ_τ, $2 \leqslant \tau \leqslant 3$,

$$\xi = (\xi_+, \xi_-, \xi_0) \in B_\varepsilon \mapsto ((3-\tau)\xi_+, \xi_-, \xi_0) \in B_\varepsilon$$

carries $(0_B \cup B^{0-}_\varepsilon, B^{0-}_\varepsilon)$ into $(0_B \cup (D_- \oplus D_0)^{0-}, (D_- \oplus D_0)^{0-})$.

To show that the inclusion of the product $(0_{D_0} \cup D^{0-}_0, D^{0-}_0) \times (D_-, D^{0-}_-)$ in the pair $(0_B \cup (D_0 \oplus D_-)^{0-}, (D_0 \oplus D_-)^{0-})$ is an isomorphism in homology, we define a map

$$\Psi : (0_B \cup (D_0 \oplus D_-)^{0-}, (D_0 \oplus D_-)^{0-}) \to (0_{D_0} \times D^{0-}_0, D^{0-}_0) \times (D_-, D^{0-}_-)$$

as follows. If $\xi = (\xi_0, \xi_-) \in (D_0 \oplus D_-)^{0-}$ and $E_0(\xi_0) < 0$ put $\Psi(\xi) = \xi$. Otherwise, put $\Psi(\xi) = (0_{D_0}, \xi_-)$. The induced map in homology is then an isomorphism. Thus we have proved (†).

To complete the proof of (4.2.3) it remains to show that $H_*(0_0 \cup D^{0-}_0, D^{0-}_0)$ has a finite base. To see this we observe that our previous arguments show that

$$H_*(0_0 \cup D^{0-}_0, D^{0-}_0) = H_*(Z_{0,\varepsilon}, Z^{0-}_{0,\varepsilon}) = H_*(Z_{0,\varepsilon}, Z^{-\delta}_{0,\varepsilon}).$$

Here, $0 < \delta \leqslant \varepsilon$, and $Z_{0,\varepsilon}$ is defined as a neighborhood of $0_0 \in D_0$ with the help of $E_0 := E|D_0$ and the induced Riemannian metric on D_0, just as Z_ε was defined above. The upper index $-\delta$ denotes the subset satisfying $E_0 \leqslant -\delta$.

We now use the well-known fact (cf. [ST]), that the function $E_0 : (D_0, 0_0) \to (\mathbb{R}, 0)$ can be approximated by a non-degenerate function $E^*_0 : (D_0, 0_0) \to (\mathbb{R}, 0)$, which has all its critical points in a small neighborhood of 0_0 inside the domain $E_0^{-1}(]-\frac{\delta}{2}, \frac{\delta}{2}[)$. Thus, the rank of the homology of $(Z_{0,\varepsilon}, Z^{-\delta}_{0,\varepsilon})$ is bounded from above by the indices of the finitely many non-degenerate critical points of E^*_0. □

We return to the case of a prime closed geodesic c_0 and assume that, for all $m = 1, 2, \ldots$, the orbits $S \,.\, c_0^m$ are isolated sets of critical points in ΛM. Put $E(c_0^m) = m^2 E(c_0) = \kappa_m$. By fixing a certain m, we also write c instead of c_0^m. Let $\mu = \mu(S \,.\, c)$ be the normal bundle over S induced from $z \in S \mapsto (z \,.\, c_0)^m = z^{\frac{1}{m}} \,.\, c_0^m \in \Lambda$ and let

$$\mu(S \,.\, c) = \mu^+(S \,.\, c) \oplus \mu^-(S \,.\, c) \oplus \mu^0(S \,.\, c)$$

be the splitting according to the sign of the eigenvalues, introduced earlier.

For every $z \in S$, consider the function $E_z : ((D_z(S \,.\, c_0^m), 0_z(S \,.\, c_0^m)) \to (\mathbb{R}, \kappa_m)$. On $D_z(S \,.\, c_0^m)$ we have the Riemannian metric $\langle\, , \rangle_{m,1}$. From (4.2.1) we then obtain the well-defined function

$$E_{0,z} := E_z \circ W_{ca} : (D_{0,z}(S \,.\, c_0^m), 0_{0,z}(S \,.\, c_0^m)) \to (\mathbb{R}, \kappa_m).$$

Since all constructions are made equivariantly with respect to the S-action on $D(S \,.\, c_0^m)$, the homology groups and type numbers defined according to

(4.2.3) and (4.2.4) are independent of the choice of $z \in S$. Thus, we denote them by $b_i(c_0^m)$ and $b_i^0(c_0^m)$, respectively.

We can also define the fundamental system of neighborhoods $B_{z,\varepsilon}(c_0^m)$ and $Z_{z,\varepsilon}(c_0^m)$ of $0_z(c_0^m)$. The union of these neighborhoods, $z \in S$, will be denoted by $B_\varepsilon(S \,.\, c_0^m)$ and $Z_\varepsilon(S \,.\, c_0^m)$, respectively. $B_\varepsilon(S \,.\, c_0^m)$ and $Z_\varepsilon(S \,.\, c_0^m)$ are S-invariant neighborhoods of the 0-section $0(S \,.\, c_0^m)$ of the bundle $\mu(S \,.\, c_0^m)$. $B_\varepsilon(S \,.\, c_0^m)$ and $Z_\varepsilon(S \,.\, c_0^m)$ may also be viewed as the total space of a fibration over S by restricting $\mu(S \,.\, c_0^m): N(S \,.\, c_0^m) \to S$ to a neighborhood of $0\,(S \,.\, c_0^m)$.

From (4.2.3) it follows that $\left(Z_\varepsilon(S \,.\, c_0^m),\ B_\varepsilon(S \,.\, c_0^m)^{\kappa_m -}\right)$ can be retracted equivariantly onto $\left(0(S \,.\, c_0^m) \cup B_\varepsilon(S \,.\, c_0^m)^{\kappa_m -},\ B_\varepsilon(S \,.\, c_0^m)^{\kappa_m -}\right)$. Denote by $b_i(S \,.\, c_0^m)$ the rank of the $\mathbb{Z}_p$-vector space $H_i\left(0(S \,.\, c_0^m)\right) \cup B_\varepsilon(S \,.\, c_0^m)^{\kappa_m -},\ B_\varepsilon(S \,.\, c_0^m)^{\kappa_m -}\right)$. $b_i(S \,.\, c_0^m)$ is called the *i-th type number* (mod $\mathbb{Z}_p$) of the orbit $S \,.\, c_0^m$. It is clearly independent of the choice of ε, cf. [ST].

4.2.5 Lemma. *Let c_0 be a prime closed geodesic and assume that, for $m = 1, 2, \ldots$, the orbits $S \,.\, c_0^m$ are isolated critical sets. Take an homology with $\mathbb{Z}_p$-coefficients. Let $\kappa_m = E(c_0^m)$, $k_m = \text{index } c_0^m$.*

Claim. (i) *For all $\varepsilon > 0$ sufficiently small, the group*

$$H_{*+k_m}\left(0\,(S \,.\, c_0^m) \cup B_\varepsilon(S \,.\, c_0^m)^{\kappa_m -},\ B_\varepsilon(S \,.\, c_0^m)^{\kappa_m -}\right)$$

is a subgroup of

$$H_*(S) \oplus H_*\left(c_0^m \cup B_{z,\varepsilon}(c_0^m)^{\kappa_m -},\ B_{z,\varepsilon}(c_0^m)^{\kappa_m -}\right).$$

Hence

$$b_{i+k_m}(S \,.\, c_0^m) \leqslant b_i^0(c_0^m) + b_{i-1}^0(c_0^m).$$

(ii) *Assume $m = q m_0$, with q, m_0 integers and such that nullity c^{m_0} = nullity c^m.* Then $b_i^0(c_0^{m_0}) = b_i^0(c_0^m)$.

Proof. From (4.2.3) we know that the homology of a fibre of

$$Z_\varepsilon(S \,.\, c_0^m)^{\kappa_m} \bmod B_\varepsilon(S \,.\, c_0^m)^{\kappa_m -}$$

coincides with the homology of

$$c_0^m \cup B_{z,\varepsilon}(c_0^m)^{\kappa_m -} \bmod B_{z,\varepsilon}(c_0^m)^{\kappa_m -}.$$

The inclusion of the fibre in the total space might kill some of the homology. In fact, two non-homologous cycles of the fibre might become homologous when considered in the total space by being the boundary of a chain of the type $e^{2\pi i r} \,.\, u$, $0 \leqslant r \leqslant l/m$, l an integer. On the other hand, $S \,.\, u$ will certainly be a cycle of the total space whenever u is a cycle of the fibre. These are clearly the only new cycles of the total space. Thus (i) follows.

To prove (ii), we consider the map

$$(\dagger)^q \qquad e \in D(S\,.\,c_0^{m_0}) \mapsto e^q \in D(S\,.\,c_0^{qm_0}),\ qm_0 = m,$$

where we identify $D(S\,.\,c^r)$ with a neighborhood of $S\,.\,c^r$ in Λ. Since we are using the metric $\langle\ ,\ \rangle_{m_0,1}$ on $D(S\,.\,c_0^{m_0})$ and the metric $\langle\ ,\ \rangle_{m,1}$ on $D(S\,.\,c_0^m)$, we see that $(\dagger)^q$ is an isometric embedding. We claim that $\mathrm{grad}_{m_0} E(e)$ is carried under $(\dagger)^q$ into $\mathrm{grad}_m E(e^q)$ (up to a constant factor). Indeed, $E(e^q) = q^2 E(e)$ and $\mathrm{grad}_m E(e^q)$, being invariant under the isotropy group of e^q, is tangent to the image of $D(S\,.\,c_0^{m_0})$ in $D(S\,.\,c_0^{qm_0})$. Finally, $\mu^0(S\,.\,c_0^{m_0})$ is carried into $\mu^0(S\,.\,c_0^m)$. We thus have that the canonically defined characteristic submanifold from (4.2.1)

$$W_{ca}(c_0^{m_0}) : (D_0(S\,.\,c_0^{m_0}),\ 0_0(S\,.\,c_0^{m_0})) \to (D(S\,.\,c_0^{m_0}), 0(S\,.\,c_0^{m_0}))$$

is transformed into the characteristic submanifold

$$W_{ca}(c_0^m) : (D_0(S\,.\,c_0^m),\ 0_0(S\,.\,c_0^m)) \to (D(S\,.\,c_0^m),\ 0(S\,.\,c_0^m)).$$

Thus, the functions $E_0 = E \circ W_{ca}(c_0^{m_0})$ on $D_0(S\,.\,c_0^{m_0})$ and $E_0 = E \circ W_{ca}(c_0^m)$ on $D_0(S\,.\,c_0^m)$ are, up to a constant factor, the same functions. This completes the proof of (ii). □

4.2.6 Proposition. *Let c be a prime closed geodesic.*

Claim. *There exists a sequence $m_1,\dots,m_s$ of positive integers, $s \leqslant 2^n$, and for each integer $j \in [1,s]$ a strictly increasing sequence $\{q_{ji}, i=1,2,\dots\}$ of positive intergers such that the sets $N_j := \{m_j q_{ji}, i=1,2,\dots\}$ form a partition of $N^* = N - \{0\}$ and*

$$(*)_j \qquad \text{nullity } c^{m_j q_{ji}} = \text{nullity } c^{m_j}.$$

Proof. Recall from (3.2.9) that nullity $(c^m) = \sum_\rho \rho$-nullity(c), $\rho^m = 1$. According to (3.2.12), ρ-nullity (c) = dimension of the eigenspace of the Poincaré map P, associated to c, for the eigenvalue ρ.

Consider those roots $\rho = e^{2\pi i a}$, $\bar\rho = e^{-2\pi i a}$ of P, $0 < a \leqslant \frac{1}{2}$, where $a = p/q$ is rational, with p,q relatively prime. Denote by D the (possibly empty) set of denominators of these a. For each subset $E \subset D$, let $m(E)$ denote the least common multiple of all elements in E. In case $E = \emptyset$ put $m(\emptyset) = 1$. Let $m_1,\dots,m_s$ be the set of different numbers obtained in this manner. Clearly, $s \leqslant 2^n$. We may assume that $m_1 = 1$.

For each j, $1 \leqslant j \leqslant s$, we consider the maximal subsequence $\{q_{ji}, i=1,2,\dots\}$ of positive integers such that none of the m_k, $k \neq j$, divides $m_j q_{ji}$.

We see at once that $(*)_j$ is satisfied. Moreover, every positive integer m can be written as $m = m_j q$, q a positive integer, whereby j is determined by the condition that it be the maximal divisor of m among the elements $m_1,\dots m_s$. □

4.2.7 Proposition. *Let c be a closed geodesic and assume that index $c^m > 0$ for some $m \geqslant 1$.*

Then there exist $\alpha > 0$, $\beta \geqslant 0$ such that for all integers $m_0, m_1 \geqslant 1$

$$\text{index } c^{m_0+m_1} \geqslant \text{index } c^{m_0} + m_1 \alpha - \beta.$$

Proof. From the hypothesis it follows that the number α_c defined in (3.2.15) is > 0, since otherwise $I_{c,j} = 0$, for all j, and thus $I_c(\rho) = 0$, for all ρ. Now apply (3.2.15) to obtain $\alpha = \alpha_c, \beta = 2\beta_c$. □

We can now prove the crucial lemma of Gromoll and Meyer [GM 2]. We would like to view it as a semi-local result since it concerns the topology of a neighborhood of a single closed geodesic c and all of its coverings $c^m, m = 1, 2 \dots$.

4.2.8 Lemma. *Let c be a closed geodesic and assume that all its coverings $c^m, m = 1, 2, \dots$, give rise to an isolated orbit $S \,.\, c^m$ of critical points. Then the sum $\sum_m b_i(c^m)$ of the i-th type numbers with $\mathbb{Z}_p$-coefficients is bounded above by a constant $R(c)$ uniformly for all $i \geqslant$ some i_0.*

Proof. From (4.2.6) and (4.2.5) we know that for every m there is a m_j in a finite collection $\{m_1, \dots, m_s\}$ such that $b_i^0(c^m) = b_i^0(c^{m_j})$. Now, $b_i^0(c^{m_j}) = 0$ for all $i >$ some $i(m_j)$. Let $i_0 = \sup_j i(m_j)$. Then, if $i > i_0$, $b_i^0(c^m) = 0$ for all m. Note that in the case in which all c^m are non-degenerate we have $b_i^0(c^m) = 0$, for all m and all $i > 0$.

Now fix $i > i_0$. Either $b_i(c^m) = 0$ for all m, or there exists a smallest $m_0 \geqslant 0$ such that $b_i(c^{m_0}) > 0$. From (4.2.5) it then follows that index $c^{m_0} > 0$. There exist $\alpha > 0, \beta \geqslant 0$ such that (4.2.7) holds. Hence, if $m_1 > (i_0 + \beta)/\alpha$, we have

$$\text{index } c^{m_0+m_1} > \text{index } c^{m_0} + i_0$$

i.e. $b_i(c^{m_0+m_1}) = 0$. Thus, in $\sum_m b_i(c^m)$ at most $[(i_0 + \beta)/\alpha + 1]$ summands are non-zero. From (4.2.5) and (4.2.6), we have an upper bound for the $b_i(c^m)$ for all $i > i_0$, and all m. Thus, we get the desired upper bound for $\sum_m b_i(c^m)$. □

Note. If we assume that all $c^m, m = 1, 2, \dots$, are non-degenerate, the lemma is a trivial consequence of (4.2.7), which in turn is an immediate consequence of Bott's formula (3.2.9). Thus, in the case for which all closed geodesics on a manifold M are non-degenerate, the following theorem possesses a very simple proof.

In order to formulate the subsequent theorem we denote by *hypothesis* $(GM)_p$ the property that the sequence $\{b_i \Lambda M\}$ of $\mathbb{Z}_p$-Betti numbers of the manifold ΛM is unbounded.

4.2.9 Theorem. *Let M be a compact Riemannian manifold and assume that the hypothesis $(GM)_p$ is satisfied for some p. Then there are infinitely many prime closed geodesics on M.*

Remark. This theorem is due to Gromoll and Meyer [GM 2] for the case $p=\infty$, i.e. rational Betti numbers. Also note that we make no assumption concerning the fundamental group of ΛM.

Proof. From (4.2.9) we know that, for a fixed $j, 1 \leqslant j \leqslant s$, and for large i, the sum $\sum_m b_i(c_j^m)$ is bounded above by some $R(c_j)$. Thus, $R=\sum_j R(c_j)$ is an upper bound of $\sum_c b_i(c)$.

Now take k such that $b_k(\Lambda M)>R$, $k>\dim M$. There exists a non-critical value $\kappa>0$ such that the finitely many critical orbits $S.c$ with $b_k(S.c)>0$ all belong to $\Lambda^\kappa M$. Combining (2.1.2) and (4.2.3) with the proof of the Morse inequalities (2.4.12) we get a contradiction

$$R<b_k(\Lambda,\Lambda^\varepsilon)=b_k(\Lambda^\kappa,\Lambda^\varepsilon)\leqslant \Sigma b_k(S.c)\leqslant R,$$

where the last sum is taken over the critical orbits in $\Lambda M-\Lambda^0 M$. □

Remark. This theorem leads to the question as to which compact manifolds M satisfy the hypothesis $(GM)_p$. Clearly, $(GM)_p$ involves the homotopy type of M only.

As far as $p=\infty$ is concerned, Sullivan and Vigué [VS] gave a complete characterization of those M for which $(GM)_\infty$ is satisfied. See the first part of (4.2.10).

We recall that a truncated rational polynominal ring $T_{d,h}(x)$ of degree d and height h is defined as the quotient ring $\mathbb{Q}[x]/(x^h)$ with x of degree d, cf. [Sp]. For $h=2$, $T_{d,h}(x)$ is the cohomology ring of the sphere of dimension d. For $d=2,4,8$ and $h\geqslant 3$ arbitrary, except $d=8$, where one must take $h=3$, $T_{d,h}(x)$ is the cohomology ring of one of the projective spaces over the complex numbers, the quaternions or the Cayley numbers.

4.2.10 Theorem. (i) *Let M be a simply connected compact differentiable manifold. The hypothesis $(GM)_\infty$ is satisfied if and only if the rational cohomology ring $H^*(M;\mathbb{Q})$ is not a truncated polynomial ring.*

(ii) *The hypothesis $(GM)_p$ is not satisfied for any p if M has the homotopy type of one of the irreducible symmetric spaces of rank* 1.

For the proof of (i) we refer to [VS]. The proof uses essentially Sullivan's theory [Su] of the rational homotopy type; see also (4.3) for more details. (ii), on the other hand, follows from the work of Švarc [Šv], who computed the integer cohomology of ΛM for the symmetric spaces of rank 1. □

We wish to stress that there are many other spaces besides the symmetric spaces of rank 1, for which the rational cohomology ring is a truncated polynomial ring; see also (4.2.11) below.

Prior to Sullivan and Vigué, Klein [Kle] and Sacks [Sa] obtained various partial conditions which implied the hypothesis $(GM)_\infty$.

Sullivan [Su] has shown that for every compact, simply connected manifold M there exists a sequence $\{w_r\}$ of rational cohomology classes of ΛM with $\dim w_r =2ar+b$, a,b integers, $a>0, b\geqslant 0, r=1,2,\ldots$, see (4.3) for more details. From this we conclude that if M has the homotopy type of the product $M' \times M''$ of two simply connected, compact manifolds M' and M'', then M satisfies $(GM)_\infty$.

Indeed, observe that

$$H_k\Lambda(M' \times M'') = H_k(\Lambda M' \times \Lambda M'') = \sum_i H_i\Lambda M' \otimes H_{k-i}\Lambda M''.$$

$H^*\Lambda M'$ contains classes w'_r of dimension $2a'r+b'$, and $H^*\Lambda M''$ contains classes w''_r of dimension $2a''r+b''$. Hence, the Betti number of $\Lambda(M' \times M'')$ in dimension $k=2a'a''m+b'+b''$ is $\geqslant m+1$, since $H_i\Lambda M'$ and $H_{k-i}\Lambda M''$, for $i=2a'a''j+b'$, contain $w'_{ja''}$ and $w''_{(m-j)a'}$.

Ziller [Zi] has determined the $\mathbb{Z}_2$-homology of ΛM for all compact, symmetric spaces M. As a consequence, he finds that all symmetric spaces besides those of rank 1 satisfy $(GM)_p$ either for $p=\infty$ or for $p=2$. Those for which $(GM)_2$ holds but $(GM)_\infty$ does not are

$$SU(3)/SO(3);\ SO(2n+1)/SO(2n-1) \times SO(2);\ G_2/SO(4)$$

Thus, using (4.2.9), Ziller [l.c.] has proved:

4.2.11 Theorem. *If the underlying differentiable manifold of a Riemannian manifold M has the homotopy type of a symmetric space of rank* >1, *then there exist infinitely many prime closed geodesics on M.* □

4.3 The Existence of Infinitely Many Closed Geodesics

We saw in (4.2) that there are many compact differentiable manifolds with the property that, for any Riemannian metric on such a manifold, there exists an infinite number of closed geodesics, cf. in particular (4.2.10) and (4.2.11). However, as shown in (4.2.10), there are some manifolds, for instance the spheres, to which these results do not apply. The reason is that the topological structure of the associated space of closed curves is not sufficiently complicated.

In this section we will use the finer structure of the Morse complex, in particular its equivariance with respect to the S-action $\tilde{\chi}$ and the $\mathbb{Z}_2$-action θ, to show that there always exist infinitely many prime closed geodesics on a compact Riemannian manifold with finite fundamental group, cf. (4.3.5).

The basic new element used to prove this is a divisibility relation between the multiplicity of certain pairs of closed geodesics, cf. Lemma (4.3.4). In addition, we use the homology classes of ΛM which were discovered by Sullivan as an application of his theory of the minimal model; see [Su]. Actually, in the light of the result (4.2.10(i)) of Vigué and Sullivan [VS], we could restrict ourselves to simply connected compact manifolds for which the rational cohomology ring is a truncated polynomial ring. For such manifolds, the Sullivan classes are particularly simple and were already exhibited by Švarc [Šv].

Thus, if one is interested, for instance, only in the case of a sphere with an arbitrary Riemannian metric — the case with which the whole theory of closed geodesics started, with the work of Poincaré [Po 1] — then one need not know the theory of the minimal model.

We begin by stating some basic properties of the minimal model, paying particular attention to a geometric interpretation.

The *minimal model* $\mathfrak{M}(M)$ for the rational homotopy type of a simply connected countable *CW* complex M, is defined as a differential graded algebra over the rationals with product denoted by $\wedge$, having the following properties:

(i) $\mathfrak{M}(M)$ is free as an algebra (except for the relations imposed by the associativity and graded commutativity). The vector space spanned by the generators of any given degree k is finite; its dual is isomorphic to $\pi_k(M) \otimes_{\mathbb{Z}} \mathbb{Q}$;

(ii) the differential d applied to any generator is either zero or raises the degree by one and is a polynomial in generators of strictly lower degree;

(iii) $H^*(\mathfrak{M}(M)) = H^*(M)$ (over the rationals).

Given $\mathfrak{M}(M)$, one can construct a minimal model $\mathfrak{M}(\Lambda M)$ for the rational homotopy type of ΛM as follows. Each generator y of $\mathfrak{M}(M)$ is also a generator of $\mathfrak{M}(\Lambda M)$, with the same differential. Since $\pi_k \Lambda M = \pi_k M + \pi_k \Omega M$ and $\pi_k \Omega M = \pi_{k+1} M$, cf. (2.1.5), we obtain the remaining generators by associating to each generator y of $\mathfrak{M}(M)$ a generator $\bar{y}$ for $\mathfrak{M}(\Lambda M)$ of one degree less. It remains to define the differential. First, extend $\overline{}$ to all of the $\mathfrak{M}(M)$ as a derivation acting from the right; then define $d\bar{y}$ by $-\overline{dy}$.

Note. We may interpret y as a differential form and $\bar{y} := iy$ as the inner product of y with a vector field i, such that $di + id = 0$. In particular, $\overline{x \wedge y} := i(x \wedge y) = ix \wedge y + (-1)^{|x|} x \wedge iy$, where $|x| =$ degree x.

We are now ready to prove the existence of the so-called *Sullivan* (*cohomology*) *classes* in ΛM, cf. [Su]:

4.3.1 Theorem. *Let M be a compact simply connected differentiable manifold. Then there exists a sequence $\{w^*(l; s)\}$, $s = 0, 1, \ldots,$ of rational cohomology classes. Here, l is an integer ≥ 0, depending only on M. The dimension of the $w^*(l; s)$ forms an arithmetic sequence in s. Finally, if*

$$i : (\Omega M, *) \to (\Lambda M, \Lambda^0 M)$$

denotes the canonical inclusion, $i^* w^*(l; s) \neq 0$.

Remark: In case M is an odd-dimensional sphere S^n we have $l = 0$, and $w^*(0; s)$ goes, under i^*, into the $(s+1)$-th power of the $(n-1)$-dimensional basis class of $H^{n-1}(\Omega S^n)$. Similarly, if M is an even-dimensional sphere, we have $l = 1$ and $w^*(1; s)$ goes, under i^*, into the product of the generator of $H^{n-1}(\Omega S^n)$ with the s-th power of the generator of $H^{2n-2}(\Omega S^n)$.

In these and other special cases (in particular, the remaining symmetric spaces of rank 1) the Sullivan classes have been considered already by Švarc [Šv], Klein [Kle] and Sacks [Sa].

Instead of $w^*(0; s)$, we also write $w^*(s)$.

Proof. There must be generators for $\mathfrak{M}(M)$ of odd degree, since otherwise $dy^l=0$ for all powers y^l of a generator y of even degree. But then, dim $H^*(M)$ is infinite — a contradiction.

Let x be an odd generator of minimal degree. Then $dx=0$ or else dx is a polynomial in even generators, say $e_1,\ldots,e_l, de_j=0$, $1\leq j\leq l$. Put

$$w^*(0;s)\equiv w^*(s)=\bar{x}^{s+1}, \quad \text{if} \quad dx=0;$$

$$w^*(l;s)=\bar{e}_1\wedge\bar{x}^s\wedge\bar{e}_2\wedge\cdots\wedge\bar{e}_l, \quad \text{otherwise},$$

Then $dw^*=0$. This is obvious if $dx=0$. Otherwise, each summand of $d\bar{x}$ will contain one of the $\bar{e}_j\in\{\bar{e}_1,\ldots,\bar{e}_l\}$, and $\bar{e}_j\wedge\bar{e}_j=0$. Also, $d\bar{e}_j=0$.

w^* is not a boundary, since $d\mathfrak{M}(\Lambda M)$ is contained in the ideal generated by the subalgebra $\mathfrak{M}(M)\subset\mathfrak{M}(\Lambda M)$.

In order to prove the last statement we recall that the minimal model $\mathfrak{M}(\Omega M)$ of ΩM has generators $\tilde{y}$, one for each generator y of $\mathfrak{M}(M)$, with degree one less than the degree of y. The differential is given by $d\tilde{y}=0$. This corresponds to the well-known fact that the rational cohomology ring of the loop space ΩM is a free algebra with generators being a base of the rational homotopy groups.

The inclusion $i:\Omega M\to\Lambda M$ determines a map of the homotopy $\pi_*\Omega M$ into the homotopy $\pi_*\Lambda M\cong\pi_*M\oplus\pi_*\Omega M$ and is actually, an injection onto the second factor. From the duality between $\pi_k M\otimes_{\mathbb{Z}}\mathbb{Q}$ and the generators of degree k of $\mathfrak{M}(M)$ it follows that the inclusion $i:\Omega M\to\Lambda M$ determines a morphism

$$i^*:\mathfrak{M}(\Lambda M)\to\mathfrak{M}(\Omega M)$$

by which a generator of type y goes into 0 and a generator of type $\bar{y}$ goes into $\tilde{y}$. In particular, the induced homomorphism in cohomology yields $i^*(\bar{e}_1\wedge\bar{x}^s\wedge\cdots\wedge\bar{e}_l)=\tilde{e}_1\wedge\tilde{x}^s\wedge\cdots\wedge\tilde{e}_l$. This completes the proof. □

In preparation for the construction of certain cycles which represent dual homology classes to Sullivan's cohomology classes, we consider a sphere $(S^{|y|},*)$ of constant curvature equal to 1 with base point$*$, dim $S^{|y|}=|y|\geq 2$.

Consider the loop space $(\Omega S^{|y|},*)$ of $S^{|y|}$, based at$*$. From the end of (2.5) we know that the $(r+1)$-fold covered great circles c_0^{r+1} in $(\Omega S^{|y|},*)$ form a non-degenerate critical submanifold $B_{r+1}^{\Omega}\sim S^{|\bar{y}|}$. Let $W_u^{\Omega}(B_{r+1}^{\Omega})$ be its unstable bundle, cf. (2.5.6). The closure of $W_u^{\Omega}(B_{r+1}^{\Omega})$ forms, for $|y|$ odd, a base element of $H_{(2r+2)|\bar{y}|}(\Omega S^{|y|};\mathbb{Z})$, which we denote by $v_*^{|\bar{y}|}(2r+2)$.

The restriction of $W_u^{\Omega}(B_{r+1}^{\Omega})$ to the fibre over an element $c_0^{r+1}\in B_{r+1}^{\Omega}$ gives the strong unstable manifold of c_0^{r+1}

$$W_{uu}^{\Omega}(c_0^{r+1}):(\mathbb{R}^{(2r+1)|\bar{y}|},0)\to(\Omega S^{|y|},c_0^{r+1}).$$

Again, for $|y|$ odd, the closure of this map represents a basis element of $H_{(2r+1)|\bar{y}|}(\Omega S^{|y|};\mathbb{Z})$ which we denote by $v_*^{|y|}(2r+1)$.

In particular, for $r=0$ we get a spherical cycle

$$v_*^{|\bar{y}|}(1):(S^{|\bar{y}|},*)\to(\Omega S^{|y|},*).$$

An explicit description of $v_*^{|\bar{y}|}(1)$ runs as follows. Identify $(S^{|\bar{y}|},*)$ with the sphere of codimension 1 on $(S^{|y|},*)$ which in $*$ is orthogonal to the initial direction of the great circle c_0. $v_*^{|\bar{y}|}(1)$ is the map which associates to a point $\bar{p}\in S^{|\bar{y}|}$ the circle on $S^{|y|}$ which starts out from $*$ in the same half-sphere as c_0 and meets $S^{|\bar{y}|}$ at $t=\frac{1}{2}$ orthogonal in $\bar{p}$.

It follows that $v_*^{|\bar{y}|}(2)$ consists of the totality of circles on $S^{|y|}$ which start at $*\in S^{|y|}$.

Assume that we have two spherical cycles

$$u:(S^{|u|},*)\to(\Omega M,*),\qquad \text{and} \qquad v:(S^{|v|},*)\to(\Omega M,*).$$

We can then form a new cycle by taking the Pontrjagin product $u\cdot v$ of u and v, cf. Bott and Samelson [Bo S]:

$$u\cdot v:(S^{|u|}\times S^{|v|},*)\to(\Omega M,*).$$

$u\cdot v(p,q)$ is defined as the usual composition in $(\Omega M,*)$ of the path $u(p)$ with the path $v(q)$.

For $|y|$ odd, the $s-$th power $v_*^{|y|}(1)^s$ of $v_*^{|y|}(1)$ represents a base element of $H_{s|\bar{y}|}(\Omega S^{|y|};\mathbb{Z})$. Thus, $v_*^{|y|}(1)^s$ is homologous to $v_*^{|y|}(s)$.

The multiplicative structure of the ring $H^*(\Omega S^{|y|};\mathbb{Z})$ is well known, see Serre [Se]. We restrict ourselves again to $|y|=$odd. If $\bar{y}(s)$ denotes a generator for $H^{s|\bar{y}|}(\Omega S^{|y|})$, then $\bar{y}(1)^s=s!\,\bar{y}(s)$. Hence, the value of $\bar{y}(1)^s$ on $v_*(1)^s\sim v_*(s)$ is $s!$.

Now let $y:(S^{|y|},*)\to(M,*)$, $|y|\geq 2$, represent a non-trivial homotopy class of infinite order. With this we define

$$w_*^{\bar{y}}(1)=\Omega y\circ v_*^{|y|}(1):(S^{|\bar{y}|},*)\to(\Omega M,*).$$

If $|y|$ is odd, we also define, for every integer $s\geq 1$,

$$w_*^{\bar{y}}(s):=\Omega y\circ v_*^{|y|}(s):(V_*^{|y|}(s),*)\to(\Omega M,*)$$

where $V_*^{|y|}(s)$ is the domain of $v_*^{|y|}(s)$.

We are now ready to define dual homology classes to the Sullivan cohomology classes.

First consider $w^*=w^*(2r+1)=\bar{x}^{2r+1}$, $|x|$ odd. We can assume that $x:(S^{|x|},*)\to(M,*)$ is not a proper integer multiple of some other integer spherical cycle. Define

$$w_*=w_*(2r+1)\qquad \text{as} \qquad w_*^{\bar{x}}(1)\cdot w_*^{\bar{x}}(2)^r.$$

In $w^*=w^*(l;s)=\bar{e}_1\wedge\bar{x}^s\wedge\ldots\wedge\bar{e}_l$, we can assume that the $e_j:(S^{|e_j|},*)\to(M,*)$ are indivisible elements of $H_{|e_j|}(M;\mathbb{Z})$. Define

$$\begin{aligned} w_*=w_*(l;s)&=w_*^{\bar{e}_1}(1)\cdot w_*^{\bar{x}}(1)^s\cdot w_*^{\bar{e}_2}(1)\cdot\ldots\cdot w_*^{\bar{e}_l}(1)\\ &\sim w_*^{\bar{e}_1}(1)\cdot w_*^{\bar{x}}(s)\cdot\ldots\cdot w_*^{\bar{e}_l}(1)\end{aligned}$$

where the dot again denotes the Pontrjagin product.

We want to define a homotopy w'_* of certain cycles among the w_*. In order to do so we represent the sphere $S^{|y|}$ of dimension $|y|\geq 2$ as unit sphere in the Euclidean space $\mathbb{R}^{|y|+1}$ with coordinates $(x_i)_{0\leq i\leq |y|}$. Let $*=(-1,0,\ldots,0)$. $S^{|\bar{y}|}$ is the sphere $\{x_1=0\}\cap S^{|y|}$. Let ψ_τ, $0\leq\tau\leq 1$, be the positive rotation by $\tau\,.\,180^0$ in the (x_1,x_2)-plane which is extended to an isometry of $\mathbb{R}^{|y|+1}$ by taking the identity on the orthogonal complement of this plane.

We define the homotopy

$$w'(2r+1):(W_*(2r+1),*)\times[0,1]\to(\Omega M,*),$$

where

$$W_*(2r+1)=S^{|\bar{x}|}\times V_*^{|\bar{x}|}(2)^r,\quad V_*^{|\bar{x}|}(2)=\text{domain of } v_*^{|\bar{x}|}(2), \text{ by}$$

$$w'_*(2r+1)(\;;\tau)=\Omega(x\circ\psi_\tau)\circ v_*^{|\bar{x}|}(1)\circ w_*^{\bar{x}}(2).$$

For $w_*(l;s)$, we define

$$w'_*(l;s):(W_*(l;s),*)\times[0,1]\to(\Omega M,*),$$

where

$$W_*(l,s)=S^{|\bar{e}_1|}\times(S^{|\bar{x}|})^s\times\ldots\times S^{|\bar{e}_l|},\quad \text{by}$$

$$w'_*(l;s)(\;;\tau)=\Omega(e_1\circ\psi_\tau)\circ v_*^{|\bar{e}_1|}(1)\cdot w_*^{\bar{x}}(1)^s\cdot w_*^{\bar{e}_2}(1)\cdot\ldots\cdot w_*^{\bar{e}_l}(1)$$

4.3.2 Proposition.
(i) $w'_*(2r+1)$ *is a restriction of* $w_*(2r+3)$;
(ii) $w'_*(l;s)$ *is a restriction of* $w_*(l;s+1)$.

Proof. (i) follows if we show that $v'_*(1)(\;;\tau)=\Omega\psi_\tau\circ v_*^{|\bar{x}|}(1)$ is a restriction of $v_*^{|\bar{x}|}(2)$. But $v_*^{|\bar{x}|}(2)$ is the unstable manifold of the family of great circles in $(\Omega S^{|x|},*)$.

In order to prove (ii) we consider the relation $dx=$ polynomial in $(e_1,\ldots,e_l)$. This polynomial has neither constant nor linear terms. Thus, $|e_1|\leq|\bar{x}|$.

Let $k_1:(D^{|e_1|},\partial D^{|e_1|})\to(S^{|e_1|},*')$ be the canonical mapping where $*'$ is the antipodal point of the base point $*$ of $S^{|e_1|}$. Then we can embed the disc $(D^{|e_1|},*)$ $\subset(S^{|x|},*)$ such that $x|D^{|e_1|}=e_1\circ k_1$.

Denote by $v_*^{|e_1|}(2)$ the totality of circles on $S^{|e_1|}$ which start at $*\in S^{|e_1|}$. $v_*^{|\bar{e}_1|}(2)$ contains in particular the elements of the homotopy $\psi_\tau\circ v_*^{|\bar{e}_1|}(1)$. Consider the circles $v_*^{|\bar{x}|}(1)|D^{|e_1|}$ on $S^{|x|}$. By rotating these circles around $*\in S^{|\bar{x}|}\subset S^{|x|}$ by $\pi/2$ they become circles lying on $D^{|e_1|}\subset S^{|e_1|}\subset S^{|\bar{x}|}$. That is to say, since $x|D^{|e_1|}=e_1\circ k_1$,

$$\Omega x\circ v_*^{|\bar{x}|}(1)|D^{|e_1|}=w_*^{\bar{x}}(1)|D^{|e_1|}$$

is homotopic to $\Omega e_1\circ v_*^{|\bar{e}_1|}(2)$ which contains the elements $\Omega(e_1\circ\psi_\tau)v_*^{|\bar{e}_1|}(1)$. Thus, $w'_*(l;s)(\;;\tau)$ can be obtained from $w_*(l;s+1)$ by restricting the domain $W_*(l;s+1)$ to

$$\{*\}\times D^{|e_1|}\times(S^{|\bar{x}|})^s\times\ldots\times S^{|e_l|}$$

This completes the proof of (4.3.2). □

4.3.3 Proposition.

(i) $\partial w'_*(2r+1) = -\theta w_*^{\bar{x}}(1) \cdot w_*^{\bar{x}}(2)^r - w_*^{\bar{x}}(1) \cdot w_*^{\bar{x}}(2)^r$;

(ii) $\partial w'_*(1; s) = -\theta w_*^{\bar{e}_1}(1) \cdot w_*^{\bar{x}}(1)^s \cdot \ldots \cdot w_*^{\bar{e}_1}(1)$

$- w_*^{\bar{e}_1}(1) \cdot w_*^{\bar{x}}(1)^s \cdot \ldots \cdot w_*^{\bar{e}_1}(1)$.

Proof. Note that ψ_1 carries the great circle $c_0 \in (\Omega S^{|z|}, *)$ in the (x_0, x_1)-plane into θc_0. Also, the circles near c_0 of $v_*^{|\bar{z}|}(1)$ are carried into circles of $\theta \circ v_*^{|\bar{z}|}(1)$ under ψ_1. The sign follows from the fact that $\theta \circ \psi_1$, restricted to the circles on the subsphere $\{x_0^2 + x_1^2 + x_2^2 = 1\}$, is the reflection at the circle $c_0 = \{x_0^2 + x_1^2 = 1\}$, whereas $\theta \circ \psi_1$ is the identity on the orthogonal complement. □

We now can prove our fundamental lemma. In order to simplify notation we use $w_*(r)$ to denote either $w_*(2r+1)$ or $w_*(l; r)$. We also write $w'_*(r)$ instead of $w'_*(2r+1)$ and $w'_*(l; r)$.

4.3.4 Lemma. *Let M be a compact, simply connected Riemannian manifold. Fix an $r^* \geq 0$ and choose κ^* such that it is non-critical and image $w'_*(r^*) \subset \Lambda^{\kappa^*} M$. Assume that the closed geodesics in $\Lambda^{\kappa^*} M$ are non-degenerate.*

Then there exist, for every $r = 0, \ldots, r^*$, closed geodesics $c(r)$ and $c'(r)$ of index dim $w_*(r)$ *and* dim $w'_*(r)$, *respectively, such that the homotopy classes of $w_*(r)$ and $w'_*(r)$ are locally represented by the strong unstable manifold of $c(r)$ and $c'(r)$, respectively.*

Moreover, if $E(c(r)) = \kappa(r), E(c'(r)) = \kappa'(r)$,

(i) $\kappa(r) < \kappa'(r) < \kappa(r+1)$.

If $m(r)$ and $m'(r)$ are the multiplicity of $c(r)$ and $c'(r)$, respectively, then $m'(r)$ divides $m(r)$, i.e.

(ii) $m'(r) \mid m(r)$.

Proof. We fix r. Instead of $w_*(r)$, $w'_*(r)$, $c(r)$, $c'(r)$, $\kappa(r)$, $\kappa'(r)$, $m(r)$, $m'(r)$ etc. we also write w, w', c, c', κ, κ', m, m', etc.

Denote by $\mathscr{M}^{\kappa^*}M, \mathscr{M}^{\kappa^*}$, or simply $\mathscr{M}$, the Morse complex, formed by the unstable manifolds $W_u(S \,.\, c), E(c) < \kappa^*$, cf. (2.5).

From condition (C) it follows that $\phi_s w'$ will belong to any prescribed neighborhood of $\mathscr{M}$, if only s is sufficiently large. By applying, if necessary, a non-standard deformation which does not increase the E-value, we can bring an element of the homotopy class of w' into any prescribed tubular neighborhood of the subcomplex of $\mathscr{M}$ which is formed by the unstable and strong unstable manifolds of dimension $\leq \dim w'$. In retracting the tubular neighborhood onto the Morse complex we obtain a representation of the homotopy class of w' in $\mathscr{M}$; we denote this representation again by w'. $w' | \{\tau = 0\}$ is the representation of the homotopy class of w in $\mathscr{M}$. We denote it again by w.

Actually, we may assume that $w' : W' \to \mathscr{M}$ is a sum with integer coefficients of strong unstable manifolds $W_{uu}(c')$ and of (partial) stable manifolds $\{z \,.\, W_{uu}(c'');\ z \in$

some interval of positive length on $S\}$, both of dimension dim w'. That is to say, if $W_{uu}(c')$ occurs in w', then there exists a disc-like neighborhood D' of a point O' of W' such that $w'|(D',O')$ coincides, up to a diffeomorphism, with $W_{uu}(c')$. Similar properties hold for the (partial) stable manifolds. The corresponding properties hold for w.

Note. If we take the quotient $\tilde{\pi}\circ w'$ of w' with respect to the S-action, then the stable manifolds lose one dimension and hence can be neglected.

The existence of strong unstable manifolds in w and w' will be proved below.

Observe that the deformation of the original map $w': W'\to \Lambda M$ into the map $w': W'\to \mathcal{M}$ carried S-orbits into S-orbits. We use this observation to introduce the following extensions of the maps w and w':

$$S.w: S.W\to \Lambda M; \quad z.p\mapsto z.w(p)$$

$$S.w': S.W' = S.W\times[0,1]\to \Lambda M; \quad (z.p,\tau)\mapsto z.w'(p;\tau).$$

Here, the domain $S.W$ is the completion of W to full S-orbits. On the range ΛM we have the S-action $\chi\tilde{}$. $S.w$ and $S.w'$ are thus equivariant with respect to the canonical S-action on $S.W$ and the S-action $\chi\tilde{}$. Also, the deformations of w and w' into maps $w: W\to\mathcal{M}$ and $w': W'\to\mathcal{M}$ can be extended so as to give equivariant maps

$$S.w: S.W\to\mathcal{M}; \quad z.p\mapsto z.w(p),$$

$$S.w': S.W'\to\mathcal{M}; \quad (z.p,\tau)\mapsto z.w'(p;\tau).$$

We first consider the case $w=w_*(0):(S^{|\bar{y}|},*)\to(\Lambda M,*)$. We claim that the representation of $w_*(0)$ in the Morse complex contains a $W_{uu}(c)$ of dimension $|\bar{y}|=\dim w_*(0)$ with odd coefficient, whereas $\theta W_{uu}(c)=W_{uu}(\theta c)$ occurs with even coefficient only.

To prove our claim we observe that $w=w_*(0)$ determines a map $y: S^{|y|}\to M$ which is homotopic to the original map of $S^{|y|}$ into M denoted by the same letter. Let

$$i:(D^{|\bar{y}|},\partial D^{|\bar{y}|})\to(S^{|\bar{y}|},*)\subset(\Omega S^{|y|},*)$$

be the canonical mapping. In (2.1.7) we associated with each $p\in D^{|\bar{y}|}$ a circle a_p on $S^{|y|}$, starting from the point p on the half-equator $D^{|\bar{y}|}\subset S^{|y|}$. Thus, each $q\in S^{|y|}$ can be written as $q=a_p(t)$ with uniquely determined p and also uniquely determined $t\in S$ if $p\notin\partial D^{|\bar{y}|}$. Under i, to each circle a_p on $S^{|y|}$, there corresponds a circle on $S^{|y|}$ which starts at the base point $*$ and at half-way passes through $i(p)\in S^{|\bar{y}|}$. Now define

$$y: S^{|y|}\to M; \quad q\mapsto w(i(p))(t), \quad \text{if} \quad q=a_p(t).$$

Assume that our claim is false, i.e.

$$\theta\circ w_*(0)=w_*(0)\circ\theta=w_*(0) \bmod 2.$$

For our mapping y this has the following consequence. Consider the involution

$$\overline{\theta}: S^{|y|} \to S^{|y|}; \qquad q = a_p(t) \mapsto \theta q = (\theta a_p)(t) = a_p(1-t).$$

Then

$$y(\overline{\theta} q) = \theta \circ w(i(p))(t) = w(\theta i(p))(t) = w(i(p))(t) = y(q),$$

i.e. y can be factored mod 2 by the involution $\overline{\theta}$. But this clearly violates our assumption that y represents a non-zero element of $H_{|y|}(M; \mathbb{Z}_2)$. We thus have proved that $-\tilde{\pi} \circ \partial w_*(0)' = \tilde{\pi} \circ \theta w_*(0) + \tilde{\pi} \circ w_*(0) = \pi \circ w_*(0) \neq 0$ mod 2. Here,

$$\tilde{\pi}: \Lambda M \to \tilde{\Pi} M; \qquad \pi: \Lambda M \to \Pi M$$

are the projections onto the space of unparameterized oriented and non-oriented closed curves, cf. (2.2).

Consider now the general case $w = w_*(r)$. Again we claim that

$$(\dagger) \qquad -\tilde{\pi} \circ \partial w' = \tilde{\pi} \circ w \circ (\theta \times \mathrm{id}) + \tilde{\pi} \circ w \circ (\mathrm{id} \times \mathrm{id}) \neq 0 \text{ mod} 2.$$

Here we have written the domain W of w as the product $S^{|\overline{y}|} \times \tilde{W}$, where $y = x$ in the case $w_*(r) = w_*(2r+1)$, and $y = e_1$ in the case $w_*(r) = w_*(1; r)$. The base points of $S^{|\overline{y}|}$ and $\tilde{W}$ will be denoted by $\bar{*}$ and $\tilde{*}$, respectively.

If $(\dagger)$ were false, it would mean that, for every $\tilde{c} \in \tilde{W}$ and with i as above,

$$\tilde{\pi} \circ w | \theta i D^{|\overline{y}|} \times \{\tilde{c}\} = \tilde{\pi} \circ w | i D^{|\overline{y}|} \times \{\tilde{c}\} \text{ mod } 2.$$

This relation commutes with deformations of $\tilde{c}$ on $\tilde{W}$. Consider in particular the deformation into $\tilde{*} \in \tilde{W}$. Since $w_*(r) | S^{|\overline{y}|} \times \{\tilde{*}\}$ coincides with $w_*(0)$, we obtain

$$\tilde{\pi} \circ \theta \circ w_*(0) = \tilde{\pi} \circ w_*(0) \circ \theta = \tilde{\pi} \circ w_*(0) \text{ mod } 2.$$

But we have shown above that this is false.

$(\dagger)$ implies the existence of some $W_{uu}(c)$ in $w = w_*(r)$ of dimension dim w such that $\tilde{\pi} \circ W_{uu}(c)$ occurs in $\tilde{\pi} \circ w = \tilde{\pi} \circ w \circ (\mathrm{id} \times \mathrm{id})$ with odd coefficient, whereas in $\tilde{\pi} \circ w \circ (\theta \times \mathrm{id})$ it occurs with even coefficient only. Hence, there exists $W_{uu}(c')$ in w' of dimension dim w' such that $W_{uu}(c) \subset \partial W_{uu}(c')$. Since $W_{uu}(c')$ is invariant under the isotropy group $I^{\sim}(c')$ of c', also the full $I^{\sim}(c')$-orbit $\{W_{uu}(z_0 \cdot c); z_0 \in I^{\sim}(c')\}$ will belong to w – note that w and the homotopy w' of w are defined equivariantly with respect to the S-action on ΛM.

We shall show that such a pair (c, c') of geodesics (with c' possibly being replaced by a geodesic again denoted by c') possesses the properties (i) and (ii) of (4.3.7). Taking (c, c') as $(c(r), c'(r))$, we shall have thus proved the lemma.

We first prove (ii). Consider the chain $\partial W_{uu}(c') \cap w$. It contains $W_{uu}(c)$. In the simplest case it may happen that $\partial W_{uu}(c) = 0$. Thus, $w_0 = W_{uu}(c)$ is a cycle. It is contained in $\partial W_{uu}(c') \cap w \subset \partial w' \cap w$. Denote by w_0' the restriction to w_0 of the homotopy w'. To w_0 and w_0', apply all the previous arguments. In particular, w_0 and w_0' consist of a sum of strong stable manifolds plus possibly some (partial) S-orbits of such manifolds of one lower dimension. There will again be a $W_{uu}(c') \subset w_0'$ with $W_{uu}(c) = w_0 \subset \partial W_{uu}(c') \cap w_0$. The right-hand side is invariant under $\tilde{I}(c')$ as will be the left-hand side, i.e. $\tilde{I}(c')$ is a subgroup of $\tilde{I}(c)$ and (ii) holds.

In general, we shall have $\partial W_{uu}(c) \neq 0$. If none of the $W_{uu}(c_1) \subset \partial W_{uu}(c)$ also occurs (with opposite sign) in $\partial W_{uu}(z_0 \,.\, c)$ where $z_0 \,.\, c$ is an element of the $\tilde{I}(c')$-orbit of c different from c, then we can find a cycle $w_0 \subset \partial w' \cap w$ which contains only the element $W_{uu}(c)$ and none other of the $\tilde{I}(c')$-orbit of $W_{uu}(c)$. By restricting the homotopy w' of w to w_0, we obtain the homotopy w_0'. Just as before, we find a $W_{uu}(c')$ in w_0' having $W_{uu}(c)$ in its boundary. The chain $\partial W_{uu}(c') \cap w_0$ is invariant under $\tilde{I}(c')$. It therefore contains the full $\tilde{I}(c')$-orbit of $W_{uu}(c)$. But only $W_{uu}(c)$ belongs to w_0. Thus, $z_0 \,.\, c = c$ for $z_0 \in \tilde{I}(c')$, i.e. (ii) holds.

Finally assume that $W_{uu}(c_1) \subset \partial W_{uu}(c)$ occurs with opposite sign in the boundary $\partial W_{uu}(z_0 \,.\, c)$ of some $W_{uu}(z_0 \,.\, c) \neq W_{uu}(c)$ of the $\tilde{I}(c')$-orbit of $W_{uu}(c)$. This implies that

$$W_{uu}(c_1) - W_{uu}(\bar{z}_0 \,.\, c_1) \subset \partial W_{uu}(c).$$

Consider the chain

$$(\dagger\dagger) \qquad W_{uu}(c_1)\,[0, -r_0] = \{e^{-2\pi i r} \,.\, W_{uu}(c_1); \quad 0 \leqslant r \leqslant r_0\}$$

with $z_0 = e^{2\pi i r_0}$. The boundary of the chain

$$W_{uu}(c) + W_{uu}(c_1)\,[0, -r_0]$$

contains neither $W_{uu}(c_1)$ nor $-W_{uu}(\bar{z}_0 \,.\, c_1)$. Proceed in the same manner with the remaining elements of $\partial W_{uu}(c)$.

In this way we can construct a cycle w_0 which contains $W_{uu}(c)$ but no other element of the $\tilde{I}(c')$-orbit of $W_{uu}(c)$. For instance, if $z_0^j \,.\, c$, $1 \leqslant j \leqslant k-1$, are the elements in the $\tilde{I}(c')$-orbit of c different from c, then we replace the chain

$$\sum_1^{k-1} W_{uu}(z_0^j \,.\, c)$$

in $\partial W_{uu}(c') \cap w$ by the chain $(\dagger\dagger)$. Both chains have the same boundary; their domain of definition in each case is a cell. $\tilde{\pi} \circ w_0$ belongs mod 2 to $\tilde{\pi} \circ (\partial w' \cap w)$.

We define a homotopy w_0' of w_0 as follows. On $W_{uu}(c) \subset w_0$ it is the restriction of w', and the same holds for the deformation of the boundary of $W_{uu}(c)$, e.g. $W_{uu}(c_1)$. The homotopy of one of the $\bar{z} \,.\, W_{uu}(c_1)$, $\bar{z} = e^{-2\pi i r}$, $0 \leqslant r \leqslant r_0$, is obtained by applying $\bar{z}$ to the already defined homotopy of $W_{uu}(c_1)$.

Just as before we conclude the existence of a $W_{uu}(c') \subset w_0'$ containing $W_{uu}(c) \in w_0$ in its boundary. The invariance of $\partial W_{uu}(c') \cap w_0$ under $\tilde{I}(c')$ and the fact that w_0 contains only the element $W_{uu}(c)$ of the $I^{\sim}(c')$-orbit imply that $I^{\sim}(c')$ leaves c invariant. This completes the proof of (ii).

To prove (i), we obviously have $\kappa(r)=E(c(r))<E(c'(r))=\kappa'(r)$. To prove also $\kappa'(r)<\kappa(r+1)$, we recall from (4.3.3) that $w'_*(r)$ can be viewed as a restriction of $w_*(r+1)$.

We can choose $c(r+1)$ such that $\pi\circ W_{uu}(c(r+1))$ has dimension dim $w_*(r+1)$ and occurs in $\pi\circ w_*(r+1)$, and $\pi\circ W_{uu}(c'(r))$ is obtained from a restriction of the closure of $\pi\circ W_{uu}(c(r+1))$. Indeed, $\pi\circ w'_*(r)$ may also be viewed as a restriction of $\pi\circ w_*(r+1)$. □

Remark. In the proof of (4.3.4) it was vitally important that $\tilde{\pi}\circ\partial w_*(0)'\neq 0 \bmod 2$. This was due to the fact that the cycle $\bar{y}:S^{|\bar{y}|}\to\Omega M\subset\Lambda M$ came from a cycle $y:S^{|y|}\to M$ which represented an indivisible integer homotopy class of M of infinite order. To make this point more clear, for $M=S^2$ we take y to be the Hopf map $y:S^3\to S^2$. y does not have the aforementioned properties. Choose on $S^{|y|}=S^3$ the sphere $S^{|\bar{y}|}=S^2$ in such a way as to be orthogonal to the fibre $y^{-1}(*)$ over the base point $*$ of S^3. Then all the circles on S^2 passing through $*$ are contained twice in the image of

$$\bar{y}:(S^{|\bar{y}|},*)=(S^2,*)\to(\Omega M,*)=(\Omega S^2,*).$$

It follows that $\theta\circ\bar{y}$ has the same image as $\bar{y}$. Thus, if we define, for $w_*(0)=\bar{y}$, the homotopy

$$\bar{y}'=w_*(0)':(S^{|y|},*)\times[0,1]\to(\Omega S^2,*)$$

as above, we obtain $\partial w_*(0)'=0 \bmod 2$.

We are now ready for the proof of our main theorem.

4.3.5 Theorem. *Let M be a compact Riemannian manifold with finite fundamental group. Then there exist on M infinitely many prime unparameterized closed geodesics.*

Note. It suffices to prove the theorem for simply connected manifolds, since under the projection map $\tilde{M}\to M$ of the universal covering $\tilde{M}$ of M onto M, different prime closed geodesics on $\tilde{M}$ cannot cover the same prime closed geodesic on M.

We do not claim that the prime closed geodesics have no self-intersections. Indeed, as the example of a 2-dimensional ellipsoid with pairwise different axes of approximately the same length shows, there need not exist more than three closed geodesics without multiple points. In this case these are the three principal ellipses. See also (5.1.2) for this example.

Proof. In order to bring out the basic idea of the proof more clearly, we first assume that all closed geodesics on M are non-degenerate. The case for degenerate closed geodesics will be handled with the help of the Gromoll-Meyer Theorem.

So assume that there are only finitely many prime closed geodesics, say c_j, $1\leqslant j\leqslant s$. We consider only those geodesics c_j for which there exists an integer $m>0$ with index $c_j^m>0$. Denote these geodesics again by c_j, $1\leqslant j\leqslant s$.

From (3.2.15) and (4.2.7) we have the existence of real numbers $\alpha>0$, $A>0$, $\beta\geq 0$ such that, for all $j\in\{1,\ldots,s\}$,

$$(*)\qquad \begin{aligned}&\text{(i) } \alpha(m'-m)-\beta\leq \operatorname{index} c_j^{m'}-\operatorname{index} c_j^m,\\ &\text{(ii) } \operatorname{index} c_j^m/A\leq m.\end{aligned}$$

Fix an integer $r_0\geq 0$ and consider the sequence $w'_*(r_0+i)$, $0\leq i\leq s$, of $s+1$ subsequent homotopies of the Sullivan classes $w_*(r_0+i)$. From (4.3.4), there follows the existence of geodesics $c,c'\in\{c_1,\ldots,c_s\}$ and integers $a,b,0\leq a<b\leq s$, such that the geodesics $c(r_0+b)$, $c'(r_0+a)$, $c'(r_0+b)$ determined by $w(r_0+b)$, $w'(r_0+a)$, $w'(r_0+b)$ are of the form c^m, c'^{m_a}, c'^{m_b}. Here, $m=qm_b$ for some integer $q\geq 1$. Moreover,

$$E(c'^{m_a})<E(c^m)<E(c'^{m_b}).$$

Put $E(c)=\kappa$, $E(c')=\kappa'$. Then we can write

$$m_a^2\kappa'<q^2m_b^2\kappa<m_b^2\kappa',$$

and hence

$$0<(\kappa'/\kappa-q^2)<(m_b^2/m_a^2-1)q^2<(m_b^2/m_a^2-1)\,(\kappa'/\kappa).$$

We shall obtain the desired contradiction to our assumption that there are only finitely many prime closed geodesics if we can show that $m_b^2/m_a^2\to 1$, as $r_0\to\infty$. Note that there are only finitely many quotients $\kappa'/\kappa>1$.

Now, (*) (i) yields the estimate

$$\begin{aligned}\alpha(m_b-m_a)-\beta&\leq \operatorname{index} c'^{m_b}-\operatorname{index} c'^{m_a}\\ &=\dim w'(r_0+b)-\dim w'(r_0+a)\leq d(s+1)\end{aligned}$$

where $d=2|\bar{x}|$, x the generator of minimal odd degree in the minimal model $\mathfrak{M}(M)$ of M. Thus

$$m_b/m_a\leq (d(s+1)+\beta)/\alpha m_a+1.$$

But $m_a\to\infty$ as $r_0\to\infty$. This follows from (*) (ii) and the fact that $\dim w'(r_0+a)$ $=\operatorname{index} c'^{m_a}\geq \dim w'(r_0)$ grows linearly in r_0.

Let us now consider the general case, i.e., there are only finitely many prime closed geodesics $\{c_1,\ldots,c_s\}$ on M and all coverings c_j^m of these c_j lie on isolated critical S-orbits $S\,.\,c_j^m$. Again we assume that for every j there is an $m\geq 1$ such that index $c_j^m>0$.

For each j, $1\leq j\leq$s, there exists a finite set $M^*(c_j)=\{m_1^*,\ldots,m_{s(j)}^*\}$ of positive integers with the property (cf. (4.2.5), (4.2.6)) that for every covering c_j^m of c_j there exists a well-determined $m^*\in M^*(c_j)$ such that $m=qm^*$, q an integer ≥ 1; moreover, the characteristic manifold $W_{ca}(c_j^{m^*})$ of $c_j^{m^*}$ is isometric to the characteristic manifold $W_{ca}(c_j^m)$ of c_j^m.

In fact, the map $e \mapsto e^q$, defined on a neighborhood $D(S\,.\,c_j^{m*})$ of $S\,.\,c_j^{m*}$, maps $D(S\,.\,c_j^{m*})$ into a neighborhood $D(S\,.\,c_j^m)$ of $S\,.\,c_j^m$. Taking appropriate metrics in these neighborhoods, $e \mapsto e^q$, restricted to $W_{ca}(m_j^{m*})$, is an isometry onto $W_{ca}(c_j^m)$. In particular, $b_i^0(c_j^{m*}) = b_i^0(c_j^m)$.

Let $m^*(c_j)$ be the least common multiple of the elements of the set $M^*(c_j)$.

Now fix a non-critical $\kappa^* > 0$. Denote the finitely many critical orbits in $\Lambda^{\kappa^*}M - \Lambda^0 M$ by $S\,.\,c_\alpha$, $1 \leq \alpha \leq \alpha_0$. Put $E(c_\alpha) = \kappa_\alpha$.

We consider open neighborhoods $\mathscr{U}$ in Λ^{κ^*-} of the critical set of $\Lambda^{\kappa^*}M$ of the form

$$\mathscr{U} = \Lambda^{\varepsilon_0} \cup M \bigcup_\alpha \mathscr{U}(S\,.\,c_\alpha).$$

Here, $\varepsilon_0 > 0$ is so small that $\Lambda^{\varepsilon_0}M$ can be retracted onto $\Lambda^0 M$ by ϕ_s, $s \to \infty$, cf. (1.4.15). $\mathscr{U}_\alpha = \mathscr{U}(S\,.\,c_\alpha)$ is an open S-invariant neighborhood of $S\,.\,c_\alpha$ such that $\theta\mathscr{U}(S\,.\,c_\alpha) = \mathscr{U}(S\,.\,\theta c_\alpha)$. The $\mathscr{U}_\alpha$ are pairwise disjoint. The multiplicity of an element of $\mathscr{U}_\alpha$ is a divisor of the multiplicity $m(c_\alpha)$ of c_α.

Clearly, such neighborhoods $\mathscr{U}$ exist, and they form a fundamental system of neighborhoods for the critical set in $\Lambda^{\kappa^*}M$.

We wish to consider certain modifications $\tilde{g}$ of the Riemannian metric g on M. That is to say, $\tilde{g}$ shall be an element in a certain neighborhood of g in the space $\mathscr{G}M$ of Riemannian metrics on M, cf. (3.3). The underlying differentiable manifold, endowed with the metric $\tilde{g}$, shall be denoted by $\tilde{M}$. $\tilde{E}$ denotes the energy integral on $\Lambda\tilde{M}$.

Fix a neighborhood $\mathscr{U}$ of the type described above and a $\varepsilon > 0$. We consider modifications $\tilde{g} = \tilde{g}(\mathscr{U}, \varepsilon)$ of g satisfying the following conditions:

(a) κ^* is not a critical value of $\tilde{E}$ on $\Lambda\tilde{M}$;

(b) for every α, the critical orbit $S\,.\,c_\alpha$ in ΛM splits into a finite number of non-degenerate critical orbits, all of which belong to $\mathscr{U}_\alpha = \mathscr{U}(S\,.\,c_\alpha)$. Thereby, the $\tilde{E}$-values of the new critical orbits in $\mathscr{U}(S\,.\,c_\alpha)$ differ by less than ε in absolute value from $E(c_\alpha) = \kappa_\alpha$. Moreover,

$$H_*(\Lambda^{\kappa_\alpha -}M \cup \mathscr{U}_\alpha, \Lambda^{\kappa_\alpha -}M) = H_*(\Lambda^{\kappa_\alpha -}\tilde{M} \cup \mathscr{U}_\alpha, \Lambda^{\kappa_\alpha -}\tilde{M});$$

(c) if c_α is of the form c_j^m, then the multiplicity $\tilde{m}$ of any of the non-degenerate closed geodesics $\tilde{c}$ of $\Lambda\tilde{M}$ in $\mathscr{U}_\alpha$ is such that the integer $m/\tilde{m}$ divides the least common multiple $m^*(c_j)$ of the elements of $M^*(c_j)$. Thus, in particular, $m/\tilde{m} \leq m^*(c_j)$.

We refer to the methods and results of (3.3) to show that there actually exist modifications $\tilde{g} = \tilde{g}(\mathscr{U}, \varepsilon)$ having properties (a), (b), (c). Indeed, (a) and (b) follow at once from (3.3.7). To see that (c) can also be satisfied, we fix j. From the definition of $m^*(c_j)$ it follows that the characteristic manifold $W_{ca}(c_j^{m^*(c_j)})$ of $c_j^{m^*(c_j)}$ has maximal dimension among all the characteristic manifolds $W_{ca}(c_j^{m_i^*})$, $m_i^* \in M^*(c_j)$.

By an appropriate modification $\tilde{g}$ of g we can replace the critical orbit $S\,.\,c_j^{m^*(c_j)}$ by a certain number of non-degenerate critical orbits $S\,.\,\tilde{c}$, all of which belong to a prescribed $\mathscr{U}(S\,.\,c_j^{m^*(c_j)})$. We also say that $W_{ca}(c_j^{m^*(c_j)})$ has been dissolved into the $S\,.\,\tilde{c}$.

Fix some integer $q_0 \geqslant 1$. As long as $\tilde{g}$ is sufficiently near g, the $W_{ca}(c_j^{qm^*(c_j)})$, for all $q \leqslant q_0$, will be dissolved into a number of critical non-degenerate orbits near $S \,.\, c_j^{qm^*(c_j)}$. Actually, these orbits will be the images $S \,.\, \tilde{c}^q$ under the map $e \to e^q$ of the non-degenerate orbits $S \,.\, \tilde{c}$ near $S \,.\, c_j^{m^*(c_j)}$.

Now if $m_i^* \in M^*(c_j)$, we can write $m^*(c_j) = pm_i^*$, where p is a positive integer. Consider the map of a neighborhood $D(S \,.\, c_j^{m_i^*})$ of $S \,.\, c_j^{m_i^*}$ into a neighborhood $D(S \,.(c_j^{m^*(c_j)})$given by $e \to e^p$. Under this map, $W_{ca}(c_j^{m_i^*})$ goes into a submanifold of $W_{ca}(c_j^{m^*(c_j)})$. Thus, if under the modification $\tilde{g}$ of g, $W_{ca}(c_j^{m^*(c_j)})$ is dissolved into a number of non-degenerate critical orbits $S \,.\, \tilde{c}$ near $S \,.\, c_j^{m^*(c_j)}$, we can assume that at the same time $W_{ca}(c_j^{m_i^*})$ is dissolved into non-degenerate critical orbits $S \,.\, \tilde{c}(m_i^*)$ near $S \,.\, c_j^{m_i^*}$ which under $e \mapsto e^p$ become orbits of the type $S \,.\, \tilde{c} \in \mathscr{U}(S \,.\, c_j^{m^*(c_j)})$. In particular, $p \cdot$ multiplicity of $\tilde{c}(m_i^*)$ divides $m^*(c_j)$. Thus, for all c_j^m, $m \leqslant$ some m_0, where $m = m_i^* q$, the non-degenerate orbits of $\Lambda\tilde{M}$ in $\mathscr{U}(S \,.\, c_j^m)$, which are obtained by dissolving $W_{ca}(c_j^m) \cong W_{ca}(c_j^{m_i})$, have a multiplicity $\tilde{m}$ such that $m/\tilde{m}$ divides $m^*(c^j)$.

This completes the proof that modifications $\tilde{g}$ of g satisfying (a), (b), (c) exist.

According to (4.2.5) and (4.2.6) there exists, for each $j \in \{1, \ldots, s\}$, an $i_0(j)$ such that $i > i_0(j)$ implies that $b_i^0(c_j^m) = 0$, for all m. Put $i_0 = \sup_j i_0(j)$. From (4.2.4) it follows, if $i > i_0$ and $b_i(c_j^m) \neq 0$, that

$$(**) \qquad 0 < \text{index } c_j^m \leqslant i \leqslant \text{index } c_j^m + i_0 \,.$$

Consider the finite set K of real numbers of the form $(m_k^{*2}\kappa_k)/(m_l^{*2}\kappa_l)$ which are >1, where m_k^* divides $m^*(c_k)$ and m_l^* divides $m^*(c_l)$. Choose $\underline{\lambda} \in K$, $\underline{q}$ an integer $\geqslant 1$ such that $\underline{\lambda} - \underline{q}^2$ is >0 and minimal among the $\lambda - q^2 > 0, \lambda \in K$. Let $\bar{\lambda} = \max \{\lambda \in K\}, d = 2|\bar{x}|$ and $\alpha > 0$, $\beta > 0$ as above. Choose an integer $m_0 > 0$ such that

$$(\dagger) \qquad [((d(s+1) + i_0 + \beta)/(\alpha m_0) + 1)^2 - 1]\bar{\lambda} < \underline{\lambda} - \underline{q}^2 .$$

There exists $i_* \geqslant i_0$ such that $i > i_*$ and $b_i(c_j^m) \neq 0$ implies $m \geqslant m_0$. Indeed, $(**)$ and $(*)$ (ii) yield $0 < (i - i_0)/A \leqslant m$. Choose $i_* \geqslant Am_0 + i_0$. There also exists an integer $r_0 \geqslant 0$ such that $r \geqslant r_0$ implies that dim $w_*(r) > i_*$.

Consider $w'_*(r_0 + s)$, $0 \leqslant i \leqslant s$. Choose $\kappa^* > 0$ to be non-critical and such that the $w'_*(r_0 + s)$ have their carrier in $\Lambda^{\kappa^*-}M$. With respect to this κ^*, take a neighborhood $\mathscr{U}$ of the critical set of Λ^{κ^*} as above and consider, for small $\varepsilon > 0$, modifications $\tilde{g} = \tilde{g}(\mathscr{U}, \varepsilon)$ of the Riemannian metric g on M having properties (a), (b), (c).

As a first consequence, we obtain from (4.3.4) for each of the $w'_*(r_0 + i)$, $w_*(r_0 + i)$, $0 \leqslant i \leqslant s$, closed geodesics $\tilde{c}'(r_0 + i)$, $\tilde{c}(r_0 + i)$ on $\tilde{M}$ satisfying conditions (i) and (ii) of (4.3.4). Each of these geodesics belongs to a well-determined neighborhood $\mathscr{U}(S \,.\, c_\alpha)$ of one of the critical orbits $S \,.\, c_\alpha$ of $\Lambda^{\kappa^*}M - \Lambda^0 M$.

There exist integers a and b, $0 \leqslant a < b \leqslant$s, and $c' \in \{S \,.\, c_1, \ldots, S \,.\, c_s\}$ such that

$$\tilde{c}'(r_0 + a) \in \mathscr{U}(S \,.\, c'^{m_a}); \qquad \tilde{c}'(r_0 + b) \in \mathscr{U}(S \,.\, c'^{m_b}).$$

Let $\tilde{c}(r_0 + b) \in \mathscr{U}(S \,.\, c^m)$, $c \in \{S \,.\, c_1, \ldots, S \,.\, c_s\}$. Put $E(c') = \kappa'$, $E(c) = \kappa$.

The multiplicities $\tilde{m}_a$, $\tilde{m}_b$, $\tilde{m}$ of $\tilde{c}'(r_0+a), \tilde{c}'(r_0+b), \tilde{c}(r_0+b)$ satisfy the relations

$$m_a = \tilde{m}_a m''^*, \qquad m_b = \tilde{m}_b m'^*, \qquad m = \tilde{m} m^*.$$

Here, m''^* and m'^* belong to the set $M^*(c')$, and thus divide $m^*(c')$. m^* belongs to $M^*(c)$, thus $m^* | m^*(c)$. Property (i) of (4.3.4) states that

$$(\dagger\dagger) \qquad E(\tilde{c}'(r_0+a)) < E(\tilde{c}(r_0+b)) < E(\tilde{c}'(r_0+b))$$

and according to (ii) of (4.3.4) there exists an integer $q \geqslant 1$ such that $\tilde{m} = q\tilde{m}_b$.

Since we can choose ε in $\tilde{g}(\mathscr{U}, \varepsilon)$ arbitrarily small, we can conclude from (††) that

$$E(c'^{m_a}) = m_a^2 \kappa' < E(c^m) = m^2 \kappa = q^2 \tilde{m}_b^2 m^{*2} \kappa = q^2 m_b^2 (m^{*2}/m'^{*2}) \kappa <$$
$$E(c'^{m_b}) = m_b^2 \kappa' = \tilde{m}_b^2 m'^{*2} \kappa'.$$

Thus

$$1 \leqslant q^2 < m'^{*2}\kappa'/m^{*2}\kappa; \qquad m'^{*2}\kappa'/m^{*2}\kappa < q^2 (m_b^2/m_a^2).$$

Hence

$$(\S) \qquad 0 < \underline{\lambda} - \underline{q}^2 \leqslant (m'^{*2}\kappa')/(m^{*2}\kappa) - q^2 < (m_b^2/m_a^2 - 1) q^2$$
$$< ((m_b^2/m_a^2) - 1)\,(m'^{*2}\kappa'/m^{*2}\kappa) \leqslant (m_b^2/m_a^2 - 1)\bar{\lambda}.$$

$w'_*(r_0+a)$ defines locally at $\tilde{c}'(r_0+\mathrm{a})$ a non-trivial cycle of dimension dim $w'_*(r_0+a)$. From property (b) of the modified metric $\tilde{g}$ it follows that the type number $b_i(c'^{m_a})$ of c'^{m_a} in dimension $i = \dim w'_*(r_0+a)$ is not zero. Similarly, $b_j(c'^{m_b}) \neq 0$ for $j = \dim w'_*(r_0+a)$. Thus (**) implies that

$$\dim w'_*(r_0+a) \leqslant \text{index } c'^{m_a} + i_0; \qquad \text{index } c'^{m_b} \leqslant \dim w'_*(r_0+b).$$

From (*) (i) we therefore obtain

$$\alpha(m_b - m_a) - \beta - i_0 \leqslant \text{index } c'^{m_b} - \text{index } c'^{m_a} - i_0 \leqslant \dim w'_*(r_0+b)$$
$$- \dim w'_*(r_0+a) \leqslant d(s+1).$$

Hence, since $m_a \geqslant m_0$,

$$m_b/m_a \leqslant (d(s+1) + i_0 + \beta)/(\alpha m_0) + 1.$$

Together with (†), this gives a contradiction to (§).

This completes the proof of the theorem. □

4.3 Appendix. The Minimal Model for the Rational Homotopy Type of ΛM

By Jonathan Sacks

A differential graded-commutative algebra (over $\mathbb{Q}$) (abbreviated DGA) is a graded algebra

$$\mathscr{A} = \bigoplus_{k \geqslant 0} \mathscr{A}^k$$

together with a differential

$$d : \mathscr{A} \to \mathscr{A}$$

such that

(i) $xy = (-1)^{kl} yx \qquad (x \in \mathscr{A}^k,\ y \in \mathscr{A}^l)$,

(ii) $d(xy) = dx \cdot y + (-1)^k x \cdot dy \qquad (x \in \mathscr{A}^k)$, and

(iii) $d^2 = 0$.

Given a DGA $\mathscr{A}$, let $I(\mathscr{A}) = \bigoplus_{k \geqslant 1} \mathscr{A}^k$ be the ideal of positive degree elements.

A DGA $\mathfrak{M}$ is said to be *minimal* if

(i) $\mathfrak{M}$ is free as a graded-commutative algebra (thus $\mathfrak{M}$ is a tensor product of a polynomial algebra on even degree generators with an exterior algebra on odd degree generators), and

(ii) $dI(\mathfrak{M}) \subseteq I(\mathfrak{M}) \,.\, I(\mathfrak{M})$.

Let $\{t, dt\}$ denote $\mathbb{Q}[t] \otimes \mathbb{Q}[t]dt$, i.e. polynomial forms in one variable t. Two maps of DGA's

$$\mathscr{A} \xrightarrow{f_i} \mathscr{B} \qquad (i = 0,1)$$

are called *homotopic* if there exists a map of DGA's

$$\mathscr{A} \xrightarrow{F} \mathscr{B} \otimes \{t, dt\}$$

such that

$$f_i = F|_{t=i} \quad (i = 0,1).$$

(The notation $F|_{t=0}$ means compose F with the DGA map on $\mathscr{B} \otimes \{t, dt\}$ given by setting $t = 0 = dt$).

4.3.A.1 Theorem. *The relation of homotopy between maps of DGA's is an equivalence relation.*

Given a simple homotopy type X, i.e. one having the homotopy type of a countable CW complex in which the fundamental group acts trivially on homology and homotopy, one can choose a rational polyhedron representing the homotopy type of X. On this rational polyhedron one can define the DGA over $\mathbb{Q}$ consisting of rational $P.L.$ forms, denoted by $\mathscr{A}_{P.L.}(X)$. The equivalence class of $\mathscr{A}_{P.L.}(X)$ under the equivalence relation of homotopy of DGA's contains a unique (up to isomorphism) representative which is minimal. This minimal DGA is called the minimal model for $\mathscr{A}_{P.L.}(X)$ and is denoted by $\mathfrak{M}(X)$.

A map $f: X \to Y$ induces $f^*: \mathscr{A}_{P.L.}(Y) \to \mathscr{A}_{P.L.}(X)$ by pull-back of $P.L.$ differential forms. In turn, f^* induces a map $\bar{f}: \mathfrak{M}(Y) \to \mathfrak{M}(X)$ whose homotopy class depends only on the homotopy class of f.

The main theorem on this subject is the following:

4.3.A.2 Theorem (*Quillen, Sullivan*). *There is an equivalence of categories*

$$\{\textit{simple rational homotopy types}\} \leftrightarrow \{\textit{DGA's over } \mathbb{Q} \textit{ modulo homotopy equivalence}\}.$$

In this equivalence of categories, the assignments are
$\{\textit{rational space } X\} \leftrightarrow \mathfrak{M}(X)$;
$\{\textit{maps } X \xrightarrow{f} Y\} \leftrightarrow \{\bar{f}: \mathfrak{M}(Y) \to \mathfrak{M}(X)\}$
where $\{\ \}$ denotes homotopy class. (For a proof, see "Homotopy theory & differential forms", R. Friedlander, P.A. Griffiths, J. Morgan, Seminario di Geometria, Florence (1972)). □

Given two DGA's $\mathscr{A}$ and $\mathscr{B}$, we can make their tensor product (as vector spaces) into a DGA as follows.

We begin by providing $\mathscr{A} \otimes \mathscr{B}$ with a grading by letting $\deg a \otimes b = \deg a + \deg b$, where $a \in \mathscr{A}$ and $b \in \mathscr{B}$ are generators.

To define multiplication on $\mathscr{A} \otimes \mathscr{B}$ it is sufficient to do so on its generators and extend it in the obvious way to all of $\mathscr{A} \otimes \mathscr{B}$. So let $a_1, a_2 \in \mathscr{A}$ and $b_1, b_2 \in \mathscr{B}$ be generators. Define

$$(a_1 \otimes b_1) \cdot (a_2 \otimes b_2) = (-1)^{\deg b_1 \cdot \deg b_2} a_1 a_2 \otimes b_1 b_2 .$$

Suppose $d_{\mathscr{A}}$ and $d_{\mathscr{B}}$ are the differentials in $\mathscr{A}$ and $\mathscr{B}$, respectively. Define the differential d in $\mathscr{A} \otimes \mathscr{B}$ by

$$d = d_{\mathscr{A}} \otimes \mathrm{id} + \mathrm{id} \otimes d_{\mathscr{B}} .$$

It is easy to check that with all these definitions $\mathscr{A} \otimes \mathscr{B}$ satisfies all the axioms for a DGA.

Let M be a simply connected manifold and suppose $\mathfrak{M}(M)$ is the DGA over $\mathbb{Q}$ with generators $\{x_i\}_{i \in \mathbb{N}}$ and differential d. Let $\mathscr{A}$ be the DGA over $\mathbb{Q}$ with generators $\{x_i, \bar{x}_i\}_{i \in \mathbb{N}}$, where degree $\bar{x}_i$ is one less than degree x_i. Extend $\bar{\ }$ to all of $\mathfrak{M}(M)$ as a derivation acting from the right, and define the differential on $\mathscr{A}$ by letting it act the same way on unbarred generators as in $\mathfrak{M}(M)$ and defining $d\bar{x}_i = \overline{dx_i}$.

4.3.A.3 Theorem. *$\mathscr{A}$ is isomorphic as a* DGA *to* $\mathfrak{M}(\Lambda M)$.

The theorem is proved by means of the following:

4.3.A.4 Proposition *Let $\{e\}$ be the minimal model for the rational homotopy type of S^1, i.e. $\{e\}$ is the* DGA *over $\mathbb{Q}$ generated by one element e of degree one with $de=0$. Then there is a map of* DGA's

$$\Phi: \mathfrak{M}(M) \to \mathscr{A} \otimes \{e\}$$

such that for every DGA $\mathscr{B}$, *and for every map*

$$\Psi: \mathfrak{M}(M) \to \mathscr{B} \otimes \{e\}$$

of DGA'*s, there exists a unique map*

$$\tilde{\Psi}: \mathscr{A} \to \mathscr{B}$$

of DGA'*s which makes the following diagram commute*

$$\begin{array}{ccc} \mathfrak{M}(M) & \xrightarrow{\quad \Phi \quad} & \mathscr{A} \otimes \{e\} \\ & \searrow^{\Psi} & \downarrow{\scriptstyle \tilde{\Psi} \otimes \mathrm{id}} \\ & & \mathscr{B} \otimes \{e\} \end{array}$$

Proof of Proposition. Define ϕ on generators by

$$\phi(x_i) = x_i \otimes 1 + \bar{x}_i \otimes e$$

and extend to all of $\mathfrak{M}(M)$ as an algebra map in the obvious way.

We must first show that Φ is a map of DGA'*s*, i.e. that Φ commutes with the differentials. It is easy to show that it suffices to prove that Φ commutes with the differentials on generators. Suppose $x \in \mathfrak{M}(M)$ is a generator and let

$$dx = \sum_{(i_1, \ldots, i_r)} a_{i_1 \ldots i_r} x_{i_1} \ldots x_{i_r} \quad \text{where} \quad a_{i_1 \ldots i_r} \in \mathbb{Q}.$$

Then

$$\begin{aligned} d(\Phi(x)) &= d(x \otimes 1 + \bar{x} \otimes e) = dx \otimes 1 + \overline{dx} \otimes e \\ &= \Big(\sum_{(i_1, \ldots, i_r)} a_{i_1 \ldots i_r} x_{i_1} \ldots x_{i_r}\Big) \otimes 1 \\ &\quad + \Big(\sum_{(i_1, \ldots, i_r)} a_{i_1 \ldots i_r} \sum_{j=1}^{r} (-1)^{(\deg x_{i_{j+1}}) \ldots (\deg x_{i_r})} x_{i_1} \ldots x_{i_{j-1}} \bar{x}_{i_j} x_{i_{j+1}} \ldots x_{i_r}\Big) \otimes e \end{aligned}$$

since $\bar{\ }$ acts as a derivation from the right. Also,

$$\begin{aligned} \Phi(dx) &= \Phi\Big(\sum_{(i_1, \ldots, i_r)} a_{i_1 \ldots i_r} x_{i_1} \ldots x_{i_r}\Big) \\ &= \sum_{(i_1, \ldots, i_r)} a_{i_1 \ldots i_r} \Phi(x_{i_1}) \ldots \Phi(x_{i_r}) \\ &= \sum_{(i_1, \ldots, i_r)} a_{i_1 \ldots i_r} (x_{i_1} \otimes 1 + \bar{x}_{i_1} \otimes e) \ldots (x_{i_r} \otimes 1 + \bar{x}_{i_r} \otimes e) \end{aligned}$$

$$= \sum_{(i_1,\ldots,i_r)} a_{i_1\ldots i_r}(x_{i_1}\ldots x_{i_r})\otimes 1$$

$$+\Big(\sum_{(i_1,\ldots,i_r)} a_{i_1\ldots i_r}$$

$$\sum_{j=1}^{r}(-1)^{(\deg x_{i_{j+1}})\ldots(\deg x_{i_r})} x_{i_1}\ldots x_{i_{j-1}}\bar{x}_{i_j}x_{i_{j+1}}\ldots x_{i_r}\Big)\otimes e$$

since $e^2=0$. Note that the factor $(-1)^{(\deg x_{i_{j+1}})\ldots(\deg x_{i_r})}$ comes from the definition of multiplication in $\mathscr{A}\otimes\{e\}$. Thus $d\Phi(x_i)=d\Phi(x_i)$ for each generator x_i, as claimed. Next we show that the pair $(\mathscr{A}\otimes\{e\},\Phi)$ has the universal property described in the statement of the proposition.

Suppose $\Psi:\mathfrak{M}(M)\to\mathscr{B}\otimes\{e\}$ is a map of DGA's. For each generator $x_i\in\mathfrak{M}(M)$ we can write $\Psi(x_i)$ uniquely in the form

$$\Psi(x_i)=b_i\otimes 1+\tilde{b}_i\otimes e$$

where $b_i\otimes 1$ is the component of $\Psi(x_i)$ without an e factor and $\tilde{b}_i\otimes e$ is the component of $\Psi(x_i)$ with an e factor.

Define $\tilde{\Psi}:\mathscr{A}\to\mathscr{B}$ on generators by

$$\tilde{\Psi}(x_i)=b_i,\quad\text{and}$$

$$\tilde{\Psi}(\bar{x}_i)=\tilde{b}_i$$

and extend as an algebra map to all of $\mathscr{A}$. Clearly, $\tilde{\Psi}$ is the unique algebra map such that $(\tilde{\Psi}\otimes\mathrm{id})\circ\Phi=\Psi$.

Also, for each generator $x_i\in\mathfrak{M}(M)$, $\Psi(dx_i)=d(\Psi x_i)$, since Ψ is a map of DGA's. But

$$\begin{aligned}\Psi(dx_i)&=(\tilde{\Psi}\otimes\mathrm{id})\,(\Phi(dx_i))\\&=(\tilde{\Psi}\otimes\mathrm{id})\,(d(\Phi(x_i))),\ \text{since }\Phi\text{ is a DGA map,}\\&=(\tilde{\Psi}\otimes\mathrm{id})\,(dx_i\otimes 1+d\bar{x}_i\otimes e)\\&=\tilde{\Psi}(dx_i)\otimes 1+\tilde{\Psi}(d\bar{x}_i)\end{aligned}$$

and

$$\begin{aligned}d(\Psi(x_i))&=d((\tilde{\Psi}\otimes\mathrm{id})\,(\Phi(x_i)))\\&=d((\tilde{\Psi}\otimes\mathrm{id})\,(x_i\otimes 1+\bar{x}_i\otimes e))\\&=d(\tilde{\Psi}(x_i)\otimes 1+\tilde{\Psi}(\bar{x}_i)\otimes e)\\&=d\tilde{\Psi}(x_i)\otimes 1+d(\tilde{\Psi}(\bar{x}_i))\otimes e\end{aligned}$$

from which it follows that

$$\tilde{\Psi}(dx_i)=d\tilde{\Psi}(x_i)\quad\text{and}\quad\tilde{\Psi}(d\bar{x}_i)=d\tilde{\Psi}(\bar{x}_i)$$

for all generators $x_i,\bar{x}_i$ in $\mathscr{A}$, i.e. $\tilde{\Psi}$ is a map of DGA's. □

Proof of theorem. Let $ev : \Lambda M \times S^1 \to M$ be the evaluation map

$$ev(c,t) = c(t) \quad \text{for} \quad c \in \Lambda M \quad \text{and} \quad t \in S^1.$$

The pair $(\Lambda M \times S^1, ev)$ has the following universal property.

Given any space X and any continuous map $f : X \times S^1 \to M$, there exists a unique continuous function $\tilde{f} : X \to \Lambda M$ so that the following diagram commutes

$$
\begin{array}{ccc}
M & \xleftarrow{\ ev\ } & \Lambda M \times S^1 \\
& \nwarrow f & \Big\uparrow \tilde{f} \times \mathrm{id} \\
& & X \times S^1
\end{array}
$$

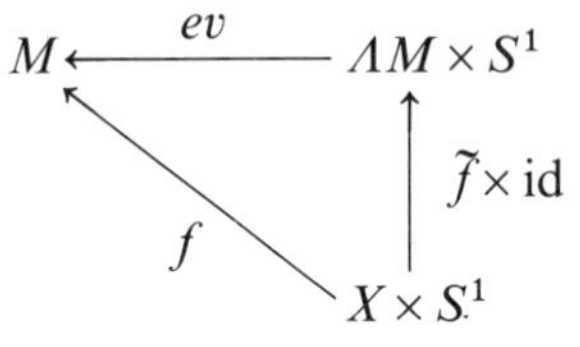

$\tilde{f}$ is defined by $\tilde{f}(x)(t) = f(x,t)$.

Now let $X = X_{\mathscr{A}}$, the rational homotopy type corresponding to $\mathscr{A}$ of the proposition just proved, and let $f = f_{\Phi}$ be any continuous map which is sent to $\Phi : \mathfrak{M}(M) \to \mathscr{A} \otimes \{e\}$ under the equivalence of categories in the theorem of Quillen and Sullivan. By this same theorem, the above commutative diagram gives rise to a diagram in the category of DGA's over $\mathbb{Q}$ which homotopy commutes

$$
\begin{array}{ccc}
\mathfrak{M}(M) & \xrightarrow{\ \Psi_{ev}\ } & \mathfrak{M}(\Lambda M) \otimes \{e\} \\
& \searrow \Phi & \Big\downarrow \tilde{\Phi} \otimes \mathrm{id} \\
& & \mathscr{A} \otimes \{e\}
\end{array}
$$

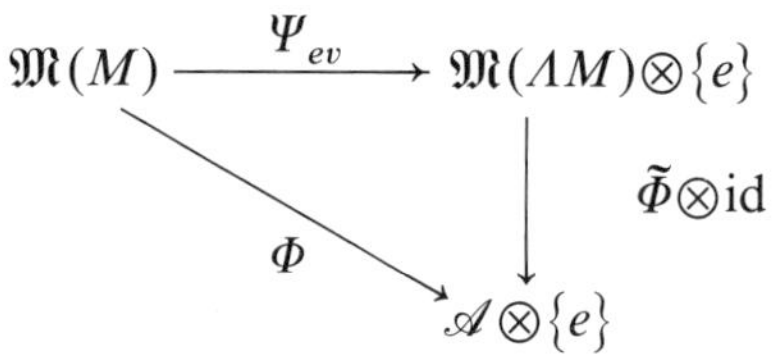

where Ψ_{ev} is the DGA map coming from ev and $\tilde{\Phi} \otimes \mathrm{id}$ is the DGA map coming from $\tilde{f} \times \mathrm{id}$. Consider the diagram

$$
\begin{array}{ccc}
\mathfrak{M}(M) & \xrightarrow{\ \Phi\ } & \mathscr{A} \otimes \{e\} \\
& \searrow \Psi_{ev} & \Big\downarrow \tilde{\Psi}_{ev} \otimes \mathrm{id} \\
& & \mathfrak{M}(\Lambda M) \otimes \{e\} \\
& \searrow \Phi & \Big\downarrow \tilde{\phi} \otimes \mathrm{id} \\
& & \mathscr{A} \otimes \{e\}
\end{array}
$$

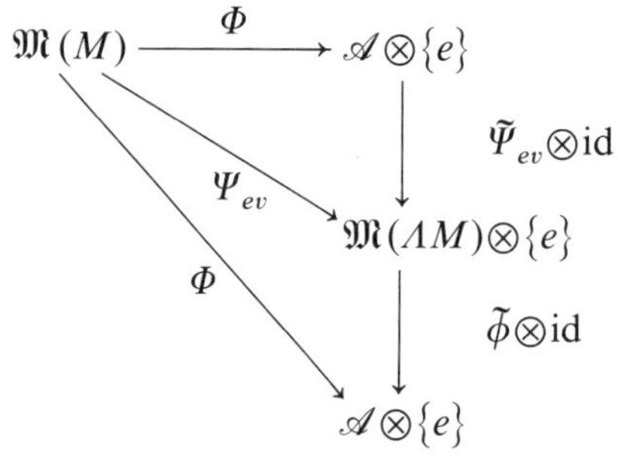

where $\tilde{\Psi}_{ev}$ is the unique DGA map (whose existence is guaranteed by the proposition) which makes the upper triangle commute. Since the lower triangle homotopy commutes, the diagram

$$
\begin{array}{ccc}
\mathfrak{M}(M) & \xrightarrow{\;\Phi\;} & \mathscr{A}\otimes\{e\} \\
 & \overset{\Phi}{\searrow} & \big\downarrow{\scriptstyle(\tilde{\Phi}\circ\tilde{\Psi}_{ev})\otimes \mathrm{id}} \\
 & & \mathscr{A}\otimes\{e\}
\end{array}
$$

also homotopy commutes.

By the proposition, this implies that $(\tilde{\Phi}\circ\tilde{\Psi}_{ev})\otimes\mathrm{id}$ is homotopic to $\mathrm{id}\otimes\mathrm{id}$, which in turn implies that $\tilde{\Psi}_{ev}:\mathscr{A}\to\mathfrak{M}(\Lambda M)$ is a homotopy equivalence. Since $\mathfrak{M}(\Lambda M)$ is the unique (up to isomorphism) minimal DGA in its homotopy class, and $\mathscr{A}$ is minimal by construction, $\mathscr{A}$ is isomorphic as a DGA to $\mathfrak{M}(\Lambda M)$. □

4.4 Some Generic Existence Theorems

In (4.3) we proved the existence of infinitely many closed geodesics by making a detailed analysis of the Morse complex of ΛM. No use at all was made of the geodesic flow, except for Bott's formula (3.2.9) for the index of an iterated closed geodesic in terms of the ρ-index.

Here we want to show how the generic properties of the geodesic flow which were derived in (3.3) can be employed to prove that generically there exist infinitely many closed geodesics.

Of course, this is less than (4.3.5). But the proof is somewhat simpler and, in addition, we obtain under certain conditions information on the existence of infinitely many closed geodesics near a given one of elliptic type. Such a result was envisioned by Poincaré and Birkhoff in their early papers.

Again, a crucial role is played by the divisibility relation (ii) of our Fundamental Lemma (4.3.4). Other tools are the Sullivan classes and, most important, the Birkhoff-Lewis Fixed Point Theorem (3.3.3). At the very end, we prove for even-dimensional manifolds of non-negative curvature and with non-trivial fundamental group the existence of non-hyperbolic closed geodesics. This gives support to our conjecture that, in general, on a manifold with finite fundamental group not all closed geodesics can be hyperbolic. Also, the proof is very simple and avoids the use of the delicate Fundamental Lemma.

The results of this section (except Theorem (4.4.6)) were presented in [Kl 16].

We consider the dual homology classes $w_*(r)$, $r=0,1,\dots$, of certain Sullivan cohomology classes introduced in (4.3). Recall that $w_*(r)$ is dual either to the class $\bar{x}\wedge\bar{x}^{2r}$ or to $\bar{e}_1\wedge\bar{x}^r\dots\wedge\bar{e}_l$.

4.4.1 Lemma. *Let M be a compact, simply connected Riemannian manifold. Assume that all closed geodesics on M are hyperbolic, cf.* (3.2). *Then there exists, for each $r=0,1,\dots$, a prime closed geodesic $c'(r)$ of index* $\dim w'_*(r)$, *$c'(r)$ constructed with the help of $w'_*(r)$, as in* (4.3.4).

Note. It seems unlikely that on a compact simply connected Riemannian manifold M all closed geodesics are hyperbolic. In fact, Poincaré [Po 1] has shown that there exists an open set of riemannian metrics g on S^2 in the neighborhood of the standard metric g_0 such that (S^2,g) possesses an elliptic closed geodesic of length approximately equal to 2π. See also (4.4.6). Thorbergsson recently has generalized these results.

On the other hand, if M has negative sectional curvature or, more generally, if the geodesic flow on the tangent bundle of M is of Anosov type, then all closed geodesics are hyperbolic. Note, however, that in this case the fundamental group is very large; actually, it has exponential growth, cf. [An], [Kl 12], (5.3).

Proof. By our hypothesis, all closed geodesics on M are non-degenerate. Thus, we can construct, as in (4.3.4), for every $r=0,1,\ldots$ closed geodesics $c(r)$ and $c'(r)$ of index $\dim w_*(r)$ and $\dim w'_*(r)$, respectively, such that the multiplicity $m'(r)$ of $w'_*(r)$ divides the multiplicity $m(r)$ of $w_*(r)$.

From (3.2.13) we have a simple formula for the index of the hyperbolic closed geodesics $c(r)=c_0(r)^{m(r)}$ and $c'(r)=c'_0(r)^{m'(r)}$. This yields

$$m(r)\ \mathrm{index}\ c_0(r)+1=m'(r)\ \mathrm{index}\ c'_0(r).$$

Thus, $m'(r)|m(r)$ implies that $m'(r)=1$, and we have proved (4.4.1). □

On the space $\mathscr{G}=\mathscr{G}M$ of Riemannian metrics on a compact, simply connected differentiable manifold we consider the following subsets:

$\mathscr{G}_H$ = set of those g for which (M,g) has only hyperbolic closed geodesics.

$\mathscr{G}_T$ = set of those g for which (M,g) has at least one closed geodesic of twist type, i.e. the Poincaré map of this closed geodesic is of twist type, cf. (3.3). Clearly, $\mathscr{G}_T$ is open.

As already stated above, $\mathscr{G}_H$ is likely to be empty. Denote by $\mathring{\mathscr{G}}_H$ the interior of $\mathscr{G}_H$.

4.4.2 Theorem. *Let M be a compact, simply connected differentiable manifold. Then the set $\tilde{\mathscr{G}}=\mathring{\mathscr{G}}_H\cup\mathscr{G}_T$ is an open dense set in $\mathscr{G}=\mathscr{G}_M$. In particular, if $\mathscr{G}_H$ is empty, there exists an open dense set $\tilde{\mathscr{G}}$ in $\mathscr{G}$ such that, for each $\tilde{g}\in\tilde{\mathscr{G}}$, $\tilde{M}=(M,\tilde{g})$ has at least one prime closed geodesic $\tilde{c}_0$ such that there exist infinitely many prime closed geodesics $\tilde{c}_0(k)$, $k=1,2,\ldots$, with length $\tilde{c}_0(k)$ tending to infinity, in any neighborhood of the immersed circle $t\in S\mapsto\tilde{c}_0(t)\in\tilde{M}$ on $\tilde{M}$.*

Note. The interesting feature of this theorem is its second part, i.e. the existence of infinitely many prime closed geodesics on $\tilde{M}$ which cluster around a closed geodesic $\tilde{c}_0$ on $\tilde{M}$. This phenomenon occurs whenever we have a closed geodesic of twist type. One may hope that some day one will be able to show that generically there exist so many closed geodesics of twist type that the closure of the $X_0\in T_1M$ which give rise to a periodic orbit of the geodesic flow on T_1M has positive measure. Take (3.3.4) together with the theorem (4.4.6) below, applied to the projective plane with a metric of positive curvature, as a partial result in this direction.

Proof. $\tilde{\mathscr{G}}$ clearly is open, and is also dense. Indeed, the closure of $\tilde{\mathscr{G}}$ contains the set of all those metrics $g\in\mathscr{G}$ for which (M,g) has at least one closed geodesic c with a non-hyperbolic Poincaré map P_c, cf. (3.3.7) and (3.3.8). Thus, the complement of the closure of $\tilde{\mathscr{G}}$ is an open set in the complement of $\mathring{\mathscr{G}}_H$, consisting of those g' for which (M,g') has only hyperbolic closed geodesics, i.e. this complement is empty. The remainder now follows from (3.3.3) □

We continue this section by pointing out still another residual set $\mathscr{G}^*$ in $\mathscr{G}_M$ which might be useful in future work on generic properties of closed geodesics. We used this set, in the paper [Kl 16] to show that, at least generically, there always exist infinitely many prime closed geodesics.

Given $g \in \mathscr{G}$, we consider for each closed geodesic c on (M,g) the linear Poincaré map P_c. Write the eigenvalues of P_c of modulus 1 in the form $e^{\pm 2\pi i a}, 0 \leqslant a \leqslant {}^1\!/\!_2$. Recall that $a=0$ means that c is degenerate, cf. (3.2.12).

Let $A(M,g)$ be the set of the numbers a defined thus, for all closed geodesics c on (M,g). We consider the following property of the metric g on M:

(α) the set $A(M,g) \cup \{1\}$ is free over the rationals.

In other words, none of the $a \in A(M,g)$ is rational and any finite set of elements of $A(M,g)$ is independent over the rationals.

Note that a Riemannian metric g satisfying (α), has only non-degenerate closed geodesics. Abraham [Ab] called such metrics bumpy. Thus we are tempted to call a metric satisfying (α) *super bumpy*.

4.4.3 Lemma. *The set $\mathscr{G}(\alpha)$ of Riemannian metrics satisfying (α) contains a residual subset.*

Proof. (cf. [Kl 16]). For each integer $m>0$ we consider the property:

(α, m) given any family $\{k_a,\ a \in A(M,g)\}$ of integers satisfying

$$1 \leqslant \sum_a |k_a| \leqslant m, \quad \text{one has} \quad \sum_a k_a a \notin \mathbb{Z}.$$

Note that (α, m) implies, in particular, that none of the eigenvalues of a Poincaré map P_c is a mth root of unity.

Clearly, (α) is the intersection of the properties (α, m), for all $m \in \mathbb{N}$. Thus it suffices to show that the set $\mathscr{G}(\alpha, m)$ of Riemannian metrics satisfying (α, m) contains a residual set.

To see this we proceed as in the proof of (3.3.9). Consider the property:

(α, m, κ) All closed geodesics c with $E(c) \leqslant \kappa$ satisfy (α, m).

Then we show that

(i) the set $\mathscr{G}(\alpha, m, \kappa) \subset \mathscr{G}$, satisfying (α, m, κ), is dense;

(ii) $\mathscr{G}(\alpha, m, \kappa)$ is contained in the interior of $\mathscr{G}(\alpha, m, \kappa/2)$.

To prove (i) we recall that the set of metrics with all closed geodesics non-degenerate is dense, cf. (3.3.9). For such a metric, there exist only finitely many closed geodesics in Λ^κ, for a given κ. Thus, we get from these geodesics only a finite contribution to the set $A(M,g)$. Perturbations of the type employed in the proof of (3.3.9) show that g can be approximated arbitrarily close by elements in $\mathscr{G}(\alpha, m, \kappa)$.

For the proof of (ii) we observe that sufficiently small perturbations of a metric g in $\mathscr{G}(\alpha,m,\kappa)$ will carry closed geodesics of E-value $\leqslant\kappa/2$ in closed geodesics of E-value $<\kappa$ without invalidating the open and invariant property (α,m) for these geodesics.

To complete the proof that (α,m) is residual, we need only observe that (i) and (ii) imply that $\mathscr{G}(\alpha,m,\kappa)$ contains the interior of $\mathscr{G}(\alpha,m,2\kappa)$, which in turn contains the dense subset $\mathscr{G}(\alpha,m,4\kappa)$. Since $\mathscr{G}(\alpha,m)$ is the intersection of the $\mathscr{G}(\alpha,m,\kappa)$, as κ tends to ∞ through a sequence, $\mathscr{G}(\alpha,m)$ contains a residual set. This completes the proof of (4.4.3). □

As a further preparation for our next result we introduce the *average index* $\bar{I}_c$ of a closed geodesic by

$$\bar{I}_c=\lim_{m\to\infty}\frac{1}{m}\,I_{c^m}$$

where $I_{c^m}=\operatorname{index} c^m$, cf. Bott [Bo 3].

We have to show that the limit exists. This and a formula for $\bar{I}_c$ are already contained in (3.2.15). Actually, $\bar{I}_c=\beta_c$.

Recall that if $P=P_c$ is the Poincaré map associated to the closed geodesic c, we denote by $(\rho_j,\bar{\rho}_j)=(e^{2\pi i a_j},e^{-2\pi i a_j})$, $1\leqslant j\leqslant l-1$, those roots of P_c which have modulus 1. Here

$$a_0=0\leqslant a_1<\cdots<a_{l-l}\leqslant a_l=1/2.$$

Let $I_{c,j}$ be the value of the ρ-index $I_c(\rho)$ for $\rho=e^{2\pi i a}$, $a_{j-1}<a<a_j$, $1\leqslant j\leqslant l$.

4.4.4 Proposition. *The average index of a closed geodesic c is given by*

$$\bar{I}_c=I_{c,l}-2\sum_{j=1}^{l-2}a_j(I_{c,j+1}-I_{c,j}).$$

Proof. From (3.2.15) we have

$$\bar{I}_c=\beta_c=2\sum_{j=1}^{l}I_{c,j}(a_j-a_{j-1}).$$

Set $a_l=1/2$. □

We are now ready to prove the following theorem.

4.4.5 Theorem. *Let M be a compact, simply connected differentiable manifold. $\mathscr{G}^*$ is the subset of the space $\mathscr{G}$ of Riemannian metrics g on M for which the exponents of the Poincaré maps associated to the closed geodesics on (M,g) are free over the rationals, cf. (α). $\mathscr{G}^*$ contains a residual set.*

If $g^*\in\mathscr{G}^*$ then Lemma (4.3.4), *together with Proposition* (4.4.4) *concerning the average index, implies that on* (M,g^*) *there are infinitely many closed geodesics.*

Proof. We will derive a contradiction from the assumption that there exists a sequence $\{r_i\}$ of integers with the property that if $c(r_i), c'(r_i)$ are geodesics constructed according to Lemma (4.3.4) with the cycles $w_*(r_i)$ and their homotopies $w'_*(r_i)$, then $c(r_i)$ and $c'(r_i)$ are all coverings of just two prime closed geodesics c and c', respectively. Clearly, this will imply our theorem.

To derive the contradiction, we consider $c(r_i)=c^{m_i}$, $c'(r_i)=c'^{m'_i}$ with $m_i=q_i m'_i$, q_i an integer, and $E(c(r_i))=m_i^2 E(c)<E(c'(r_i))=m_i'^2 E(c')$.

Clearly, $c \neq c'$. We claim that the $\{q_i\}$ are bounded. To see this we recall from (4.2.7) (see also (3.2.15)) that there exist $\alpha>0$ and $\beta \geqslant 0$ such that

$$q_i m'_i \alpha - \beta + 1 \leqslant \text{index } c^{q_i m'_i} + 1 = \text{index } c'^{m'_i}.$$

On the other hand, (3.2.15) implies the existence of $\alpha', \beta' \geqslant 0$ such that index $c'^{m'_i} \leqslant \alpha' m'_i + \beta'$. Thus, $q_i \leqslant (\alpha' + \beta' + \beta)/\alpha$, and is bounded.

Hence, by going to a subsequence, which we denote again by $\{i\}$, we can assume that $m_i = q m'_i$. From index $c'(r_i) = \text{index } c(r_i) + 1$, we therefore obtain $q\bar{I}_c = \bar{I}_{c'}$, i.e., using (4.4.4),

$$2q \sum_j a_j \Delta I_j - q I_l = 2 \sum_{j'} a'_{j'} \Delta I'_{j'} - I'_{l'},$$

where ΔI_j, I_l, $\Delta I'_{j'}$, $I'_{l'}$ are integers. The hypothesis on the rational independence of the exponents implies that all the ΔI_j and $\Delta I'_{j'}$ are zero. Thus, $\bar{I}_c = I_c(\rho)$, for ρ not a root of P_c, and $\bar{I}_{c'} = I_{c'}(\rho)$, for ρ not a root of $P_{c'}$.

Since neither P_c nor $P_{c'}$ have eigenvalues which are roots of unity, we obtain, for all i,

$$\text{index } c^{m_i} + 1 = m_i \bar{I}_c + 1 = m'_i \bar{I}_{c'}, \quad m_i = q m'_i,$$

which is a contradiction.

This completes the proof of (4.4.5). □

We conclude with a special result about the existence of a non-hyperbolic closed geodesic. Recall that $\mathcal{G}_T$ is the open subset in the space $\mathcal{G} = \mathcal{G}_M$ of Riemannian metrics on a compact differentiable manifold M with the property that, for $g^* \in \mathcal{G}_T$, (M, g^*) possesses a closed geodesic of twist type. According to (3.3.3), (M, g^*) has infinitely many prime closed geodesics.

4.4.6 Theorem. *Let (M,g) be a non-simply connected Riemannian manifold of even dimension and with curvature $K \geqslant 0$. Let c be a non-null-homotopic closed geodesic of minimal E-value in its free homotopy class. Such geodesics exist according to* (2.1.3). *Then c is non-hyperbolic. If $K>0$, the metric g belongs to the closure of the set $\mathcal{G}_T$. In particular, g can be approximated by Riemannian metrics g^* such that (M,g^*) has infinitely many prime closed geodesics.*

Proof. Since E has a minimal value on c, index $c=0$. Let $\tilde{c}=c^2$ be the 2-fold covering of c, i.e. $\tilde{c}(t)=c(2t)$, $t \in S$. $\tilde{c}$ is orientable. That is to say, the parallel translation of the normal fibre over $0 \in S$ of the immersion $t \in S \mapsto \tilde{c}(t) \in M$ along $\tilde{c}(t)$ into the fibre over $1=0 \in S$ is an element of the special orthogonal group

$SO(2m-1)$, $2m=\dim M$. Hence, there exists a parallel periodic vector field $\xi(t)$ along $\tilde{c}(t):|\xi(t)|=1$, $\langle\xi(t),\tilde{c}(t)\rangle=0$, $\nabla\xi(t)=0$.

We find, for the index form, that

$$D^2E(\tilde{c})\,(\xi,\xi)=-\int_0^1\langle R(\xi(t),\dot{\tilde{c}}(t),\dot{\tilde{c}}(t)),\xi(t)\rangle dt\leqslant 0$$

since $\quad\langle R(\xi(t),\dot{\tilde{c}}(t),\dot{\tilde{c}}(t)),\xi(t)\rangle=K(\xi,\tilde{c})\,(t)|\dot{\tilde{c}}(t)|^2\geqslant 0.$

Equality holds only if $K(\xi,\dot{\tilde{c}})\,(t)\equiv 0$, i.e. if $\xi(t)$ is a periodic Jacobi field, $\tilde{c}$ is degenerate.

Assume now that c is hyperbolic. From (3.2.13) we then have index $\tilde{c}=$ 2. index $c=0$. Thus, $\tilde{c}$ is degenerate, which is impossible since along with c, $\tilde{c}$ is hyperbolic. □

Remark. Consider a compact manifold M of arbitrary dimension and $K>0$. Then a sufficiently high covering c^m of a closed geodesic c will contain conjugate points and hence have positive index. Thus, if index $c=0$, c cannot be hyperbolic. This shows the existence of a non-hyperbolic closed geodesic on M, provided M is not simply connected.

Chapter 5. Miscellaneous Results

In this chapter we collect additional results about closed geodesics. We begin with a proof of the Theorem of the Three Closed Geodesics (5.1.1). This is a generalization for arbitrary compact simply-connected Riemannian manifolds of the classical result of Lusternik and Schnirelmann [LS] on the existence of three closed geodesics on a surface of the type of the sphere, cf. the Appendix.

The gist of the theorem is that there always exist at least three relatively short closed geodesics – note that in the existence theorems of Chapter 4, nothing was said about the length of the geodesics compared to other geometric quantities of the manifold.

In the second section, we collect a number of results on special compact simply-connected manifolds such as symmetric spaces, spaces with all geodesics closed and of the same length, or manifolds with some additional integrals for the geodesic flow, e.g. Liouville's surfaces.

In section three we give a detailed exposition of the geodesic flow on manifolds of hyperbolic type. Typical examples of such manifolds are those which allow a metric of strictly negative curvature.

We conclude with a result on manifolds which can be covered by a torus.

5.1 The Theorem of the Three Closed Geodesics

Lusternik and Schnirelmann [LS] proved in 1929 that on an arbitrary surface of the topological type of the 2-sphere there always exist three closed geodesics without self-intersections. Full details of their proof were not given until much later, by Lysternik [Ly]. Compare the Appendix for a completely elementary proof.

The proof of the theorem of Lusternik and Schnirelmann uses, crucially, the Jordan Curve Theorem for the 2-sphere. We will present here a generalization of this theorem to arbitrary compact simply-connected manifolds. The proof is based on a divisibility relation of the type we established in (4.3.4).

In our generalization we do not claim that the geodesics have no self-intersections. Their main feature is that they are comparatively short. At the end of the section we derive Morse's result of the existence of only "few" short closed geodesics in general.

We begin by constructing certain chains in the space of circles on the sphere S^k. From (2.3) we recall that we have in the spaces AS^k and ΓS^k of parameterized and unparameterized circles on S^k a bundle structure

$$\alpha: AS^k - A^0S^k \to BS^k \cong V(2,k-1),$$

$$\gamma: \Gamma S^k - \Gamma^0 S^k \to \Delta S^k \cong G(2,k-1),$$

where the fibre is a $(k-1)$-disc. Here, we had to remove from AS^k and ΓS^k the trivial circles A^0S^k and $\Gamma^0 S^k$.

The maps α and γ associate to a non-trivial circle the great circle in the plane parallel to the plane carrying the circle.

As in (2.3), we think of S^k as being isometrically embedded in the Euclidean space $\mathbb{R}^{k+1}$ with coordinates $(x_0, \ldots, x_k)$.

In (2.3) we introduced the basis $[a_1, a_2]$, $0 \leqslant a_1 \leqslant a_2 \leqslant k-1$, of the $\mathbb{Z}_2$-homology of $\Delta S^k = G(2,k-1)$. Denote by $v(a_1, a_2)$ the cycle of ΓS^k mod $\Gamma^0 S^k$ which corresponds to $[a_1, a_2]$ under the Thom isomorphism

$$H_*(G(2,k-1)) \to H_{*+(k-1)}(\Gamma S^k, \Gamma^0 S^k).$$

That is to say, $v(a_1, a_2)$ is the restriction of the $(k-1)$-disc bundle γ to the cycle $[a_1, a_2]$ of the base space of γ modulo $\Gamma^0 S^k$. Note that dim $[a_1, a_2] = a_1 + a_2$, and hence dim $v(a_1, a_2) = a_1 + a_2 + k - 1$.

We begin with the $(k-1)$-cycle $v(0,0)$. $v(0,0)$ can be written in the form $\pi \circ u(0,0)$. Here

$$\pi = \pi | AS^k : (AS^k, A^0S^k) \to (\Gamma S^k, \Gamma^0 S^k)$$

is the projection and

$$u(0,0): (D^{k-1}, \partial D^{k-1}) \to (AS^k, A^0S^k)$$

consists of the parameterized circles parallel to the parameterized great circle $(\cos 2\pi t, \sin 2\pi t, 0, \ldots, 0)$.

Our next goal is to find chains $u(a_1, a_2)$ for the remaining (a_1, a_2), $0 \leqslant a_1 \leqslant a_2 \leqslant k-1$ in AS^k mod A^0S^k such that $\pi \circ u(a_1, a_2) = v(a_1, a_2)$. This is done by constructing certain homotopies.

Let $\psi_\tau^{1,2}$, $0 \leqslant \tau \leqslant 1$, be the positive rotation by $\tau \cdot 180°$ in the (x_1, x_2)-plane which is extended to the orthogonal $(k-1)$-plane by the identity. Define

$$u(0,1): (D^{k-1}, \partial D^{k-1}) \times [0,1] \to (AS^k, A^0S^k)$$

by

$$u(0,1)\,(c;\tau) := \psi_\tau^{1,2} \circ u(0,0)\,(c).$$

We have

$$\partial u(0,1) = -\theta u(0,0) - u(0,0),$$

i.e. $u(0,1)$ is a homotopy from $u(0,0)$ to $-\theta u(0,0)$. Indeed, $\theta u(0,1)$ $(0;1)$ operates on the part $x_3 = \cdots = x_k = 0$ of the fibre as reflection on its midpoint.

Moreover, $\pi \circ u(0,1) = v(0,1)$.

Let $\psi_\tau^{2,3}$, $0 \leqslant \tau \leqslant 1$, be the positive rotation by τ . 180° in the (x_2, x_3)-plane which extended by the identity to the orthogonal $(k-1)$-plane. Define, for $k-1>1$,

$$u(0,2): (D^{k-1} \times [0,1], \partial D^{k-1} \times [0,1]) \times [0,1] \rightarrow (AS^k, A^0 S^k)$$

by

$$u(0,2)\,(c;\tau_1,\tau) = \psi_\tau^{2,3} \circ u(0,1)\,(c;\tau_1)$$

Clearly, $\pi \circ u(0,2) = v(0,2)$.

Moreover, $\partial u(0,2) = \theta u(0,1) - u(0,1)$, where we neglect, of course, chains of dimension $<k$.

Indeed, $\theta u(0,2)$ $(\;;,1)$ operates as a reflection at the (x_0, x_1)-plane on the great circles of $S^2 = \{x_0^2 + x_1^2 + x_2^2 = 1\}$ which start at $x_0 = 1$. Thus, the orientation is changed on the chain formed by these great circles which form the basis of $u(0,1)$ in AS^k. On the fibre over the great circle in the (x_0, x_2)-plane, the operation is the reflection on its center.

Continuing in this manner, we finally construct a homotopy

$$u(0,k-1): (D^{k-1} \times [0,1]^{k-2}, \partial D^{k-1} \times [0,1]^{k-2}) \times [0,1] \rightarrow (AS^k, A^0 S^k)$$

by

$$u(0,k-1)\,(c;\tau_1,\ldots,\tau_{k-2},\tau) = \psi_\tau^{k-1,k} \circ u(0,k-2)\,(c;\tau_1,\ldots,\tau_{k-2}).$$

Here, $\psi_\tau^{k-1,k}$ is the positive rotation by τ . 180° in the (x_{k-1}, x_k)-plane which is extended by the identity to the orthogonal complement.

Clearly, $\pi \circ u(0,k-1) = v(0,k-1)$.

Moreover, $\partial u(0,k-1) = (-1)^{k-1} \theta u(0,k-2) - u(0,k-2)$.

We now continue by constructing chains $u(i,k-1)$ with $\pi \circ u(i,k-1) = v(i,k-1)$, $0 \leqslant i \leqslant k-1$. We start with

$$u(1,k-1): (D^{k-1} \times [0,1]^{k-1}, \partial D^{k-1} \times [0,1]^{k-1}) \times [0,1] \rightarrow (AS^k, A^0 S^k)$$

where

$$u(1,k-1)\,(c;\tau_1,\ldots,\tau_{k-1},\tau) = \psi_\tau^{0,1} \circ u(0,k-1)\,(c;\tau_1,\ldots,\tau_{k-1}).$$

Here, $\psi_\tau^{0,1}$ is the positive rotation by τ . 180° in the (x_0, x_1)-plane, extended as usual by the identity to all of $\mathbb{R}^{k+1}$.

We claim that

$$\partial u(1,k-1) = -\theta e^{i\pi} \,.\, u(0,k-1) - u(0,k-1).$$

To see this, we observe that $e^{i\pi}$. $u(1,k-1)$ (;..., 1) operates on the great circles starting at $x_0 = 1$ as a reflection on the (x_0, x_1)-plane. Hence, $\theta e^{i\pi}$. $u(1,k-1)$(,...,1) is for these great circles the reflection on the $(x_2, \ldots, x_k)$-space; it therefore reverses the orientation of the base of $u(0,k-1)$. On the fibre over a great circle belonging to this $(k-1)$-space, this operation also reverses the orientation. Hence our claim holds.

Clearly, $\pi \circ u(1,k-1) = v(1,k-1)$. Actually, $\pi|u(1,k-1)$ is 1:1 except on the boundary $\partial u(1,k-1)$.

We continue these constructions to define, for $k-1>1$,

$$u(2,k-1)\,(c;\, \tau_1, \ldots, \tau_k, \tau)$$

by

$$\psi_\tau^{0,2} \circ u(1,k-1)\,(c;\, \tau_1, \ldots, \tau_k),$$

where $\psi_\tau^{0,2}$ is the positive rotation by τ . 180° in the (x_0, x_2)-plane with its usual extension. Again,

$$\partial u(2,k-1) = \pm\theta e^{i\pi} \,.\, u(1,k-1) - u(1,k-1)$$

and

$$\pi \circ u(2,k-1) = v((2,k-1).$$

Thus we can proceed and finally define $u(k-1,k-1)$ with $\pi \circ u(k-1,\, k-1) = v(k-1,k-1)$.

Actually, by stopping the first set of deformations with $u(0,a_2)$ and then starting with the second set we see that, for every pair (a_1, a_2), $0 \leqslant a_1 \leqslant a_2 \leqslant k-1$, we can construct a chain $u(a_1, a_2)$ which under π projects onto $v(a_1, a_2)$ such that $\pi|u(a_1, a_2)$ is 1:1, except at the boundary.

We now formulate the so-called *Theorem of the Three Closed Geodesics*.

5.1.1 Theorem. *Let (M,g) be a compact simply-connected Riemannian manifold. Here, M denotes the underlying differentiable manifold and g the Riemannian metric on M. Let*

$$f\colon (S^k, g_0) \to (M,g)$$

be a map of the standard k-sphere which represents an indivisible integer homology class of infinite order. It is well known that such maps exist and $k \geq 2$.

Then there exist on (M,g) $2k-1\geqslant 3$ prime closed geodesics which are short in the following sense. Let $\lambda(f)$ be the supremum of the length of curves on (M,g) which are images under f of circles on (S^k,g_0). Then these $2k-1$ prime closed geodesics have length $\leqslant\lambda(f)$.

Note. In the case $M=S^2$, Lusternik and Schnirelmann [LS] use the Jordan Curve Theorem crucially for the proof of this theorem. Their proof, therefore, cannot be generalized. Our proof of the theorem uses a divisibility property, similar to Lemma (4.3.4).

Observe that we do not claim that the $2k-1$ geodesics have no self-intersections. A result in this direction for $M=S^n$ and under certain restrictions on the curvature was proved in (2.3.8).

Proof. Instead of (M,g) we also simply write M if the metric g is kept fixed.

Let $\kappa^*>0$ be a non-critical value of $E:\Lambda M\to\mathbb{R}$ such that $\Lambda f(\Lambda S^k)\subset\Lambda^{\kappa^*}M$.

We first assume that there are only non-degenerate closed geodesics in Λ^{κ^*}. We then have the Morse complex $\mathscr{M}=\mathscr{M}^{\kappa^*}M$ of $\Lambda^{\kappa^*}M$, cf. (2.5). Using the same arguments as in the proof of (4.3.4) we see that the homotopy class of each of the chains $u(a_1, a_2)$ can be represented in $\mathscr{M}$ by a sum, with integer coefficients, of unstable and strong unstable manifolds of dimension $\leqslant\dim u(a_1, a_2)$. We denote this representation in $\mathscr{M}$ by $u_f(a_1, a_2)$.

We consider the maps (cf. (2.2.))

$$\tilde{\pi}:\Lambda M\to\tilde{\Pi}M=\Lambda M/S;\quad \pi:\Lambda M\to\Pi M=\Lambda M/O(2).$$

The hypothesis on $f:S^k\to M$ implies that

$$\pi\circ u_f(0,0):(D^{k-1}, \partial D^{k-1})\to(\Pi M,\Pi^0 M)$$

is a non-trivial cycle mod 2, cf. the proof of (4.3.4). Thus, there occurs in $u_f(0,0)$ a strong unstable manifold $W_{uu}(c)$ of dimension $\dim u(0,0)$ such that $\pi\circ W_{uu}(c)$ in $\pi\circ u_f(0,0)$ occurs with odd coefficient. Fix such a c and call it $c(0,0)$.

We claim that there exists in $u_f(0,1)$ a $W_{uu}(c')$ of dimension $\dim u(0,1)$ such that $\pi\circ W_{uu}(c')$ occurs with odd coefficient in $\pi\circ u_f(0,1)$ and $W_{uu}(c(0,0))$ belongs to the boundary $\partial W_{uu}(c')$ of $W_{uu}(c')$.

To see this we first observe that $\pi\circ W_{uu}(c(0,0))\in\pi\circ u_f(0,0)$ mod 2 implies that the integer coefficient of $\tilde{\pi}\circ W_{uu}(c(0,0))$ in $\partial\tilde{\pi}\circ u_f(0,1)=\tilde{\pi}\circ\partial u_f(0,1)=-\tilde{\pi}\circ\theta u_f(0,0)-\tilde{\pi}u_f(0,0)$ is odd, since the integer coefficients of $\tilde{\pi}\circ W_{uu}(c(0,0))$ and $\tilde{\pi}\circ W_{uu}(\theta c(0,0))$ in $\tilde{\pi}\circ u_f(0,0)$ differ by 1 mod 2. Thus, since $W_{uu}(c(0,0))\subset u_f(0,0)$, we must have a $W_{uu}(c')\subset u_f(0,1)$ of dimension $\dim u(0,1)$ with $W_{uu}(c(0,0))\subset\partial W_{uu}(c')$. Moreover, the coefficient of $\tilde{\pi}\circ W_{uu}(c')$ in $\tilde{\pi}\circ u_f(0,1)$ differs by 1 mod 2 from the coefficient of $\tilde{\pi}\circ W_{uu}(\theta c')$ in $\tilde{\pi}\circ u_f(0,1)$, since the corresponding fact is true for the coefficients of $\tilde{\pi}\circ W_{uu}(c(0,0))\subset\tilde{\pi}\circ\partial W_{uu}(c')$ and $\tilde{\pi}\circ W_{uu}(\theta c(0,0))\subset\tilde{\pi}\circ\partial W_{uu}(c')$ in $\tilde{\pi}\circ u_f(0,0)$. This completes the proof of the claim that $\pi\circ W_{uu}(c')\in\pi\circ u_f(0,1)$ mod 2.

We do not claim that the non-vanishing $\mathbb{Z}_2$-cycle $\pi\circ u_f(0,1)$ is non-homologous to zero. All that we used for showing that $\pi\circ u_f(0,1)\neq 0$ mod 2 was that $\pi\circ u_f(0,0)\neq 0$ mod 2. Hence, we can employ the same argument to show that also $\pi\circ u_f(0,2)$

is a non-zero cycle of ΠM mod $\Pi^0 M$ (if $k-1>1$), and so on, until $\pi \circ u_f(k-1,k-1)$. Only for $\pi \circ u_f(0,0)$, $\pi \circ u_f(0,k-1)$ and $\pi \circ u_f(k-1,k-1)$ are we actually sure that the $\mathbb{Z}_2$-cycles of ΠM mod $\Pi^0 M$ are non-homologous to zero, cf. [Kl 6].

We also consider the extension of the maps $\Lambda f \circ u(a_1, a_2)$, $0 \leqslant a_1 \leqslant a_2 \leqslant k-1$, to full S-orbits:

$$S . \Lambda f \circ u(a_1, a_2) : S \times D^{k-1} \times [0,1]^{a_1+a_2} \to \Lambda M$$

$$(z, c, \tau_1, \ldots, \tau_{a_1+a_2}) \mapsto z . \Lambda f \circ u(a_1, a_2)\,(c; \tau_1, \ldots, \tau_{a_1+a_2})$$
$$= \Lambda f \circ z . u(a_1, a_2)\,(c; \tau_1, \ldots, \tau_{a_1+a_2}).$$

On the domain of this map we have the canonical S-action on the factor S. On the range ΛM we have the S-action $\chi^{\sim}$. $S . \Lambda f \circ u(a_1, a_2)$ is equivariant with respect to these S-actions. Also, the deformations from $\Lambda f \circ u(a_1, a_2)$ into $u_f(a_1, a_2)$ were such as to carry S-orbits into S-orbits. Hence, we may view these deformations as deformations bringing $S . \Lambda f \circ u(a_1, a_2)$ into the equivariant map

$$S . u_f(a_1, a_2) : S \times D^{k-1} \times [0,1]^{a_1+a_2} \to \mathcal{M}$$

$$(z, c, \tau_1, \ldots, \tau_{a_1+a_2}) \mapsto z . u_f(a_1, a_2)\,(c; \tau_1, \ldots, \tau_{a_1+a_2}).$$

We go back to $W_{uu}(c(0,0))$ in $u_f(0,0)$ and $W_{uu}(c')$ in $u_f(0,1)$ with $W_{uu}(c(0,0))$ occurring in $\partial W_{uu}(c')$, as described above. Instead of c' we now write $c(0,1)$.

The essential step in our proof is to show that the multiplicity $m(0,1)$ of $c(0,1)$ divides the multiplicity $m(0,0)$ of $c(0,0)$.

To see this we consider the chain $\partial W_{uu}(c(0,1)) \cap u_f(0,0)$. It contains $W_{uu}(c(0,0))$. Since $W_{uu}(c(0,1))$ is invariant under the isotropy group $I^{\sim}(c(0,1))$ of $c(0,1)$ the chain $\partial W_{uu}(c(0,1)) \cap u_f(0,0)$ contains the full $I^{\sim}(c(0,1))$-orbit $\{z_0 . W_{uu}(c(0,0)) = W_{uu}(z_0 . c(0,0));\ z_0 \in I^{\sim}(c(0,1))\}$ of $W_{uu}(c(0,1))$. Our claim is that one can construct from $u_f(0,0)$ a cycle $u_f^*(0,0)$ containing only $W_{uu}(c(0,0))$ and no other element of the $I^{\sim}(c(0,1))$-orbit of $W_{uu}(c(0,0))$. Moreover, $u_f^*(0,0)$ possesses a homotopy $u_f^*(0,1)$ containing a strong unstable manifold, denoted by $W_{uu}(c(0,1))$, such that $W_{uu}(c(0,0)) \subset \partial W_{uu}(c(0,1))$. Since $\partial W_{uu}(c(0,1)) \cap u_f^*(0,0)$ must contain the full $I^{\sim}(c(0,1))$-orbit of $W_{uu}(c(0,0))$ it follows that $c(0,0)$ is invariant under $I^{\sim}(c(0,1))$.

To construct $u_f^*(0,0)$ we proceed as in the proof of (4.3.4). In the simplest case we have $\partial W_{uu}(c(0,0)) = 0$. Then we put $u_f^*(0,0) = W_{uu}(c(0,0))$. Let $u_f^*(0,1)$ be the restriction of $u_f(0,1)$ to $W_{uu}(c(0,0)) \subset u_f(0,0)$. Just as above we conclude the existence of a strong unstable manifold, denoted again by $W_{uu}(c(0,1))$, in $u_f^*(0,1)$, such that $\tilde{\pi} \circ W_{uu}(c(0,0))$ occurs in $\partial \tilde{\pi} \circ W_{uu}(c(0,1))$ with odd coefficient, whereas $\tilde{\pi} \circ W_{uu}(\theta c(0,0))$ occurs with even coefficient only. But then

$$I^{\sim}(c(0,1)) . W_{uu}(c(0,0)) = W_{uu}(c(0,0)).$$

In general we have $\partial W_{uu}(c(0,0)) \neq 0$. If none of the $W_{uu}(c_1) \subset \partial W_{uu}(c(0,0))$ also occurs (with opposite sign) in the boundary of $W_{uu}(z_0 . c(0,0))$, $z_0 \in I^{\sim}(c(0,1))$, $z_0 . c(0,0) \neq c(0,0)$, then we can construct a cycle $u_f^*(0,0) \subset u_f(0,0)$ which contains $W_{uu}(c(0,0))$ but no other element of the S-orbit of $W_{uu}(c(0,0))$. By restricting the homotopy $u_f(0,1)$ to $u_f^*(0,0)$ we get a homotopy $u_f^*(0,1)$ of $u_f^*(0,0)$. To $u_f^*(0,0)$ and $u_f^*(0,1)$ apply the same arguments as to the pair $u_f(0,0)$ and $u_f(0,1)$. In particular, there will be a strong unstable manifold in $u_f^*(0,1)$, denoted again by $W_{uu}(c(0,1))$, having $W_{uu}(c(0,0))$ in its boundary. Since $u_f^*(0,1)$ must contain the full $I^{\sim}(c(0,1))$-orbit of $W_{uu}(c(0,0))$ it follows that this orbit consists of $W_{uu}(c(0,0))$ only, i.e. $I^{\sim}(c(0,1))$ is a subgroup of $I^{\sim}(c(0,1))$.

Finally assume that $W_{uu}(c_1)\subset\partial W_{uu}(c(0,0))$ occurs with opposite sign in $\partial W_{uu}(z_0 \,.\, c(0,0))$, $z_0 \,.\, c(0,0)\neq c(0,0)$, $z_0\in I^{\sim}(c(0,1))$. Then

$$W_{uu}(c_1)-W_{uu}(\bar{z}_0 \,.\, c_1)\subset\partial W_{uu}(c(0,0)).$$

Consider the chain

$$W_{uu}(c_1)\ [0,-r_0]=\{e^{-2\pi ir} \,.\, W_{uu}(c_1);\ \ 0\leq r\leq r_0\}$$

with $z_0=e^{2\pi ir_0}$. The boundary of the chain

$$W_{uu}(c(0,0))+W_{uu}(c_1)\ [0,-r_0]$$

contains neither $W_{uu}(c_1)$ nor $-W_{uu}(\bar{z}_0 \,.\, c_1)$. We proceed in the same manner with the remaining elements in $\partial W_{uu}(c(0,0))$.

In this way we construct, just as in the proof of (4.3.4), a cycle $u_f^*(0,0)$ containing $W_{uu}(c(0,0))$ but no other element of the S-orbit of $W_{uu}(c(0,0))$. The image mod 2 under $\tilde{\pi}$ of $u_f^*(0,0)$ completely belongs to $\tilde{\pi}\circ u_f(0,0)$. By restricting $u_f(0,1)$ to the strong unstable manifolds in $u_f^*(0,0)$ (and, in particular, to $W_{uu}(c(0,0))$), we get a homotopy $u_f^*(0,1)$. The definition of this homotopy on the chain $W_{uu}(c_1)\ [0,-r_0]$ is the same as in the proof of (4.3.4). To the pair $u_f^*(0,0)$, $u_f^*(0,1)$ apply the same arguments as to the pair $u_f(0,0)$, $u_f(0,1)$. We conclude that there exists a strong unstable manifold in $u_f^*(0,1)$, denoted again by $W_{uu}(c(0,1))$, having $W_{uu}(c(0,0))\subset u_f^*(0,0)$ in its boundary. The full $I^{\sim}(c(0,1))$-orbit of $W_{uu}(c(0,0))$ must belong to $\partial W_{uu}(c(0,1))\cap u_f^*(0,0)$. Hence, c(0,0) is left invariant under $I^{\sim}(c(0,1))$. We have shown in complete generality that the multiplicity $m(0,1)$ of $c(0,1)$ divides the multiplicity $m(0,0)$ of $c(0,0)$.

Proceeding to $u_f(0,2)$ (if $k-1>1$), we consider a $W_{uu}(c')\subset u_f(0,2)$ of dimension $\dim u_f(0,2)$ with $\pi\circ W_{uu}(c')\neq 0 \bmod 2$ and $W_{uu}(c(0,1))\subset\partial W_{uu}(c')$. We claim, possibly after some modifications of $u_f(0,1)$ and ensuing modifications of $u_f(0,2)$, that we can assume that the isotropy group $I^{\sim}(c')$ of c' leaves $c(0,1)$ invariant. Taking c' as $c(0,2)$ this implies that the multiplicity $m(0,2)$ of $c(0,2)$ divides the multiplicity $m(0,1)$ of $c(0,1)$.

To prove our claim we again start with the observation that $\partial W_{uu}(c')\cap u_f(0,1)$ contains the full $I^{\sim}(c')$-orbit of $W_{uu}(c(0,1))$. Just as above, we construct from $u_f(0,1)$ a cycle $u_f^*(0,1)$ containing $W_{uu}(c(0,1))$ and no other element of the S-orbit of $W_{uu}(c(0,1))$. $u_f^*(0,1)$ possesses a homotopy $u_f^*(0,2)$. There is a $W_{uu}(c(0,2))$ in $u_f^*(0,0)$ of maximal dimension having $W_{uu}(c(0,1))$ in its boundary. We conclude that $I^{\sim}(c(0,2))$ leaves $c(0,1)$ invariant.

We apply with the same argument to $u_f(0,k-1)$ which yields a critical point $c(0,k-1)$, the multiplicity $m(0,k-1)$ of which divides the multiplicity $m(0,k-2)$ of the previously constructed $c(0,k-2)$. From here we go on to construct $c(1,k-1),\ldots, c(k-1,k-1)$. Thus, we finally obtain a sequence of $2k-1$ closed geodesics $c(0,0)$, $c(0,1),\ldots, c(k-1,k-1)$ in which the multiplicity of each element divides the multiplicity of the previous one. Since the E-value on this sequence is strictly increasing, we see that the underlying prime closed geodesics in this sequence are pairwise different, with strictly increasing E-value.

It remains to consider the general case for which we do not assume that all closed geodesics in Λ^{κ^*-} are non-degenerate. We may assume, however, that the critical S-orbits in $\Lambda^{\kappa^*-}-\Lambda^0$ are isolated critical sets, say $S\,.\,c_\alpha$, $1\leqslant\alpha\leqslant\alpha_0$. We wish to reduce this case to the non-degenerate case by modifying the given Riemannian metric g. If $\tilde{g}$ is such a modification, we denote the underlying differentiable manifold endowed with the metric $\tilde{g}$ by $\tilde{M}$. The energy integral $\Lambda\tilde{M}$ will be denoted by $\tilde{E}$.

As in the proof of (4.3.5), we consider neighborhoods $\mathscr{U}$ of the critical set of $\Lambda^{\kappa^*-}M$ satisfying certain conditions. Fixing a sufficiently small $\varepsilon>0$, and such a $\mathscr{U}$, we take modifications $\tilde{g}=\tilde{g}(\mathscr{U},\varepsilon)$ satisfying conditions (a), (b), (c) as in the proof of (4.3.5). From our previous arguments, we obtain a sequence $\tilde{c}(0,0),\ldots,$ $\tilde{c}(k-1,k-1)$ of $2k-1$ prime closed geodesics of strictly increasing $\tilde{E}$-value.

The restrictions imposed upon $\tilde{g}$ imply that for each $\tilde{c}(a,b)$ in this sequence there is exactly one α with $\tilde{c}(a,b)\in\mathscr{U}(S\,.\,c_\alpha)$. We denote this c_α also by $c(a,b)$. By choosing $\varepsilon>0$ sufficiently small, the E-values $\kappa(0,0),\ldots, \kappa(k-1,k-1)$ of this sequence $c(0,0),\ldots, c(k-1,k-1)$ are increasing. We shall see in a moment that these values are actually strictly increasing, the reason for this being that the members of the sequence $v(0,0),\ldots, v(k-1,k-1)$ of cycles are produced from one another by rotations which make them mutually subordinated in a certain geometric sense.

Consider $c(0,1)$. From $\tilde{c}(0,1)\in\mathscr{U}(S\,.\,c(0,1))$ and from property (b) it follows that the homotopy class of $\Lambda f\circ u(0,1)$ has a local representation $u_f(0,1)$ as a cycle of dimension dim $(0,1)$ in $(\Lambda^{\kappa(0,1)-}M\cup c(0,1),\ \Lambda^{\kappa(0,1)-}M)$. That is to say, there exists a point $0(0,1)$ in $D^{k-1}\times\,]0,1[$ and an open neighborhood $D(0,1)$ of $0(0,1)$ such that

$$u_f(0,1)\colon (D(0,1),\ D(0,1)-0(0,1))\to(\Lambda^{\kappa(0,1)-}M\cup c(0,1),\ \Lambda^{\kappa(0,1)-}M)$$

$$u_f(0,1)\,(0(0,1))=c(0,1).$$

Thus, $E|\partial u_f(0,1)<\kappa(0,1)$, and hence $\kappa(0,0)<\kappa(0,1)$.

Similarly, one proves $\kappa(0,1)<\kappa(0,2)$, etc., until one reaches $\kappa(k-1,k-1)$.

Next we show that the modification $\tilde{g}$ of the Riemannian metric g can be chosen in such a way that $\tilde{c}(0,1)=c(0,1)$. Thus, the multiplicity $m(0,1)$ of $\tilde{c}(0,1)=c(0,1)$ divides the multiplicity $\tilde{m}(0,0)$ of $\tilde{c}(0,0)$. But $\tilde{m}(0,0)$ in turn divides the multiplicity $m(0,0)$ of $c(0,0)$, since $\tilde{c}(0,0)\in\mathscr{U}(S\,.\,c(0,0))$. Thus, $m(0,1)$ divides $m(0,0)$.

To show that g can be chosen with this property we take as typical example the case in which index $c(0,1)=k-1$. That is to say, we can represent the homotopy class of $u_f(0,1)$ as follows. We identify a subdisc D^- of $D=D(0,1)$ of co-dimension 1 through $0=0(0,1)$ with a disc in the negative eigenspace $T^-_{c(0,1)}\Lambda M$ at $c(0,1)$. Moreover, there are two unit vectors X_{-1}, X_{+1} in the null space $T^0_{c(0,1)}\Lambda M$ and a $\varepsilon>0$ such that the relative cycle at $c(0,1)$. defined by $u_f(0,1)$ can be written in the form

$$(X,t)\in D^-\times]-\varepsilon,\varepsilon[\mapsto\begin{cases}\exp(X+tX_{-1}), & \text{for } t\leq 0,\\ \exp(X+tX_{+1}), & \text{for } t\geq 0.\end{cases}$$

In particular, there does not exist a family X_s, $-1\leq s\leq +1$, of unit vectors in $T_{c(0,1)}\Lambda M$ with $E(\exp(tX_s))<\kappa(0,1)$, for small $t>0$.

Put $X_{-1}+X_{+1}=Y$, $X_{-1}-X_{+1}=Z$. Along with X_{-1} and X_{+1}, Y and Z are also periodic Jacobi fields along $c(0,1)$, orthogonal to $\dot c(0,1)$. Whereas $Z\neq 0$, the possibility that $Y=0$ is not excluded. Moreover, $\langle Y,Z\rangle_1=0$. Now, choose a modification of the 2nd order jet of the Riemannian metric along $c(0,1)$, such that in the new metric $\tilde g$ we have, besides properties (a), (b), (c)

$$D^2E(c(0,1))(Y,Y)\geq 0, \quad \text{and} \quad >0 \quad \text{if} \quad Y\neq 0, \quad \text{and}$$
$$D^2E(c(0,1))(Z,Z)<0.$$

This can be achieved by applying the methods described in (3.3).

Using the same arguments one shows that $m(0,2)$ divides $m(0,1)$, etc., until $m(k-1,k-1)$. Therefore one concludes – as in the non-degenerate case – that the underlying prime closed geodesics of the sequence $c(0,0),\ldots,c(k-1,k-1)$ are all different.

This completes the proof of the theorem. □

The following example, due to Morse [Mor 2], shows that, in general, there need exists only $n(n+1)/2$ closed geodesics which are short in the sense of (5.1.1).

5.1.2 Proposition. *For any $\kappa>2\pi^2$, all ellipsoids $E(a_0,\ldots,a_n)$ such that*

$$(*)\qquad \sum_i a_i^2x_i^2=1,\qquad i=0,\ldots,n,$$

in $\mathbb{R}^{n+1}$, with $0<a_0<\ldots<a_n$ and $|a_i-1|$ sufficiently small, have the following property. The only prime closed geodesics on $E(a_0,\ldots,a_n)$ of E-value $<\kappa$ are the $n(n+1)/2$ principal ellipses, obtained from the intersection of the ellipsoid with the coordinate planes.

Proof. The geodesics $(x_i(s))$ on $E(a_0,\ldots,a_n)$ with parameter equal to arc length, are the solutions of (*) satisfying

$$(i)\qquad x_i''(s)+\lambda(s)a_i^2x_i(s)=0$$

and $\sum_i x_i'(s)x_i'(s)=1$. This is simply the statement that a curve $(x_i(s))$ on

$E(a_0, \ldots, a_n)$ is a geodesic if and only if the second derivative is in the normal direction of the ellipsoid. The function $\lambda(s)$ is given by

$$\lambda(s) = \sum_i a_i^2 x_i'(s) x_i'(s) / \sum_i a_i^4 x_i(s) x_i(s).$$

$E(1, \ldots, 1)$ is the sphere with the standard metric g_0. In this case we have for every geodesic $(x_i(s))$ the following property. Let $x_k(s) \not\equiv 0$. Then $x_k(s)$ has a zero, say at $s = s_0$, and all the other zeros of $x_k(s)$ occur at $s = s_0 + q\pi$, $q \in \mathbb{Z}$.

Now fix $\kappa > 2\pi^2$. Choose an integer q^* satisfying $\sqrt{2\kappa} < q^*\pi + \pi/2$. From our previous remark we see that there exists an $\varepsilon = \varepsilon(q^*) > 0$ such that, for an ellipsoid $E(a_0, \ldots, a_n)$ with $|a_i - 1| < \varepsilon$, the following holds. For any geodesic $(x_i(s))$ consider a $k \in [0, n]$ with $x_k(s) \not\equiv 0$. Then there exists an s_0 with $x_k(s_0) = 0$ and the first q^* zeros $s_1, \ldots, s_{q^*}$ of $x_k(s)$ following s_0 satisfy

$$s_0 + q\pi - \pi/2 < s_q < s_0 + q\pi + \pi/2.$$

Now choose the a_i such that $|a_i - 1| < \varepsilon$ and $0 < a_0 < \cdots < a_n$. Let $(x_i(s))$ be a closed geodesic on $E(a_0, \ldots, a_n)$ of E-value $\leqslant \kappa$. If $x_i(s) \equiv 0$ for all i except two, the geodesic will be a covering of one of the principal ellipses. Hence, to prove the proposition it suffices to derive a contradiction from the assumption that there are three values for i, say $i = j, k, l$, such that $x_i(s) \not\equiv 0$. We assume $j < k < l$, hence, $a_j < a_k < a_l$.

For every i we put $(x_i(s), x_i'(s)) = X_i(s)$. Then $X_i(s + \omega) = X_i(s)$ where $\omega \leqslant \sqrt{2\kappa}$ is the length of the closed geodesic. Thus, $X_i(s)$ is a closed curve in $\mathbb{R}^2$ which is non-constant for $i \in \{j, k, l\}$.

$x_j(s)$ and $x_k(s)$ each have at least two zeros. If $x_j(s_0) = x_k(s_0) = 0$ then the subsequent zero of $x_j(s)$ will not be a zero of $x_k(s)$, as one sees from the Sturm Separation Theorem. Therefore, we have a r_0 with $x_j(r_0) = 0, x_k(r_0) \neq 0$.

We define

$$U_k(s) = (u_k(s), u_k'(s))$$

where $u_k(s)$ is a solution of $(i)_{i=k}$ determined by $U_k(r_0) = \pm X_j(r_0)$, the sign being chosen such that $\{X_k(r_0), U_k(r_0)\}$ is a positive frame in $\mathbb{R}^2$.

We now follow the behavior of the frame $\{X_k(s), U_k(s)\}$ for $r_0 \leqslant s \leqslant r_0 + \omega$. $u_k(s)$ will have a number $\leqslant q^*$ of zeros $> r_0$ in this interval. Compare these zeros with the zeros of $x_j(s)$. According to Sturm, the zeros of $u_k(s)$ will come earlier, since $a_j^2 < a_k^2$. Moreover, subsequent zeros are separated by $s_0 + q\pi - \pi/2$, $q = 1, 2, \ldots$. Hence, the vector $U_k(s_0 + \omega)$ will have a positive first component in the frame $\{X_k(s_0), U_k(s_0)\}$.

This means that the linear Poincaré map associated with the periodic equation $(i)_{i=k}$ has a normal form of the type

$$(\dagger) \qquad \begin{pmatrix} 1 & a \\ 0 & 1 \end{pmatrix}, \qquad a > 0.$$

By the linear Poincaré map, we mean the map which associates to $Y_k(s_0)=(y_k(s_0), y'_k(s_0))\in\mathbb{R}^2$ the vector $Y_k(s_0+\omega)$, where $y_k(s)$ is a solution of $(i)_{i=k}$.

By applying the corresponding arguments to a pair $\{X_k(s), V_k(s)\}$ of linearly independent solutions of $(i)_{i=k}$ with $V_k(s)$ determined by $V_k(t_0)=\pm X_l(t_0)$, we find, since $a_k^2<a_l^2$, that the Poincaré map for the equation $(i)_{i=k}$ is of the type (†) with $a<0$ instead of $a>0$.

We thus have a contradiction. This completes the proof of (5.1.2). □

Note. If we dont put any restrictions on the length of the axes of the ellipsoid then there may exist more than $n(n+1)/2$ relatively short prime closed geodesics, even without selfintersections. Cf. Viesel [Vi 2] for the case $n=2$.

5.2 Some Special Manifolds of Elliptic Type

For surfaces, it is well known that the genus determines, to a large degree, the macro-structure of the geodesic flow. Taking the size of the fundamental group $\pi_1 M$ as a rough generalization of the genus to arbitrary manifolds M, we shall show that such a statement could also be made for higher dimensions.

In this section, we consider manifolds of *elliptic type*, i.e. manifolds with finite fundamental group. For hyperbolic and parabolic types, see (5.3).

We begin with the symmetric spaces of rank 1. They have the property that all geodesics are closed and that the prime ones all have the same length and are non-self-intersecting. It is an outstanding problem to determine all manifolds with the property that all geodesics are closed. A forthcoming book by Besse [Bes] is devoted exclusively to this subject; therefore we restrict ourselves to quoting a few of the more important results.

The Riemannian manifolds with completely integrable geodesic flow provide another subject of great interest. This is a very classical field. We shall quote results by Clairaut, Liouville, Jacobi, Stäckel and Arnold.

Under a certain genericity condition closely related to the twist type introduced in (3.3), the periodic orbits of a completely integrable geodesic flow are dense. This condition is satisfied in particular for the ellipsoids with pairwise different axes.

On the other hand, there also exist among the completely integrable geodesic flows examples with "few" periodic orbits in the sense that their closure has small measure. We give an example due to Darboux. Finally, we would like to point out that Theorem (4.4.6) on the existence of elliptic closed geodesics should also be viewed as a result concerning manifolds of elliptic type.

First of all, consider homogeneous Riemannian manifolds. It is clear that, in this case, the study of the structure of the geodesic flow becomes amenable to methods developed for representation theory, cf. Auslander and Green [Au G], Moore [Moo]. Even more precise information can be obtained for Riemannian symmetric spaces, cf. Mautner [Mau], Chavel [Cha], Helgason [He]. Here we wish to report only on the *simply connected compact symmetric spaces of rank* 1.

These are

(i) the sphere $S^n = SO(n+1)/SO(n)$,

(ii) the complex projective space $P\mathbb{C}^n = U(n+1)/U(n) \times U(1)$,

(iii) the quaternion projective space $P\mathbb{H}^n = Sp(n+1)/Sp(n) \times Sp(1)$, and

(iv) the projective Cayley plane $P\mathbf{Ca}^2 = F_4/\mathrm{Spin}(9)$.

For these spaces one has the following result, due to Cartan [Ca 1].

5.2.1 Theorem. *On a simply connected compact symmetric space of rank one, all geodesics are closed and the prime ones all have the same length.*

More precisely, assume that the metric on such a space M has been normalized such that the supremum of the sectional curvature is equal to 1. *Then the length of all prime closed geodesics is* 2π.

If M is the sphere, all geodesics starting at a point $p \in M$ meet again at distance π.

If M is one of the projective spaces $P\mathbb{C}^n$, $P\mathbb{H}^n$ or $P\mathbf{Ca}^2$, of real dimension $2n, 4n$ *and* 16, *respectively, then the projective lines, having real dimension* 2,4 *and* 8, *respectively, are isometrically embedded spheres of constant curvature equal to* 1. *Hence, every prime closed geodesic lies as a great circle on a well-determined sphere S^2, S^4 or S^8; if $p \in M$, precisely those geodesics starting at p which belong to the same such sphere, passing through p, meet again at distance* π.

Proof. See [He]. □

One may ask whether the symmetric spaces of rank one are the only ones for which all geodesics are closed. This is not the case; however, as far as integer cohomology is concerned, the answer is affirmative, as was shown by Bott [Bo 2]. We recall that the integer cohomology of a symmetric space M of rank one is a truncated polynomial ring $T_{d,h}(x) = \mathbb{Z}[x]/(x^h)$, with degree $x = d$. Either, d is arbitrary and $h = 2$ – these are the spheres – or $d = 2, 4$ or 8 and h is arbitrary ≥ 3 except in the case $d = 8$, where one has to take $h = 3$; these are the projective spaces.

5.2.2 Theorem. *Let M be a compact Riemannian manifold of dimension $n+1$ and assume that there exists a point $p \in M$ and a number $\omega > 0$ such that the following holds. In any direction, there starts at p a closed geodesic without selfintersections of length* ω.

Claim. *Let λ be the index of one of the geodesic segments of length ω, starting at p. If $\lambda = 0$, then $\pi_1 M = \mathbb{Z}_2$ and the universal covering $\tilde{M}$ of M has the homotopy type of sphere of dimension $n+1$. If $\lambda \geq 1$, then $\pi_1 M = 0$ and M has the same integer cohomology ring as a simply-connected irreducible compact symmetric space of rank* 1.

In particular, the index λ is the same for all geodesic segments of length ω, starting at p.

Note. This theorem is due to Bott [Bo 2], modulo some facts of algebraic geometry, cf. [Mi 3]. In case $\lambda = n$ or $\lambda = 1$, M actually has the homotopy type of the sphere S^{n+1} and the complex projective space $P\mathbb{C}^m$, $2m = n+1$.

Proof. Let $c(t)$, $0 \leqslant t \leqslant a, |\dot{c}|=1$, be a geodesic segment of length $a>0$. According to the Morse Index Theorem, see [Mor 2], [Mi 2], the index of c, considered as a critical point in the space $\Omega(p,q)$ of H^1-curves from $p=c(0)$ to $q=c(a)$, is given by the formula

$$\operatorname{index}_{\Omega} c = \sum_{0<t<a} \dim D\phi_t V_v^n(0) \cap V_v^n(t).$$

Here, $V_v^n(t)$ is the vertical space over t and $D\phi_t V_v^n(0)$ the transport under the geodesic flow of the vertical space $V_v^n(0)$ over 0, cf. (3.2.10) and [Kl 12] for this interpretation.

Since the Poincaré map for any of the closed geodesics $c(t)$, $0 \leqslant t \leqslant \omega$, starting at p, is the identity, we have $D\phi_{\omega} V_v^n(0) = V_v^n(0)$. Hence, for $t \neq \omega$ and $|t-\omega|$ sufficiently small

$$D\phi_t V_v^n(0) \cap V_v^n(t) = 0.$$

We claim that given a closed geodesic c of length ω with $c(0)=p$, there exists a point $q=c(\varepsilon), \varepsilon>0$, on c such that every geodesic from p to q has its support on c and q is not conjugate to p. More precisely, such a geodesic is of the form

$$c(t),\ 0 \leqslant t \leqslant \varepsilon + k\omega,$$

or $$c(-t),\ 0 \leqslant t \leqslant -\varepsilon + \omega + k\omega$$

for some integer $k \geqslant 0$.

To prove this claim we assume the contrary, i.e. that there exists a sequence ε_i of positive numbers with $\lim \varepsilon_i = 0$ such that p and $p_i = c(\varepsilon_i)$ can be joined by a geodesic segment c_i not lying on the closed geodesic c. Since every c_i forms part of a closed geodesic, we may assume that $L(c_i) < \omega$. Since the exponential map is locally injective, we have a $\delta > 0$ such that $\delta < L(c_i) < \omega - \delta$. But then, any of the limit elements of the sequence $\{c_i\}$ represents a geodesic from p to p of length in the interval $[\delta, \omega - \delta]$, which is impossible. Finally, since the conjugate points are isolated on c, we can choose $p = c(\varepsilon)$ such that it is not conjugate to p.

If $\pi_1 M > 0$ we can find a geodesic segment c in $\Omega(p,p)$ of index 0. c must be a closed geodesic of length ω. Clearly, c is homotopic to its inverse and to every other geodesic segment representing a non-trivial element of $\pi_1 M$. Hence, $\pi_1 M = \mathbb{Z}_2$. If $\tilde{M}$ is the simply connected covering of M, there exists a point $\tilde{p} \in \tilde{M}$ such that in every direction at $\tilde{p}$ there starts a closed geodesic of length 2ω having index n. The latter statement follows simply from the fact that over $t=\omega$ we have

$$\dim \left(D\phi_{\omega} V_v^n(0) \cap V_v^n(0) \right) = n.$$

If, on such a closed geodesic $\tilde{c}$, we choose a point $\tilde{q} \neq \tilde{p}$ as above, then we have in the loop space $\tilde{\Omega} = \tilde{\Omega}(\tilde{p}, \tilde{q})$ only geodesic segments with index kn, $k = 0,1,2,\ldots$, as follows from the description of these segments and the index formula. Hence, since $\pi_{i+1} \tilde{M} = \pi_i \tilde{\Omega}$, $\pi_i \tilde{M} = 0$ for $0 < i < n+1$, i.e. M has the homotopy type of S^{n+1}.

Consider now the case $\pi_1 M = 0$. Then we may assume that our closed geodesic c of length ω passing through p has index $\lambda > 0$. Indeed, otherwise we would have in the loop space $\Omega = \Omega(p,q)$, with $q = c(\varepsilon)$ as above, two geodesic segments of index 0 and then only geodesics of index $\geqslant n$. For $n > 1$ this is impossible since the two geodesics of index 0 must be homotopic via a geodesic of index 1. For $n = 1$, i.e. M being a surface of the type of the 2-sphere S^2, we observe that on the cut locus of p there must be conjugate points, i.e. there must be closed geodesics of length ω through p which have index > 0, cf. [Po 1] and [My].

Let $\Omega = \Omega(p,q)$ with $q = c(\varepsilon)$, c a closed geodesic of index $\lambda > 0$. From the description of the geodesic segments in Ω and the index formula, we see that Ω has the structure of a CW-complex with one 0-cell, one λ-cell, one $(n+\lambda)$-cell, plus cells of dimension $\geqslant n + 2\lambda$.

The conclusions of the theorem now follow using standard methods in algebraic topology. Consider the space $E = E(p,M)$ of continuous curves $g(t)$, $0 \leqslant t \leqslant 1$, on M with $g(0) = p$. E is contractible and the map $g \in E \mapsto g(1) \in M$ is a fibration with fibre Ω. As we just saw, Ω is homotopically, up to dimension $\lambda + n \geqslant n + 1 = \dim M$, the sphere S^λ. Thus, the Gysin sequence applied to this fibration yields

$$H^i(M;\mathbb{Z}) = 0 \quad \text{for} \quad i \not\equiv 0 \bmod (\lambda+1)$$

and $$H^i(M;\mathbb{Z}) = \mathbb{Z} \quad \text{for} \quad i = k(\lambda+1),\ k = 0, \ldots, m > 1$$

with $$m(\lambda+1) = n+1. \quad \square$$

We can strengthen the hypothesis by requiring that, for all $p \in M$, the geodesics starting at p be closed and of constant length. One does not know much more about what happens under this additional hypothesis. Here we mention only the following result of Weinstein [We 4]. See also [Bes].

5.2.3 Theorem. *Assume that all orbits of the geodesic flow on the unit tangent bundle of a compact Riemannian manifold M are periodic of period 2π. Then the volume of M is an integer multiple of the volume of the sphere S^n of constant curvature* 1, $n = \dim M$. $\square$

It is important to note, however, that there are examples of Riemannian manifolds with all geodesics closed and of constant length which are not isometric to one of the symmetric spaces of rank 1. The first examples of such spaces were constructed by Zoll [Zo], who refined an idea of Darboux [Da].

5.2.4 Lemma. *On the differentiable 2-sphere S^2 there exists a 1-parameter family $g_t, 0 \leqslant t \leqslant \varepsilon$, of analytic metrics with g_0 being the standard metric of constant curvature* 1 *such that, for every $t > 0, (S^2, g_t)$ is not isometric to (S^2, g_0); however, (S^2, g_t) is a surface of revolution on which all geodesics are closed. The prime ones have no self-intersections and have length 2π. For sufficiently large t, (S^2, g_t) contains areas with negative curvature.*

Proof. See [Zo], also Berger [Be] and Besse [Bes]. $\square$

For the sphere (S^2, g_0) with the standard metric, the first conjugate locus of a point p is of course the antipodal point p'. Following Blaschke [Bla 1], we call a surface M of the type of the sphere a *Wiedersehensfläche* if, for every point $p\in M$, the first conjugate locus is a point $p' \neq p$. We denote by $w: M \to M$ the map which associates to a point its conjugate locus.

For a long time it was an open question whether the spheres are the only Wiedersehensflächen; see the different editions of [Bla 1] for a history of this problem. Funk [Fu] was able to prove that there is no non-trivial analytic 1-parameter family of Wiedersehensflächen containing a sphere of constant curvature. It was not until 1963 that Green proved:

5.2.5 Theorem. *A Wiedersehensfläche is isometric to a sphere of constant curvature.*

Proof. We refer to [Gre 3], [Be], [Bes]. Here we show only that, on a Wiedersehensfläche M, all geodesics are closed and the prime ones have no self-intersections and are of constant length.

We begin with the observation that the cut locus $C(p)$ of a point $p\in M$ coincides with p', i.e. all prime closed geodesics have no self-intersection. Here, the cut locus $C(p)$ is formed by the set of points on the geodesic rays emanating from p which describe the moment at which these rays cease to be minimizing curves, cf. [GKM] for more details. To see that $C(p)=p'$, we use the fact that a conjugate point of p with minimal distance from p belongs to $C(p)$, for every metric on S^2; this was proved by Myers [My].

Let c be a geodesic with $|\dot{c}|=1$, $c(0)=p$, $c(t_0)=w(p)$, $c(t_1)=p$, $0<t_0<t_1$, being the first values having this property. For $\varepsilon>0$ small, $w\circ c(\varepsilon)=c(t_0+\varepsilon_0(\varepsilon))$ and $c(\varepsilon)=w^2\circ c(\varepsilon)=c(t_1+\varepsilon_1(\varepsilon))$ with $\varepsilon_0,\varepsilon_1>0$ small, due to the standard properties of conjugate points (Sturm Separation Theorem). Hence, since this is true for all small $\varepsilon\geqslant 0$, $c(t_1+t)=c(t)$, i.e. c is closed with length t_1.

Consider $p\in M$ and a geodesic c_0 starting at p, $|\dot{c}_0|=1$. Let $c_0(t_0)=w(p)$ be the first conjugate point on c_0. The hypothesis "Wiedersehensfläche" implies that, for all geodesics c starting at p with $|\dot{c}|=1$ in a direction sufficiently near $\dot{c}_0(0)$, the first conjugate point will occur at the same value $t=t_0$. From continuity we see that for all geodesics c starting at a fixed point $p\in M$ with $|\dot{c}|=1$, the first conjugate point occurs at the fixed distance $t_0(p)$. Then $2t_0(p)$ is the length of the prime closed geodesics passing through p. Since every $q\in M$ lies on one of the geodesics through p, $2t_0(q)=2t_0(p)$, i.e. all prime closed geodesics on M have the same length.

This is all that we prove of (5.2.5). □

The analog of Wiedersehensfläche in higher dimensions has been considered by various authors. The best general result, it being a weakened version of Funk's result mentioned above, is due to Michel [Mic]:

5.2.6 Theorem. *Let (S^k, g_t), $0\leqslant t\leqslant 1$, be a deformation of the standard metric g_0 on the sphere, with g_t of class C^∞. Assume that this variation g_t is orthogonal to the orbit of the standard metric by the diffeomorphism group. If, for all t, (S^k, g_t) has the property that all geodesics starting at a point p meet again at distance π in a point p', then all derivatives of g_t as a function of t vanish.* □

Michel [l.c.] has obtained similar results for deformation of the standard metric of the other irreducible symmetric spaces of rank 1. See also Besse [Bes].

Note. Quite recently, Berger could show that a Wiedersehens manifold of even dimension is isometric to the standard sphere.

Besides homogeneous and, in particular, symmetric Riemannian manifolds, there is another class of manifolds for which the geodesic flow can be described in greater detail, i.e. those manifolds for which the geodesic flow is completely integrable.

To define this concept we consider a general Hamiltonian system (N,α,H), cf. (3.1). If $F:N\to\mathbb{R}$ is a differentiable function, we define the vector field ζ_F by

$$2\alpha_*\zeta_F=i_{\zeta_F}\,.\,\alpha=dF.$$

F is called an *integral* for (N,α,H) if F is constant along the flow lines of the vector field ζ_H, i.e. if

$$dF\,.\,\zeta_H=2\alpha(\zeta_F,\zeta_H)=-dH\,.\,\zeta_F=0.$$

Two integrals F,F' are said to be *in involution* if $2\alpha(\zeta_F,\zeta_{F'})=0$.

A Hamiltonian system (N,α,H), $\dim N=2n$, is called *completely integrable* if there exist, besides $F_1=H$, $n-1$ further functions $F_2,\dots,F_n$ on N such that the corresponding vector fields $\zeta_{F_i}(q)$ in T_qN span a Lagrangian subspace, for all q on an open dense subset of N; i.e. the $F_1,\dots,F_n$ are integrals which are in involution and $dF_1\wedge\dots\wedge dF_n\neq 0$ on an open dense subset of N. Cf. [AM], [AA], [Ta].

For completely integrable Hamiltonian systems (N,α,H) the following Theorem of Arnold [AA] is fundamental.

5.2.7 Theorem. *Let $F_1=H,F_2,\dots,F_n$ be functionally independent integrals in involution of the completely integrable Hamiltonian system (N,α,H), $2n=\dim N$. Consider the differentiable mapping*

$$\phi:N\to\mathbb{R}^n,\ p\mapsto(F_1(p),\dots,F_n(p)).$$

By hypothesis, ϕ is regular on an open, dense subset $N'\subset N$. If, for each $c\in$ image $\phi|N'$, the connected components of the n-dimensional manifold $(\phi|N')^{-1}(c)$ are compact, then each such component is an embedded n-torus T_c^n. Moreover, for sufficiently small $\delta=\delta(c)>0$, there exists a symplectic diffeomorphism

$$\psi:B_\delta^n\times T^n\to U=U(T_c^n)$$

of the δ-ball in $\mathbb{R}^n$ cross the n-Torus $T^n=(S^1)^n$ onto an open neighborhood U of T_c^n such that $\phi\circ\psi|\{I\}\times T^n$ is a constant.

The symplectic coordinates $(I,\phi)\in B_\delta^n\times T^n$, with $\phi=(\phi_1,\dots,\phi_n)$ periodic in 2π, are called *action-angular variables*. The Hamiltonian flow on U is given by

$$I(t)=I_0=\text{const};\ \phi(t)=\phi_0+t\omega(I_0)$$

with $\omega(I_0) = -\partial H \circ \psi(I, \phi)/\partial I|_{I=I_0}$ being the so-called *period* of the flow on the torus $I = I_0$.

For the proof we refer to [AA], [Ze]. □

We see that the Hamiltonian flow, restricted to a torus, is linear; it is also called *quasi-periodic*. Three cases can be distinguished:

(i) the period $\omega(I_0) \in \mathbb{R}^n$ is free over the rationals. In this case the flow is ergodic;

(ii) there exist for $\omega(I_0) \in \mathbb{R}^n$ r, $1 \leqslant r < n-1$, independent linear equations with rational coefficients. In this case the n-torus possesses a factorization $T^n = T^{n-r} \times T^r$ such that the flow is ergodic on each $T^{n-r} \times \{\omega^r\} \subset T^{n-r} \times T^r$;

(iii) $\omega(I_0) \in \omega_0 \mathbb{Q}^n$. In this case, all orbits are periodic.

Important information concerning closed geodesics is contained in the following theorem.

5.2.8 Theorem. *Assume that the geodesic flow $\phi_t : TM \to TM$ is completely integrable. If the period mapping given locally by $I \in B_\delta^n \mapsto \omega(I) \in \mathbb{R}^n$ is non-degenerate, i.e. if the Hessian*

$$(\partial^2 H(I, \phi)/\partial I_i \partial I_j)$$

is non-degenerate on an open dense set, then the periodic orbits are dense in TM.

Proof. We simply need observe that the sets $\omega_0 \mathbb{Q}^n$, $\omega_0 \in \mathbb{R}$, are dense in $\mathbb{R}^n$. □

The best-known example for this phenomenon is the ellipsoid.

5.2.9 Proposition. *The geodesic flow on the n-dimensional ellipsoid with pairwise different principal axes is completely integrable with non-degenerate period mapping. In particular, the periodic orbits of the geodesic flow are dense.*

Proof. We refer to Schüth [Schü] for a complete proof of the non-degeneracy of the period mapping. The fact that the geodesic flow on the ellipsoid is completely integrable was certainly known to Jacobi [Ja] for the 2-dimensional case. For the n-dimensional case it was proved by Rosochatius [Ros]. It is also contained in Stäckel [Sta]; see also Thimm [Th]. □

The complete integrability of the geodesic flow on the tangent bundle of the ellipsoid is based on the existence of the socalled elliptic coordinates, introduced by Jacobi [Ja]. In these coordinates the line element assumes a particularly simple form. In fact, the line element is then a special example of the *Stäckel line element* (see [Sta])

$$ds^2 = \sum_i g_{ii}(u)\,du^i du^i,$$

where the $g_{ii}(u)$ are given as follows. There exists a regular (n,n)-matrix $(h_{ij}(u))$ of functions $h_{ij}(u)$ where the elements $h_{ij}(u)$ in the i-th row are functions of u^i only. With this, the $g_{ii}(u)$ are given by

$$g_{ii}(u) = (-1)^{i+1} \det\left(h_{kl}(u)\right) : \operatorname{codet}\left(h_{i1}(u)\right).$$

The $h_{ij}(u)$ must be such that the $g_{ii}(u)$ become positive, of course. See also Prange [Pr], Blaschke [Bla 2], Thimm [Th].

In the case $n=\dim M=2$ the Stäckel line element becomes the *Liouville line element*:

$$ds^2=(U-V)\,(U_1^2du^2+V_1^2dv^2).$$

Here, U, U_1 and V, V_1 are functions of u and v only, cf. [Da], [Bla 1], [Vi 1]

5.2.10 Proposition. *The geodesic flow over a domain with a Liouville line element is completely integrable. In fact, consider on the set of non-vanishing tangent vectors X the function*

$$\begin{aligned}F(X)&=(U-V)\,(VU_1^2(duX)^2+UV_1^2(dvX)^2)/2H(X)\\&=U\cos^2\theta(X)+V\sin^2\theta(X).\end{aligned}$$

Here, $\theta(X)$ is the angle between X and the v-parameter line. Then F is constant on $\dot{c}(t), c(t)$ a non-trivial geodesic, and $dH\wedge dF\neq 0$.

Proof. Choose a constant A satisfying $V<A<U$, at least locally. We define new coordinates (u',v') by

$$du'=U_1\sqrt{U-A}\,du\pm V_1\sqrt{A-V}\,dv;\quad dv'=U_1du/\sqrt{U-A}\mp V_1dv/\sqrt{A-V}.$$

This yields

$$ds^2=(du')^2+(U-A)\,(A-V)\,(dv')^2,$$

i.e. (u',v') are geodesic parallel coordinates with $v'=$const. being geodesics.

The coordinates duX, dvX, of a tangent vector X to such a geodesic with $H(X)=B^2/2>0$ are given by

$$duX=B\sqrt{U-A}/(U-V)U_1;\quad dvX=\pm B\sqrt{A-V}/(U-V)V_1.$$

This yields

$$\cos^2\theta(X)=(A-V)/(U-V);\quad \sin^2\theta(X)=(U-A)/(U-V).$$

Hence, $F(X)=A$, $H(X)=B^2/2$, i.e. F is constant on the tangent vectors to a geodesic $v'=$const. This completes the proof of (5.2.10). □

Note. We see that if a surface possesses a Liouville line element, then it has a second integral F (besides H) which is quadratic in the coordinates duX, dvX of the vector X. A similar result holds for Stäckel line elements, given below.

5.2.11 Proposition. *The geodesic flow over a domain possessing a Stäckel line element is completely integrable. There exist n integrals $F_1=H, F_2,\dots,F_n$ in involution of the form*

$$F_i(X)=\sum_j k_{ij}(u)g_{jj}(u)^2(du^jX)^2$$

with $dF_1 \wedge \ldots \wedge dF_n \neq 0$ *for* $X \neq 0$. *Actually,* $(k_{ij}(u))$ *is the inverse to the regular matrix* $(h_{ij}(u))$ *defining the Stäckel line element.*

Proof. See [Sta], [Th]. □

Note. The existence of n integrals F_i in involution of the form (5.2.11) characterizes the Stäckel line element among the diagonal line elements; see Stäckel [Sta] and Thimm [Th]. For $n=2$, i.e. for the Liouville line element, this is due to Massieu [Thèse, Paris 1861], cf. Darboux [Da].

We mention another characterization of the Stäckel line element, due to Zwirner and Blaschke.

5.2.12 Proposition. *The Stäckel line element (and thus in particular the Liouville line element) is characterized by the property that, for every sufficiently small coordinate n-cube, all geodesic diagonals have equal length.*

Proof. See [Bla 2], [Th]. □

Whereas in general a Hamiltonian system (N, α, H) will not possess additional integrals besides H, there is the important class of systems allowing a group action which preserves the structure of (N, α, H); see Abraham-Marsden [AM] and Takens [Ta] for further details on these so-called conservation laws.

We shall consider here only the case of an isometric S-action on a compact Riemannian manifold and thereby determine those orbits which are closed geodesics.

5.2.13 Theorem. *Assume that there is given an isometric circle action on* M

$$\rho : S \times M \to M.$$

Let

$$R(p) = (d/ds)\rho(s, p)_{s=0}$$

be the generating vector field on M. *Then*

$$F : M \to \mathbb{R}; \quad X \mapsto \langle X, R(\tau X) \rangle$$

is constant along the geodesic flow lines, i.e. F *is an integral. Moreover, if*

$$\Phi : M \to \mathbb{R}; \quad p \mapsto \langle R(p), R(p) \rangle$$

has a stationary value at p_0 *and* $R(p_0) \neq 0$, *then the S-orbit through* p_0 *is a closed geodesic.*

5.2.14 Corollary. *Assume* dim $M=2$ *and* $dF \neq 0$ *on an open, dense subset of* TM. *Then the geodesic flow on* TM *is completely integrable.*

Note. The most important examples of surfaces satisfying the hypothesis of (5.2.14) are the surfaces of revolution. It can be shown that, conversely, a surface which possesses a second integral which is linear in X locally is isometric to a surface of revolution (Massieu, cf. Darboux [Da]).

For surfaces of revolution, the existence of the integral F is known as *Clairaut's Theorem*. $F(X)/|X|$ is also called *angular momentum*.

Proof. Let $c(t)$ be a geodesic. Then $F(\dot{c}(t))$ can be written as

$$\langle \dot{c}(t), R(c(t))\rangle = \left\langle \frac{\partial}{\partial t}\rho(s,c(t)), \frac{\partial}{\partial s}\rho(s,c(t))\right\rangle\Big|_{s=0}.$$

The derivation with respect to t yields, since $\rho(s,c(t))$ is a geodesic for s fixed,

$$\left\langle \frac{\partial}{\partial t}\rho(s,c(t)), \frac{\nabla}{\partial t}\frac{\partial}{\partial s}\rho(s,c(t))\right\rangle = \left\langle \frac{\partial}{\partial t}\rho(s,c(t)), \frac{\nabla}{\partial s}\frac{\partial}{\partial t}\rho(s,c(t))\right\rangle$$
$$=\frac{1}{2}\frac{\partial}{\partial s}\left\langle \frac{\partial}{\partial t}\rho(s,c(t)), \frac{\partial}{\partial t}\rho(s,c(t))\right\rangle = 0$$

since $|T\rho(s,\dot{c}(t))| = \text{const}$. Hence, $F(\dot{c}(t)) = \text{const}$.

Now assume that $D\Phi(p_0)=0$. Let $p(t)$, $0 \leqslant t \leqslant \varepsilon$, be an arbitrary curve on M starting at $p(0)=p_0$ with $dp(0)/dt = X$. Put $\rho(s,p_0)=c(s)$. Then $D\Phi(p_0)=0$ implies that

$$0=\frac{1}{2}\frac{d}{dt}\left\langle \frac{\partial}{\partial s}\rho(s,c(t)), \frac{\partial}{\partial s}\rho(s,p(t))\right\rangle\Big|_{t=0}$$
$$=\left\langle c'(s), \frac{\nabla}{\partial t}\frac{\partial}{\partial s}\rho(s,p(t))\right\rangle\Big|_{t=0} = \left\langle c'(s), \frac{\nabla}{\partial s}\frac{\partial}{\partial t}\rho(s,p(t))\right\rangle\Big|_{t=0}$$
$$= -\left\langle \frac{\nabla}{\partial s}c'(s),\ T\rho(s,X)\right\rangle.$$

Here we have used that $\langle c'(s),\ T\rho(s,X)\rangle = \langle T\rho(s,c'(0)),\ T\rho(s,X)\rangle$ is independent of s.

Hence, since $T\rho(s,X)$ can be any vector, $\nabla c'(s)/ds = 0$. This completes the proof of (5.2.13). The corollary (5.2.14) is immediate. □

We conclude with an example of a surface of revolution where the period mapping (cf. (5.2.8)) is degenerate. That is to say, the periods are the same on all the invariant 2-tori; thus, the geodesic flow is either ergodic on each 2-torus or it is periodic. The S-orbit of maximal length is a closed geodesic, cf. (5.2.14). It represents a degenerate 2-torus in this picture.

We introduce the ratio

$$\pi(M) = \text{vol}\,(\text{Per}\ T_1 M) : \text{vol}\,(T_1 M).$$

Here, Per $T_1 M$ denotes the set of those unit tangent vectors X for which the orbit $\phi_t X$ is periodic. vol (Per $T_1 M$) is the measure of the closure of Per $T_1 M$ in $T_1 M$.

The following example also serves to illustrate the erratic behavior of the number $\pi(M)$. On the sphere, $\pi(M)=1$. Also, for ellipsoids, we have $\pi(M)=1$.

5.2.15 Proposition. *There exist, in every C^0-neighborhood of the standard sphere S^2, convex surfaces of revolution for which the ratio $\pi(M)$ becomes arbitrarily small.*

Proof. (cf. [Kl 8]). Our example is based upon a construction of Darboux [Da]. See also Weinstein [We 2] for a similar argument.

The surfaces in question are constructed as follows. Let a be a real number satisfying $0<a\leqslant 1$. Then

$$f_a(u,v)=\left(a\cos u\cos v,\ a\cos u\sin v,\ \int_0^u\sqrt{1-a^2\sin^2 t}\,dt\right),$$

with $|u|<\pi/2$, represents a non-complete surface of revolution of constant curvature equal to 1. We choose a small $\delta>0$. Denote by $S^2(a,\delta)$ the surface covered by $f_a(u,v)$ when $|u|\leqslant\pi/2-\delta$, v arbitrary. By closing off $S^2(a,\delta)$ with convex caps we obtain a convex surface of revolution denoted by $M(a,\delta)$.

We shall be interested in the geodesic flow only for those $X\in T_1M(a,\delta)$ where the corresponding geodesic $c(t)=\tau\phi_tX$ stays on $S^2(a,\delta)$, for all t. From Clairaut's Theorem (5.2.14) we know that this amounts to considering only those $X\in T_1S^2(a,\delta)$ where

$$F(X)=\left\langle X,\ \frac{\partial f_a}{\partial v}\big(u(\tau X),v(\tau X)\big)\right\rangle\geqslant a\sin\delta.$$

Denote the set of these $X\in T_1M(a,\delta)$ by $T_1'S^2(a,\delta)$ or, briefly, T_1'.

To obtain a description of the geodesic flow on T_1' we first look at the geodesic flow on the domain $T_1'S^2(1,\delta)$ of the unit tangent bundle T_1S^2 of the standard sphere S^2. The flow lines in this domain project into a geodesic of the form

$$c(t)=\cos(t-t_0)E_1+\sin(t-t_0)B_\alpha,$$

with $E_1=(1,0,0)$, $B_\alpha=(0,\cos\alpha,\sin\alpha)$, $0\leqslant\alpha\leqslant\pi/2-\delta$.

If we fix $\alpha, 0<\alpha$, then the set of tangent vectors $\dot c(t)$ to these geodesics is a 2-torus T_α^2, invariant under the geodesic flow. For $\alpha=0$, we obtain the degenerate torus formed by the tangent vectors to the equator. T_α^2 is characterized by $F(X)=\cos\alpha$.

To obtain the Poincaré map we describe the orbits on T_α^2 by their initial values $\{c(0),\dot c(0)\}$. Using complex notation we see that

$$\begin{aligned} c(t)+i\dot c(t)&=e^{i(t-t_0)}(E_1-iB_\alpha)\\ &=e^{it}\big(c(0)+i\dot c(0)\big).\end{aligned}$$

Now, observe that $S^2(a,\delta)$ is isometric to the part $\{0\leqslant v<2\pi a\}$ of $S^2(1,\delta)$. Thus, the Poincaré map for the periodic orbit lying over the equator of $M(1,\delta)$ is given by

$$\mathscr{P}: c(0)+i\dot c(0)\mapsto e^{i2\pi a}\big(c(0)+i\dot c(0)\big).$$

That is to say, $\mathscr{P}$ coincides with its linear part $P = D\mathscr{P}$, cf. (3.3). The equator represents an elliptic closed geodesic.

It follows that, if and only if a is rational, all orbits of the geodesic flow on the invariant subset $T_1'S^2(a,\delta)$ are periodic. If a is irrational, only the equator yields a periodic orbit. Now, since the measure of $T_1'S^2(a,\delta)$ differs arbitrarily little from the measure of $T_1M(a,\delta)$, we obtain (5.2.15). □

Note. A behavior of the geodesic flow near an elliptic periodic orbit as we have just described is certainly exceptional. By arbitrarily small appropriate deformations of the Riemannian metric the Poincaré map will be of twist type, cf. (3.3.8). In this case, the elliptic orbit will not be isolated; rather, the closure of the periodic orbits will have positive measure in every neighborhood of that orbit.

This is a consequence of the Kolmogoroff-Arnold-Moser Theorem on Invariant Tori, together with the Birkhoff-Lewis Fixed Point Theorem (3.3.3). See Zehnder [Ze] for a complete proof. For dim $M = 2$, we mentioned this already in (3.3.4).

What happens is that the geodesic flow on the invariant tori in the Kolmogoroff-Arnold-Moser Theorem is ergodic. However, these tori are limits of periodic orbits, as the Birkhoff-Lewis Theorem shows. Observe now that the closure of the invariant tori has positive measure. An example of this situation is provided by the ellipsoids with pairwise different axes and all sufficiently nearby Riemannian metrics.

It may very well be that this is the generic behavior everywhere for the geodesic flow over a manifold of elliptic type. To get anywhere close to proving something like this, one first has to find many elliptic closed geodesics. The result of (4.4.6) is a first step in this direction; others are likely to follow.

5.3 Geodesics on Manifolds of Hyperbolic and Parabolic Type

Up to now we were mainly concerned with simply connected manifolds or manifolds with finite fundamental group. In this section we want to report on some results for manifolds which possess a cell as universal covering.

For a long time it has been known that properties of the geodesic flow associated to a closed surface depend, to a large degree, on the sign of the curvature K, see e.g. [He 2]. The main reason for this is that for $K \equiv 0$ the fundamental group is essentially infinite Abelian, whereas for $K < 0$ it has exponential growth; see [Mi 4] for this concept.

By considering manifolds with fundamental group of exponential growth, we restrict ourselves to what we call manifolds of hyperbolic type. These are manifolds permitting a Riemannian metric which induces on the unit tangent bundle a geodesic flow of Anosov type. The main examples of such manifolds (and, maybe, the only ones) are provided by manifolds permitting a Riemannian metric of strictly negative curvature. Our main result is that, for any metric on a manifold of hyperbolic type, there exists in the unit tangent bundle a relatively large invariant closed subset on which the geodesic flow has many properties in common with a geodesic flow of Anosov type, cf. (5.3.6).

We will say that a compact manifold is of parabolic type if it has an Abelian fundamental group or, more generally, if it has a finite covering with Abelian fundamental group. Of the properties of such manifolds we give only one example, due to Busemann [Bu], concerning the existence of many closed geodesics if the Riemannian metric on such a manifold has no conjugate points, cf. (5.3.7).

Recall from (3.2) that the tangent bundle τ_{T_1M}: $TT_1M \to T_1M$ of the unit tangent bundle of a Riemannian manifold M contains a distinguished subbundle of fibre codimension 1. If dim $M=n+1$, we denote this bundle by

$$\tau^{2n}: T^{2n} = T^{2n}T_1M \to T_1M.$$

The fibre $(\tau^{2n})^{-1}(X_0) = T^{2n}_{X_0}$ over $X_0 \in T_1M$ consists of those vectors which are orthogonal to the geodesic spray.

The restriction of the horizontal subbundle of τ_{TM}: $TTM \to TM$ to τ^{2n} will be denoted by τ^n_h: $T^n_h = T^n_h T_1M \to T_1M$. In the same way we obtain by restriction the vertical subbundle of τ^{2n}: τ^n_v: $T^n_v = T^n_v T_1M \to T_1M$.

Thus, every element $\tilde{Y} \in T^{2n}_{X_0}$ has a splitting $\tilde{Y} = (\tilde{Y}_h, \tilde{Y}_v) \in T^n_{X_0 h} \oplus T^n_{X_0 v}$, where $\tilde{Y}_h, \tilde{Y}_v$, considered as elements of $T_{\tau X_0}M$, satisfy

$$\langle \tilde{Y}_h, X_0 \rangle = \langle \tilde{Y}_v, X_0 \rangle = 0.$$

We denote the orbit $\phi_t X_0$ of the geodesic flow, starting at $t=0$, with the element $X_0 \in T_1M$, also by X_t. Let $c(t) := \tau \circ X_t$ be the underlying geodesic on M. Hence, $X_t = \dot{c}(t)$.

From (3.1.6) and (3.1.7) we know that, given $\tilde{Y} \in T^{2n}_{X_0}$, the vector field $D\phi_t \tilde{Y}$ is of the form $\tilde{Y}(t) = \big(Y(t), \nabla Y(t)\big) \in T^n_{X_t h} \oplus T^n_{X_t v}$ where $Y(t)$ is a Jacobi field

$$\nabla^2 Y(t) + R\big(Y(t), \dot{c}(t), \dot{c}(t)\big) = 0$$

and

$$\langle Y(t), \dot{c}(t) \rangle = \langle \nabla Y(t), \dot{c}(t) \rangle = 0.$$

This shows in particular that the flow $T\phi_t: TT_1M \to TT_1M$ induced from the geodesic flow $\phi_t: T_1M \to T_1M$, leaves the subbundle τ^{2n} invariant.

Following [Kl 12], we say that the geodesic flow ϕ_t on T_1M is of *Anosov type* if τ^{2n} permits a splitting $\tau^{2n} = \tau^n_s \oplus \tau^n_u$ into two $T\phi_t$-invariant subbundles, the so-called *stable* and *unstable bundles*.

$$\tau^n_s: T^n_s := T^n_s T_1M \to T_1M; \qquad \tau^n_u: T^n_u := T^n_u T_1M \to T_1M,$$

of fibre dimension n, each of which has the property that $T\phi_t | T^n_s$ operates as a contraction and $T\phi_t | T^n_u$ operates as an expansion, i.e. there exist positive numbers $a \leqslant 1$ and k such that, for all real $t_1 \leqslant t_2$ and all $\tilde{Y}_s \in T^n_s$, $\tilde{Y}_u \in T^n_u$ the following relations hold:

$$|D\phi_{t_2}\tilde{Y}_s|^2 \leqslant a^2 |D\phi_{t_1}\tilde{Y}_s|^2 e^{-2k(t_2-t_1)} \quad \text{and}$$

$$(\dagger) \qquad |D\phi_{t_2}\tilde{Y}_u|^2 \geqslant a^2 |D\phi_{t_1}\tilde{Y}_u|^2 e^{2k(t_2-t_1)}.$$

Note. If the geodesic flow on T_1M is of Anosov type, all closed geodesics on M are hyperbolic, i.e. the linear Poincaré map associated to a periodic orbit of the geodesic flow has no eigenvalue of modulus 1. Manifolds with geodesic flow of Anosov type are the only known examples having this property.

Eberlein [Eb 2] has given various characterizations of manifolds with geodesic flow of Anosov type. The most important examples of manifolds having this property are given by the following lemma.

5.3.1 Lemma. *Assume that the sectional curvature of a compact Riemannian manifold is strictly negative. Then the geodesic flow is of Anosov type.*

More precisely, assume that there exist numbers $k_1 \geqslant k_2 > 0$ such that, for all $X_0 \in T_1M$ and for all $Y \in TM$ with $\tau Y = \tau X_0$, $\langle Y, X_0 \rangle = 0$ we have

$$-k_1^2 |Y|^2 \leqslant \langle Y, R(Y, X_0, X_0) \rangle \leqslant -k_2^2 |Y|^2.$$

Then the constants in the relations ($\dagger$) *can be chosen such that $k = (k_2 - \varepsilon)$, $a = k_1/(k_2 - \varepsilon)$ for every small $\varepsilon > 0$. Hence, we may even take $k = k_1$, $a = k_1/k_2$.*

Proof. (cf. [An], [AnS]). Recall from (3.1.3) that on each fibre $T^{2n}_{X_0}$ of $\tau^{2n}: T^{2n} \to T_1M$ we have a symplectic structure given by

$$2\alpha(\tilde{Y}, \tilde{Y}') = \langle \tilde{Y}_h, \tilde{Y}'_v \rangle - \langle \tilde{Y}'_h, \tilde{Y}_v \rangle.$$

Let $V \subset T^{2n}_{X_0}$ be a Lagrangian subspace, i.e. a totally isotropic subspace of dimension n. If V does not intersect $T^n_{X_0 v}$, then we know from the remark following (3.2.9) that there exists a uniquely determined symmetric (n,n)-matrix S such that V can be represented as

$$V = \{(Y, SY) \in T^n_{X_0 h} \oplus T^n_{X_0 v}, \quad \text{all } Y \text{ satisfying } \langle Y, X_0 \rangle = 0\}.$$

Conversely, such a V is Lagrangian. (In (3.2) we had interchanged the roles of the vertical and the horizontal space).

Consider an orbit $X_t = \phi_t X_0$, $X_0 \in T_1M$, of the geodesic flow. Let $c(t) = \tau \circ X_t$ be the underlying geodesic, i.e. $\dot{c}(t) = X_t$. For an arbitrary $\tau \in \mathbb{R}$ we denote by S_τ a positive definite symmetric (n,n)-matrix with spectrum in $[k_2 - \varepsilon, k_1]$. Let V^n_τ be the corresponding Lagrangian subspace in $T^{2n}_{X_\tau}$. Then $V^n_\tau(t) := D\phi_{t-\tau} V^n_\tau$ is a Lagrangian subspace of $T^{2n}_{X_t}$. If we assume that $V^n_\tau(t)$ does not intersect the vertical space $T^n_{X_t v}$, then this subspace determines a symmetric matrix which we denote by $S_\tau(t)$.

Since, for all $\tilde{Y}(t) = (Y(t), \nabla Y(t)) \in V^n_\tau(t)$, we have $\nabla Y(t) = S_\tau(t)\, Y(t)$, with $Y(t)$ being a Jacobi field, $S_\tau(t)$ satisfies the Riccati equation

$$(*) \qquad \nabla S_\tau(t) + S_\tau(t) S_\tau(t) + R(\quad, \dot{c}(t), \dot{c}(t)) = 0.$$

Claim (i). For all $t \geqslant \tau$, the spectrum of $S_\tau(t)$ lies in $[k_2-\varepsilon, k_1]$. In particular, the Lagrangian space $V_\tau(t)$ does not meet $T^n_{X_t v}$, for all $t \geqslant \tau$.

To prove this we consider the functions

$$\alpha(t):=\max_{|Y|=1}\langle S_\tau(t)\,Y, Y\rangle, \qquad \beta(t):=\min_{|Y|=1}\langle S_\tau(t)\,Y, Y\rangle.$$

From (*) we deduce that

$$\dot{\alpha}(t) \leqslant -\alpha^2(t)+k_1^2; \qquad \dot{\beta}(t) \geqslant -\beta^2(t)+k_2^2,$$

or, to be more careful, the corresponding difference equation. In any case, this shows that $\alpha(t_0) \leqslant k_1$ implies that $\alpha(t) \leqslant k_1$, for all $t \geqslant t_0$, and $\beta(t_0) \geqslant k_2-\varepsilon$ implies that $\beta(t) \geqslant k_2-\varepsilon$, for all $t \geqslant t_0$.

Claim (ii). *Let $\tau' \leqslant \tau$ and let $S_\tau(t)$, $S_{\tau'}(t)$ be the solutions of (*) for $t \geqslant \tau'$ and $t \geqslant \tau$, respectively, which are determined by the initial values $S_{\tau'}, S_\tau$, having their spectrum in $[k_2-\varepsilon, k_1]$. Then, for all $t \geqslant \tau$,*

$$(**) \qquad |S_\tau(t)-S_{\tau'}(t)| \leqslant |S_\tau - S_{\tau'}(\tau)|e^{-2(k_2-\varepsilon)(t-\tau)}.$$

To prove this we put $S_\tau(t)-S_{\tau'}(t)=R(t)$. Then we have, from (*),

$$\nabla R(t) = -S_\tau(t)\,R(t) - R(t)\,S_{\tau'}(t).$$

A solution $R(t)$ is given by

$$R(t) = Q(t)\;(S_\tau - S_{\tau'}(\tau))\,Q'(t)$$

where $Q(t), Q'(t)$ are (n,n)-matrices determined by

$$\nabla Q(t) = -S_\tau(t) Q(t), \qquad Q(\tau)=E;$$
$$\nabla Q'(t) = -Q'(t)\,S_{\tau'}(t), \qquad Q'(\tau)=E.$$

Using the property of the spectrum of $S_\tau(t)$, we obtain

$$\frac{d}{dt}|Q(t)|^2 = -2\langle S_\tau(t) Q(t), Q(t)\rangle \leqslant -2(k_2-\varepsilon)|Q(t)|^2.$$

Hence, $|Q(t)| \leqslant e^{-(k_2-\varepsilon)(t-\tau)}$. The same argument applies to $Q'(t)$.

As a consequence of (ii), we have that $\lim_{\tau \to -\infty} S_\tau(t)$ exists and gives a symmetric matrix $S_u(t)$ with spectrum in $[k_2-\varepsilon, k_1]$, defined for all $t \in \mathbb{R}$. Moreover, the space

$$T^n_u(t) := \{(Y, S_u(t)\,Y) \in T^n_{X_t h} \oplus T^n_{X_t v}\,;\ \text{all } Y \text{ with } \langle Y, X_0\rangle = 0\}$$

is a Lagrangian subspace invariant under $D\phi_t$. That is to say, a flow-invariant vector field $\tilde{Y}(t)=(Y(t),\nabla Y(t))\in T_u^n(t)$ satisfies

$$\nabla Y(t)=S_u(t)\,Y(t).$$

Using the bounds $[k_2-\varepsilon,k_1]$ of the spectrum of $S_u(t)$, we therefore obtain for such a $\tilde{Y}(t)$

$$2k_1|Y(t)|^2\geqslant\frac{d}{dt}|Y(t)|^2=2\langle Y(t),S_u(t)\,Y(t)\rangle\geqslant 2(k_2-\varepsilon)|Y(t)|^2,$$

$$k_1^2|Y(t)|^2\geqslant|\nabla Y(t)|^2\geqslant(k_2-\varepsilon)^2|Y(t)|^2.$$

Hence

$$|Y(t_2)|^2\geqslant|Y(t_1)|^2e^{2(k_2-\varepsilon)(t_2-t_1)}$$

and

$$|\nabla Y(t_2)|^2\geqslant((k_2-\varepsilon)^2/k_1^2)|\nabla Y(t_1)|^2e^{2(k_2-\varepsilon)(t_2-t_1)},$$

i.e. $T_u^n(t)$ is the fibre over X_t of the unstable bundle τ_u^n.

In the same way one constructs the fibres $T_s^n(t)$ of the bundle τ_s^n.

This completes the proof of (5.3.1). $\square$

For a subsequent application, we consider a periodic orbit $\phi_t X_0$ of period ω in the unit tangent bundle T_1M, $\dim M=n+1$. Let $c_0(t)=\tau\circ\phi_t X_0$ be the underlying closed geodesic of length ω. Let $c=c_0(t)$ $0\leqslant t\leqslant m\omega$ be the m-fold covering of c_0. Consider the immersion

$$t\in S_{m\omega}\mapsto c(t)\in M$$

of the circle of length $m\omega$. We call the geodesic c *orientable* if the normal bundle $n_c: N_c^n\to S_{m\omega}$ of this immersion is orientable. In other words, the map

$$\|_c:(T_{c(0)}M,\dot{c}(0))\to(T_{c(0)}M,\dot{c}(0)),$$

obtained from parallel translation along the geodesic c has determinant $+1$.

We consider the induced bundle $(\dot{c})^*\tau^{2n}$

$$\begin{array}{ccc} V^{2n} & \longrightarrow & T^{2n}=T^{2n}T_1M \\ \Big\downarrow (\dot{c})^*\tau^{2n} & & \Big\downarrow \tau^{2n} \\ S_{m\omega} & \xrightarrow{\ \dot{c}\ } & T_1M \end{array}$$

and we denote it briefly by

$$\tau^{2n}: V^{2n} \to S_{m\omega}.$$

τ^{2n} possesses the splitting $\tau^{2n} = \tau_h^n \oplus \tau_v^n$, with

$$\tau_h^n : V_h^n \to S_{m\omega}, \qquad \tau_v^n : V_v^n \to S_{m\omega}.$$

The flow $T\phi_t$ on $T^{2n}T_1M$ induces a flow on the total space V^{2n} of τ^{2n}, which we also denote by $T\phi_t$ and $D\phi_t := T\phi_t|$ fibre, respectively.

Now assume that c is hyperbolic. That is to say, the fibre $V^{2n}(0)$ of τ^{2n} over $0 \in S_{m\omega}$ has a decomposition

$$V^{2n}(0) = V_s^n(0) + V_u^n(0)$$

into n-dimensional subspaces, invariant under $P^m = D\phi_{m\omega} = (D\phi_\omega)^m$ such that $P^m|V_s^n(0)$ has only eigenvalues of modulus <1 whereas $P^m|V_u^n(0)$ has only eigenvalues of modulus >1.

We can then define the *stable* and *unstable* bundle over $S_{m\omega}$,

$$\tau_s^n : V_s^n \to S_{m\omega}, \qquad \tau_u^n : V_u^n \to S_{m\omega},$$

by taking as fibre over $t \in S_{m\omega}$ the element $D\phi_t V_s^n(0)$ and $D\phi_t V_u^n(0)$, respectively. From (3.2.13) we then know that

$$\begin{aligned} \text{index } c &= \sum_{0 \leqslant t < m\omega} \dim\left(V_v^n(t) \cap V_s^n(t)\right) = \\ &= m \sum_{0 \leqslant t < \omega} \dim\left(V_v^n(t) \cap V_s^n(t)\right) = m \text{ index } c_0. \end{aligned}$$

5.3.2 Lemma. *Let $c = (c(t), 0 \leqslant t \leqslant m\omega)$, $|\dot{c}| = 1$, be a hyperbolic closed geodesic.*

Claim. *Index c is even if and only if the normal bundle $n_c : N_c^n \to S_{m\omega}$ and the stable bundle $\tau_s^n : V_s^n \to S_{m\omega}$, associated to c and its orbit $(\dot{c}(t), 0 \leqslant t \leqslant m\omega)$ in T_1M, are either both orientable or both non-orientable.*

Proof. (cf. [Kl 12]) By changing, if necessary, the initial point, we may assume that $V_v^n(0) \cap V_s^n(0) = 0$. This is equivalent to saying that the linear map $T\tau|V_s^n(0): V_s^n(0) \to N_c^n(0)$ ($=$fibre of N_c^n over $0 \in S_{m\omega}$) is a bijection.

Put $\dim\left(V_v^n(t) \cap V_s^n(t)\right) = d(t)$. Choose an orientation of $V_s^n(0)$. The bijection $T\tau|V_s^n(0)$ then determines an orientation of $N_c^n(0)$. We extend these orientations to all $V_s^n(t)$, $0 \leqslant t < m\omega$, via the map $D\phi_t$, and to all $N_c^n(t)$, $0 \leqslant t < m\omega$, via parallel translation.

Assume that $t_0 > 0$ is the first value with $d(t_0) \neq 0$. Then, for $\varepsilon > 0$ sufficiently small, $T\tau|V_s^n(t_0+\varepsilon): V_s^n(t_0+\varepsilon) \to N_c^n(t_0+\varepsilon)$ is a bijection which preserves or changes the orientation, according to whether $d(t_0)$ is even or odd. Since index $c = \sum_{0 \leqslant t < \omega} d(t)$ it follows that, for $\varepsilon > 0$ sufficiently small, $T\tau|V_s^n(m\omega-\varepsilon): V_s^n(m\omega-\varepsilon) \to N_c^n(m\omega-\varepsilon)$ is orientation-preserving if and only if index c is even. □

Let $c(t)$, $t\in\mathbb{R}$, be a geodesic, $|\dot{c}|=1$. We recall that c is said to have a *conjugate point* if there is a non-trivial Jacobi field $Y(t)$ along $c(t)$ with $Y(0)=Y(t_0)=0$, for some $t_0>0$. $c(t_0)$ is called *conjugate to* $c(0)$ *along* $c|[0,t_0]$, cf. [GKM].

An alternative definition of a conjugate point would use the orbit $\phi_t X_0$, $X_0=\dot{c}(0)$, in $T_1 M$. As above, the immersion

$$t\in\mathbb{R}\mapsto \phi_t X_0 \in T_1 M$$

induces bundles $\tau^{2n}: V^{2n}\to\mathbb{R}$, $\tau_h^n: V_h^n\to\mathbb{R}$, $\tau_v^n: V_v^n\to\mathbb{R}$, $\tau^{2n}=\tau_h^n\oplus\tau_v^n$. The existence of a conjugate point at $t=t_0>0$ then means that

$$D\phi_{t_0} V_v^n(0)\cap V_v^n(t_0)\neq 0.$$

We recall still another characterization of conjugate points due to Morse [Mor 2] (for a proof, see e.g. [GKM]):

5.3.3 Lemma. *Let* $c=c(t)$, $t\in\mathbb{R}$, $|\dot{c}|=1$, *be a geodesic on a Riemannian manifold.*

(i) *Assume there exists a* $\omega>0$ *and an* H^1*-vector field* ξ *along* $c|[0,\omega]$ *satisfying* $\xi(0)=\xi(\omega)=0$, ξ *not a Jacobi field,* ξ *not identically* 0, *and*

$$H_\omega(\xi,\xi):=\int_0^\omega \langle\nabla\xi(t),\nabla\xi(t)\rangle-\langle R(\xi(t),\dot{c}(t),\dot{c}(t)),\xi(t)\rangle dt\leqslant 0.$$

Then c has a conjugate point at some t_0, $0<t_0<\omega$

(ii) *Conversely, if c has a conjugate point at* $t_0>0$, *then, for every* $\omega>t_0$, *there exists a vector field* ξ *possessing the properties stated in* (i). □

We now prove the main result on manifolds of Anosov type (cf. [Kl 10]):

5.3.4 Theorem. *Assume that the geodesic flow* $\phi_t: T_1 M\to T_1 M$ *is of Anosov type.*

Claims. (i) *There are no null-homotopic closed geodesics on* M; *all closed geodesics are hyperbolic and have index* 0.

(ii) *There are no conjugate points on* M; *as a consequence, the universal covering* $\tilde{M}$ *of* M *is diffeomorphic to any tangent spaces via the exponential map.*

(iii) *For every* $X_0\in T_1 M$, *the stable fibre over* X_0 *and the vertical fibre over* X_0 *have only the* 0*-vector in common; the same is true for the unstable fibre.*

(iv) *Let* $k>0$ *be the exponent in the Anosov condition. There exist constants* $t_0>0$, $k'>0$, $k>k'$, *such that the following holds. Let* $Y(t)$ *be a Jacobi field along a geodesic* $c=(c(t))$, $|\dot{c}|=1$, $\langle\dot{c}(t),\ Y(t)\rangle=0$, $Y(0)=0$. *Then, for all* $t\geqslant t_0$,

$$|Y(t)|\geqslant|\nabla Y(t)|\sinh(k't).$$

(v) *Let* Λ' *be a connected component of* ΛM *which does not contain the constant curves. Then the critical set of* Λ' *consists of a single critical orbit* $S\,.\,c'$ *with* $E(c')=\inf E|\Lambda'$.

(vi) *The set* Per $T_1 M$ *of those* $X_0\in T_1 M$ *for which* $\phi_t X_0$ *is periodic, is dense in* $T_1 M$.

(vii) *The geodesic flow is transitive in the following sense. There exists a subset* Trans $T_1 M$ *of measure equal to the measure of* $T_1 M$ *such that, for* $X_0\in$ Trans $T_1 M$, *the orbit* $\phi_t X_0$, $t\geqslant 0$ (*or* $\phi_t X_0$, $t\leqslant 0$) *is dense in* $T_1 M$.

(viii) *The fundamental group* $\pi_1 M$ *of* M *has exponential growth; every non-trivial Abelian subgroup of* $\pi_1 M$ *is infinite cyclic.*

Note. The assertions (vi) and (vii) are among the main results of Anosov [An] for flows satisfying the so-called Anosov-hypothesis, provided the flow is measure preserving; we will not reproduce their proofs here. Rather, we concentrate on the proof of the assertions (i)–(v), which will make essential use of Anosov's results (vi) and (vii).

Proof. (cf. [Kl 12]) The assertion (i) is equivalent to saying that there are no closed geodesics on the universal covering $\tilde{M}$ of M. To prove this latter statement we observe that the universal covering of the stable bundle, $\tilde{\tau}_s^n: T_s^n T_1 \tilde{M} \to T_1 \tilde{M}$, is orientable. Hence, if $\tilde{c}$ is a closed geodesic on $\tilde{M}$, its associated stable bundle and its normal bundle both are orientable. Thus, according to (5.3.2), index $\tilde{c}$ is even.

Now let $\tilde{c}_0$ be a closed geodesic on $\tilde{M}$. Then, as we just saw, index $\tilde{c}_0$ is even. Looking at the proof of Fet's Theorem (4.1.8), we see that if $\tilde{c}_0$ is such a geodesic and if, in addition, $\tilde{c}_0$ has minimal index and minimal E-value among these, then there also exists a closed geodesic $\tilde{c}_1$ on $\tilde{M}$ with index $\tilde{c}_1 =$ index $\tilde{c}_0 + 1$ which is odd — a contradiction.

To prove (ii), we denote by Conj $T_1 M$ the set of $X_0 \in T_1 M$ for which the geodesic $c(t) = \tau \circ \phi_t X_0$ has a conjugate point. Clearly, Conj $T_1 M$ is open. Moreover, we see from (5.3.3) that ϕ_{-t} Conj $T_1 M \subset$ Conj $T_1 M$, for $t \geqslant 0$.

Now assume that $\phi_t: T_1 M \to T_1 M$ is of Anosov type and Conj $T_1 M$ is non-empty. From (vii), it then follows that Conj $T_1 M$ is open and dense. (vi) therefore implies the existence of a periodic orbit $\phi_t X_0$, $0 \leqslant t \leqslant \omega$, for which the underlying geodesic $c(t) = \tau \circ \phi_t X_0$, $t \in \mathbb{R}$, has a conjugate point at, say, $t_0 > 0$.

We claim that index $c > 0$, where we denote by c the prime closed geodesic $c(t)$, $0 \leqslant t \leqslant \omega$. Since c is hyperbolic and therefore index $c^m = m$ index c, this is equivalent to saying that index $c^m > 0$ for some $m \geqslant 1$.

Now choose m so large that $m\omega > t_0$. From (5.3.3), it then follows that there exists a non-trivial periodic vector field $\xi(t)$, $0 \leqslant t \leqslant m\omega$, along $c(t)$, $0 \leqslant t \leqslant m\omega$, satisfying $D^2 E(c^m)(\xi, \xi) \leqslant 0$. Since $\xi \notin$ nullspace $D^2 E(c^m)$, this proves our claim.

Denote by Λ' the connected component of ΛM containing the non-null-homotopic closed geodesic c of index $c > 0$. According to (2.1.3), there also exists a closed geodesic c' in Λ' with $E(c') = \inf E|\Lambda' : = \kappa' > 0$, index $c' = 0$.

We claim that index c and index c' have the same parity, hence, index $c = 2k > 0$. To see this, we observe that the normal bundles of c and c' are either both orientable or both non-orientable. The same is true for the stable bundles associated with c and c', since these bundles can be viewed as quotients by the same element of the fundamental group operating on the bundle $\tilde{\tau}_s^n: T_s^n T_1 \tilde{M} \to T_1 \tilde{M}$, reduced to the universal covering of c and c', respectively.

Now let c be a closed geodesic in the Λ' under consideration which has minimal positive index, say $2k$. Among these, c shall have minimal E-value. From the negative disc, or, better, the unstable manifold, at c, we obtain a non-trivial spherical cycle

$$u: (S^{2k}, *) \to (\Lambda', c').$$

Composition of this cycle with the usual projection map $\Lambda M \to M$ gives a non-trivial cycle

$$v:(S^{2k},*)\to(M,c'(0)).$$

But then we obtain from (2.1.8) a null-homotopic closed geodesics on M – a contradiction to (i). This proves (ii), except for the last statement, which is the usual extension of the Hadamard-Cartan theorem, cf. [GKM].

To prove (iii) we assume the contrary, i.e. there is a geodesic $c(t)$, $|\dot{c}|=1$ on M, such that, for $X_1=\dot{c}(1)$, we have

$$V^n_{X_1 s}\cap V^n_{X_1 v}\neq 0.$$

That is to say, there exists a non-trivial $\tilde{Y}(t)\in V^n_s(t)$ with $Y(1)=0$. Choose $\omega \gg 1$. Define a H^1-vector field ξ along $c|[0,\omega]$ as follows:

$$\xi(t)=0 \quad \text{for} \quad 0\leqslant t\leqslant 1; \quad \xi(t)=Y(t) \quad \text{for} \quad 1\leqslant t\leqslant \omega-1;$$

$$\xi(t)=(\omega-t)\,Y(\omega-1) \quad \text{for} \quad \omega-1\leqslant t\leqslant \omega.$$

We then find for $|H_\omega(\xi,\xi)|$, (5.3.3), an estimate of the form $|H_\omega(\xi,\xi)|\leqslant b|\tilde{Y}(\omega)|^2$ with $b>0$ independent of the geodesic c and the choice of ω. Define a vector field $\eta(t)$ along $c(t)$, $0\leqslant t\leqslant \omega$, by $\eta(0)=0$, $\eta(t)=0$ for $2\leqslant t\leqslant \omega$, and $H_\omega(\xi,\eta)=1$. For the vector field $\zeta(t):=\eta(t)-H_\omega(\eta,\eta)\,\xi(t)$ we then find that

$$H_\omega(\zeta,\zeta)=H_\omega(\eta,\eta)\,(-1+H_\omega(\eta,\eta)\,H_\omega(\xi,\xi)),\ |H_\omega(\xi,\xi)|\leqslant ba|\tilde{Y}(0)|e^{-2k\omega}.$$

If $H_\omega(\eta,\eta)$ is not already $\leqslant 0$, certainly, for sufficiently large ω, we have $H_\omega(\zeta,\zeta)\leqslant 0$; thus, according to (5.3.3), c has a conjugate point, in contradiction to (ii). This completes the proof of (iii).

To prove (iv), we first show that the Jacobi field $Y(t)$ under consideration satisfies the relation

$$(\dagger) \qquad |\nabla Y(t)|\leqslant k_1|Y(t)|$$

with $-k_1^2$ being a lower bound for the sectional curvature on M. To see this we observe that the vertical space $V^n_v(0)$ over $0\in\mathbb{R}$ is Lagrangian and, therefore, the same is true for all $D\phi_t V^n_v(0)$. We know, moreover, from (ii), that $D\phi_t V^n_v(0)\cap V^n_v(t)=0$ for all $t>0$. Hence, as in the proof of (5.3.1), there exists a symmetric (n,n)-matrix $S(t)$ for $t>0$, satisfying a Riccati equation. This equation, in turn, allows us to derive an upper bound for the spectrum of $S(t)$ of the form $\leqslant k_1$.

Now consider $\tilde{Y}(t)=(Y(t),\nabla Y(t))$. In the decomposition $\tilde{Y}(t)=\tilde{Y}_s(t)+\tilde{Y}_u(t)\in V^n_s(t)+V^n_u(t)$ we have $\tilde{Y}_u(0)\neq 0$, according to (iii). Hence, using (†) and the inequalities for stable and unstable fields, we obtain

$$|Y(t)|^2\geqslant|\tilde{Y}(t)|^2/(1+k_1^2)$$

and

$$|\tilde{Y}(t)| \geqslant |\tilde{Y}_u(t)| - |\tilde{Y}_s(t)| \geqslant a(|\tilde{Y}_u(0)|e^{kt} - |\tilde{Y}_s(0)|e^{-kt}) \geqslant |\tilde{Y}(0)| \sinh(k't),$$

the last inequality being valid only for sufficiently large t. Thus we have proved (iv).

(v) is now a simple consequence of (i). In fact, if there were two different critical orbits $S.c_1'$, $S.c_2'$, in a connected component Λ' of ΛM, $\Lambda^0 M \not\subset \Lambda'$, there would have to exist a homotopy in Λ' from c_1' to c_2'. Both these geodesics have index 0; the homotopy, when deformed to its critical level, would give rise to a closed geodesic of index 1, which is impossible.

For the proof of (viii) we refer to [Kl 12]. (vi) and (vii) are proved in [An].

Thus we have completed the proof of (5.3.4). □

We say that a compact differentiable manifold M is of *hyperbolic type* if there exists a Riemannian metric g^* on M such that (M, g^*) (i.e. M endowed with the Riemannian metric g^*) possesses, in its tangent bundle, a geodesic flow of Anosov type.

In the paper [Kl 10], we defined the concept 'hyperbolic' by the possibly stronger condition that there exists a metric g^* for which (M, g^*) has strictly negative curvature. Our next goal is to investigate the geodesic flow and, in particular, the periodic orbits on a Riemannian manifold where the underlying differentiable manifold is of hyperbolic type.

One can easily show that for orientable surfaces, "hyperbolic type" is equivalent with "genus >1". This case was investigated by Morse [Mor 1]. See [Kl 10] for further references.

5.3.5 Lemma. *Let g^* and g be two Riemannian metrics on a compact differentiable manifold. Assume that the geodesic flow on $T_1(M, g^*)$ is of Anosov type. Let $(\tilde{M}, g^*)$, $(\tilde{M}, g)$ be the Riemannian universal covering of (M, g^*) and (M, g), respectively.*

Claim. *There exists a constant $\rho > 0$ having the following property. Let p and q be any two points on $\tilde{M}$ and let c^* and c be minimizing geodesics from p to q in $(\tilde{M}, g^*)$ and $(\tilde{M}, g)$, respectively. Then every point on c has g-distance $\leqslant \rho$ from c^*.*

Note. $(\tilde{M}, g^*)$ has no conjugate points. Hence, there is precisely one geodesic c^* joining p and q. By g-distance and g-geodesic on $\tilde{M}$ we mean distance and geodesic on $(\tilde{M}, g)$. Similarly, we use the concept g^*-distance, etc.

Proof. (cf. [Kl 12] and [Es]). Since M is compact, there exist positive constants b, a, $b \leqslant 1 \leqslant a$, such that, for all $X \in TM$, we have the relations

$$(\dagger) \qquad b^2 g^*(X, X) \leqslant g(X, X) \leqslant a^2 g^*(X, X).$$

This implies that, on the universal covering $\tilde{M}$ of M, the length of curves measured in the g-metric and in the g^*-metric are uniformly comparable.

From (5.3.4(iv)), we conclude that there exists a constant $\sigma_0 > 0$ having the following property. Let pqr be a triangle in $(\tilde{M}, g^*)$ with the side qr having distance $\geqslant \sigma_0/2$ from p. Denote by $\tilde{M}'$ the surface of constant negative curvature $-k'^2$

with k' as in (5.3.4(iv)). There exists, up to congruence, a well-determined triangle $p'q'r'$ in M' of which the sides have the same length as in the g^*-triangle pqr. We claim that the angle $\measuredangle qpr$ at $p \leqslant$ the angle $\measuredangle q'p'r'$ at p'.

We refer to this result as the *Comparison Theorem for Large Triangles*. The proof is obtained as usual by integrating the relation (5.3.4(iv)), cf. [GKM] for similar techniques.

As a further auxiliary tool we need the following assertion.

Assertion. There exists a constant $\sigma \geqslant \sigma_0$ having the following property. Let $c^* = c^*(t)$, $t \in \mathbb{R}$, $|\dot{c}^*| = 1$, be a complete geodesic on $(\tilde{M}, g^*)$. Let $c_1 = c_1(t)$, $0 \leqslant t \leqslant 1$, be a geodesic segment on $(\tilde{M}, g)$ from $p = c_1(0)$ to $q = c_1(1)$ such that, for all $t \in [0,1]$, the g^*-distance $d^*(c_1(t), c^*) \geqslant \sigma$ and $d^*(p, c^*) = d^*(q, c^*) = \sigma$. Assume, moreover, that c_1 has g-length equal to $d(p,q)$. Then, g^*-projections p^* and q^* of p and q onto c^* have g^*-distance $d^*(p^*, q^*) < \sigma_0$.

Assume for the moment that the assertion holds. Then $\rho := 6a\sigma$, a as in (†), is a constant having the property required in the lemma. To see this, assume that there is a point $r \in c$ with $d(r, c^*) > \rho$, hence, $d^*(r, c^*) > \rho/a = 6\sigma$, cf. (†). We can find a segment $c_1 = c_1(t)$, $0 \leqslant t \leqslant 1$, on the geodesic c satisfying the hypotheses of the assertion with $r = c_1(t_0)$, $0 < t_0 < 1$, in its interior. Projections p^* and q^* of $p = c_1(0)$ and $q = c_1(1)$ onto c^* then have g-distance $< \sigma_0$ from each other. Thus, the curve going from p to p^* to q^* to q has g^*-length $< 3\sigma$, hence g-length $< \rho/2$. But then, the geodesic segment c_1, containing the point $r = c_1(t_0)$ of g-distance $> \rho$ from c, cannot be a curve of minimal g-length from p to q.

It remains to prove the assertion. To do so we derive a contradiction from the assumption that there are arbitrarily large $\sigma \geqslant \sigma_0$ for which $d^*(p^*, q^*) \geqslant \sigma_0$, with σ_0 as above.

First of all, we take a g^*-equidistant subdivision $c_0^* = p^*, c_1^*, \ldots, c_r^* = q^*$ of the segment p^*q^* into segments of length $\geqslant \sigma_0$ and $< 2\sigma_0$. We can find a subdivision $t_0 = 0 < t_1 < \ldots < t_r = 1$ of $[0,1]$ such that the g^*-projection of $c_1(t_i)$ onto c^* yields the point $c_i^*, 0 \leqslant i \leqslant r$.

For each $i > 0$, we compare the g^*-angle at c_i^* of the large triangle $c_{i-1}^* c_i^* c_1(t_i)$ with the corresponding angle of the corresponding triangle in the hyperbolic plane $\tilde{M}'$. The latter angle is $< \gamma_0 < \pi/2$, where γ_0 is the base angle in the isosceles triangle in $\tilde{M}'$ having a base of length $2\sigma_0$ and sides going to infinity. Hence, the g^*-angle at c_i^* of the g^*-triangle $c(t_{i-1}) c_i^* c(t_i)$ is bounded away from zero.

Since the points of $c_1 | [t_{i-1}, t_i]$ have g^*-distance $\geqslant \sigma \geqslant \sigma_0$ from c_i^*, we obtain from (5.3.4(iv)) a lower bound for the g^*-length of $c_1 | [t_{i-1}, t_i]$ of the form $a_0 \sinh(k_0 \sigma)$, with positive constants a_0, k_0, independent of σ and i.

Using (†), we thus obtain a lower bound for the g-length of c_1 of the form $u(\sigma) = r a_0 \sinh(k_0 \sigma)/b$. Since c_1 is a minimizing g-curve from p to q, an upper bound for the g-length of c_1 is given by the g-length of the curve pp^*q^*q, i.e. by $v(\sigma) = 2a(\sigma + r\sigma_0)$. But for large σ, $v(\sigma) < u(\sigma)$. This completes the proof of (5.3.5). □

To formulate the main implications of (5.3.5), we consider the universal Riemannian covering $(\tilde{M}, g)$ of the compact Riemannian manifold (M, g). By $\gamma : (\tilde{M}, g) \to (M, g)$ we denote the covering map, i.e. the quotient with respect to the group $\Gamma \cong \pi_1 M$ of deck transformations. We use γ also to denote the induced map from $T_1(\tilde{M},g)$ onto $T_1(M,g)$. A geodesic $\tilde{c}$, $|\dot{\tilde{c}}|=1$, on $(\tilde{M},g)$, is said to be *minimizing* if, for any two of its points, the segment of $\tilde{c}$ between these points has length equal to the g-distance between these points.

Let $\tilde{c}^*$, $|\dot{\tilde{c}}^*|=1$, be a minimizing geodesic on $(\tilde{M},g^*)$ which is left invariant under some element $\tau \neq \mathrm{id}$ of Γ, i.e. $\tau\tilde{c}^*(t)=\tilde{c}^*(t+\omega)$ for some $\omega>0$. This is equivalent to saying that the projected geodesic $c^*=\gamma\tilde{c}^*$ on (M,g^*) is periodic, i.e. $c^*(t+\omega)=c^*(t)$. Now if $\tilde{c}$, $|\dot{\tilde{c}}|=1$, is a minimizing geodesic on $(\tilde{M},g)$ which has bounded distance from $\tilde{c}^*$, then we call $\tilde{c}$ *almost-invariant*, since also all the images $\tau^k\tilde{c}$, $k=\pm 1, \pm 2, \ldots$, of $\tilde{c}$ have bounded distance from $\tilde{c}$. The projection $c=\gamma\tilde{c}$ in (M,g) will be called *almost periodic* with respect to the element τ. By $\mathrm{Per}\, T_1(M,g)$, we denote the set of tangent vectors to almost periodic geodesics on (M,g), parameterized by arc length.

5.3.6 Theorem. *Let M be a compact manifold of hyperbolic type and let g be an arbitrary Riemannian metric on M. Let g^* be a Riemannian metric on M such that the geodesic flow on $T_1(M,g^*)$ is of Anosov type.*

Claim. (i) *Given any complete geodesic $\tilde{c}^*$, $|\dot{\tilde{c}}^*|=1$, on $(\tilde{M},g^*)$ (which, in particular, is also a minimizing geodesic), there exists a complete minimizing geodesic $\tilde{c}$, $|\dot{\tilde{c}}|=1$, on $(\tilde{M},g)$, having bounded distance from $\tilde{c}^*$. If $\tilde{c}^*$ is left invariant under some element $\tau \neq \mathrm{id}$ of Γ, then one can find a closed set $\tilde{C}$ of such minimizing geodesics $\tilde{c}$ which is left invariant under τ.*

(ii) *Let $\tilde{c}^*$, $|\dot{\tilde{c}}^*|=1$, be a complete geodesic on $(\tilde{M},g^*)$ such that the set $\{\dot{c}^*(t),\, t\in\mathbb{R}\}$ of tangent vectors is dense in $T_1(M,g^*)$, i.e. $\phi_t\dot{c}^*(0)$ is a dense orbit in $T_1(M,g^*)$. Then there exists among the minimizing geodesics of $(\tilde{M},g)$ with bounded distance from $\tilde{c}^*$ one, say $|\tilde{c}|$, $\dot{\tilde{c}}=1$, having the following properties. Put $\gamma\tilde{c}=c$ and denote by $T_{1,c}(M,g)$ the closure of the set $\{\dot{c}(t),\, t\in\mathbb{R}\}$ and define* $\mathrm{Per}\, T_{1,c}(M,g)=T_{1,c}(M,g)\cap \mathrm{Per}\, T_1(M,g)$. *Then*

(a) *$T_{1,c}(M,g)$ is ϕ_t-invariant;*

(b) *$\phi_t\, T_{1,c}(M,g)$ is topologically transitive. More precisely, the orbit $\phi_t\dot{c}(0)$ is dense in $T_{1,c}(M,g)$;*

(c) $\mathrm{Per}\, T_{1,c}(M,g)$ *is dense in $T_{1,c}(M,g)$;*

(d) *for every $\tau\in\Gamma$, $\tau\neq\mathrm{id}$, there exists an almost periodic geodesic c' with respect to τ such that $c'\in T_{1,c}(M,g)$.*

Remark 1. We see that the geodesic flow $\phi_t|T_{1,c}(M,g)$ has many properties in common with the geodesic flow of Anosov type on $T_1(M,g^*)$; the main difference is that, in general, $T_{1,c}(M,g)$ will only be a proper subset of $T_1(M,g)$.

As for property (d), we observe that to every non-trivial conjugacy class in Γ there corresponds a connected component $\Lambda'(M,g)$ of $\Lambda(M,g)$, with $\Lambda'(M,g)$ not containing the point curves. From (2.1.3), we know that there exists a closed geodesic c' in $\Lambda'(M,g)$ with $E(c')=\inf E|\Lambda'(M,g)$. A lift $\tilde{c}'$ of c' into $(\tilde{M},g)$ will then have bounded distance from a complete geodesic $\tilde{c}^*$ of $(\tilde{M},g^*)$, $\tilde{c}^*$ being invariant under some element τ in the conjugacy class under consideration. Also, $\tilde{c}'$ will be invariant under τ; however, $\tilde{c}'$ need not be minimizing in general. It is for this reason that we introduced above the concepts of an almost-invariant and an almost-periodic geodesic.

Remark 2. In the case where M is a compact orientable surface, Theorem (5.3.6) can be strengthened, as was shown by Morse [Mor 1]. Whereas in general one has only almost-periodic minimizing geodesics, we claim that in this particular case the geodesics are actually periodic. Thus, the set $\tilde{C}$ in (5.3.6(i)) consists of a single minimizing geodesic, invariant under the element $\tau \neq \mathrm{id}$ of Γ. And in (5.3.6(iid)), the geodesic c' can be assumed to be closed and of minimal length in its homotopy class.

The explanation for this phenomenon comes from Corollary 2 of the Index Theorem (3.2.12). Let c be a closed geodesic of minimal length in its free homotopy class on an orientable surface. c therefore has index 0 and is orientable. According to (3.2.14), c is then either hyperbolic or degenerate. From this we can conclude that all the multiple coverings c^q of c also have index 0. This will prove our claim; indeed, this is precisely the property which guarantees that a complete geodesic $\tilde{c}$ on the universal covering $(\tilde{M}, g)$ of (M, g) which projects onto the closed geodesic c on (M, g) is minimizing between any two of its points and not only between points $\tilde{c}(t)$, $\tilde{c}(t+\omega) = \tau\tilde{c}(t)$ having distance $\omega = \text{length } c$.

Now, if c is hyperbolic of index 0, also index $c^q = q$. index $c = 0$; see (3.2.13). If, on the other hand, c is degenerate, this means that there exists a non-trivial periodic Jacobi field $Y(t)$ along $c(t)$. Since index $c = 0$ and c is orientable, $Y(t) \neq 0$, for all t. Let $c^q = \{c(t);\ 0 \leqslant t \leqslant q\omega\}$ be the q-fold covering of c. Then any periodic vector field ξ along c^q may be written as $\xi(t) = w(t)Y(t)$, $0 \leqslant t \leqslant q\omega$. It follows that $D^2E(c^q)\,(\xi, \xi) \geqslant 0$ with $= 0$ only if $w(t) = w_0 = \text{const}$ (compare the end of the proof of (3.2.12)), i.e. ξ in the null space of the index form $D^2E(c^q)$.

Proof of (5.3.6) (i). For a given geodesic $\tilde{c}^*$ in $(\tilde{M}, g^*)$, we denote by $\tilde{c}_m$ a minimizing geodesic segment in $(\tilde{M}, g)$ from $\tilde{c}^*(-m)$ to $\tilde{c}^*(m)$, m an arbitrary integer $\geqslant 1$. From (5.3.5), we conclude that the sequence $\{\dot{\tilde{c}}_m(0)\}$ of tangent vectors stays in a bounded domain of $T_1(\tilde{M}, g)$; hence, it has a convergent subsequence. A limit point of such a sequence determines a complete minimizing geodesic $\tilde{c}$ in $(\tilde{M}, g)$ of bounded distance from $\tilde{c}^*$.

If there is a $\tau \neq \mathrm{id}$ in Γ with $\tau\tilde{c}^* = \tilde{c}^*$ then, starting with a geodesic $\tilde{c}$ of the type just constructed, the sequence $\{\tau^n\tilde{c}\}$ of minimizing geodesics of $(\tilde{M}, g)$ stays in a bounded domain along $\tilde{c}^*$. Hence, this sequence has accumulation elements. The set of such accumulation elements is clearly globally invariant under τ. In general, there exist invariant subsets in this set. If such an invariant subset consists of only a single complete geodesic, the projection into (M, g) gives a closed geodesic c which, together with all its iterates, is an element of minimal E-value in its connected component.

To prove (ii) we start with a minimizing geodesic $\tilde{c}$, $|\dot{\tilde{c}}| = 1$, in $(\tilde{M}, g)$, which has bounded distance from the geodesic $\tilde{c}^*$ in $(\tilde{M}, g^*)$. Put $\gamma\tilde{c} = c$, $T_{1,c}(M, g) = \text{closure of } \{\dot{c}(t),\ t \in \mathbb{R}\}$. We claim that $T_{1,c}(M, g)$ satisfies property (d).

To see this, we recall that, given any $\tilde{X}_0^* \in T_1(\tilde{M}, g^*)$, there exists a sequence $\{t_n\}$ of real numbers $\geqslant 0$ and a sequence $\{\tau_n\}$ of elements in Γ such that $\lim \tau_n\dot{\tilde{c}}^*(t_n) = \tilde{X}_0^*$. From (5.3.5), it follows that the sequence $\{\tau_n\dot{\tilde{c}}(t_n)\}$ is bounded, hence, it contains accumulation elements $\tilde{X}_0$. If, in particular, the g^*-geodesic $\tilde{c}_0^*$ determined by $\dot{\tilde{c}}^*(0) = \tilde{X}_0^*$ is invariant under some non-trivial $\tau \in \Gamma$, then the geodesic $\tilde{c}_0$ determined by $\dot{\tilde{c}}_0(0)=\tilde{X}_0$ is almost invariant under τ. The closure of the set of tangent vectors to the geodesic $c_0=\gamma\tilde{c}_0$ on (M,g) belongs to $T_{1,c}(M,g)$.

Since Per $T_1(M,g^*)$ is dense in $T_1(M,g^*)$, there exists a sequence $\{\tau_n\}$ of non-trivial elements of Γ and a sequence $\{\tilde{c}_n^*\}$ of complete geodesics of $(\tilde{M},g^*)$ with $\tilde{c}_n^*$ invariant under τ_n, such that lim $\dot{\tilde{c}}_n^*(0)=\dot{\tilde{c}}^*(0)$, $\tilde{c}^*$ being the original geodesic with which we started. For every n, we also have a τ_n-invariant set $\tilde{C}_n$ of τ_n-almost-invariant geodesics $\tilde{c}_n$ on $(\tilde{M},g)$. The set of tangent vectors to these $\tilde{c}_n$, when projected into $T_1(M,g)$, belongs to $T_{1,c}(M,g)$. The limit set of a sequence $\{\tau_n\tilde{c}_n\}$, $\tilde{c}_n\in\tilde{C}_n$, contains a geodesic $\tilde{c}'$ of $(\tilde{M},g)$ of bounded distance from $\tilde{c}$ and $\tilde{c}^*$.

Define, as before, $T_{1,c'}(M,g)$, with the help of $c'=\gamma\tilde{c}'$. Then $T_{1,c'}(M,g) \subset T_{1,c}(M,g)$ and $T_{1,c'}(M,g)$ again satisfies property (d).

By iterating this process we obtain a sequence $\tilde{c}=\tilde{c}^{(0)}$, $\tilde{c}'=\tilde{c}^{(1)}$, $\tilde{c}^{(2)},\ldots$, of minimizing geodesics of $(\tilde{M},g)$, all having bounded distance from $\tilde{c}^*$. Thus, there must be limit geodesics. Denote such a limit again by $\tilde{c}$. The set $T_{1,c}(M,g)$, defined as above as the closure of the tangent vectors of $c=\gamma\tilde{c}$, now has not only property (d) but also property (c), as follows immediately from the construction of $\tilde{c}$ as limit of the $\tilde{c}^{(k)}$. Properties (a) and (b) are clear from the definition of $T_{1,c}(M,g)$. This concludes the proof of (5.3.6). □

The geodesic flow for Riemannian manifolds without conjugate points has been the object of a great number of investigations. The special case for which M is a closed surface of negative curvature has already been investigated by Hadamard [Had]; of the later papers we mention besides [An] and [AnS], only Cartan [Ca 2], Hedlund [He 2], Busemann [Bu], Green [Gre 1,2] and Eberlein [Eb 2].

Note that, in particular, a flat Riemannian manifold possesses no conjugate points. A fundamental result of Bieberbach, cf. [Wo], states that a compact flat Riemannian manifold has as a finite covering a torus. We take this property to define a manifold of parabolic type. More precisely, we say that a compact differentiable manifold is of *parabolic type* if M possesses a finite, and hence compact, covering with Abelian fundamental group.

Of the numerous results on Riemannian manifolds of parabolic type, we present here only the following, which is due to Busemann [Bu].

5.3.7 Theorem. *Let M be a compact Riemannian manifold without conjugate points. Assume that the fundamental group $\Gamma\equiv\pi_1 M$ is Abelian.*

Claim. *For every element $\tau\neq$ id of Γ, there exists in the universal covering $\tilde{M}$ of M, a simple covering by complete τ-invariant minimizing geodesics. That is to say, through every $\tilde{p}\in\tilde{M}$ there exists exactly one geodesic $\tilde{c}=\tilde{c}(t)$, $t\in\mathbb{R}$, with $\tilde{c}(0)=\tilde{p}$, $|\dot{\tilde{c}}|=1$, $\tau\tilde{c}(t)=\tilde{c}(t+\omega)$ with $\omega>0$ and independent of the choice of $\tilde{p}\in\tilde{M}$. Moreover, for any pair $t_0,t_1,t_0<t_1$, of real numbers, $d(\tilde{c}(t_0),\tilde{c}(t_1))=t_1-t_0$.*

As a consequence we have for every $\tau\notin$id in Γ which is prime, i.e. which can not be written as $\tau=\tau_0^m$ for some integer $m>1$, some $\tau_0\in\Gamma$, a simple covering of M by prime closed geodesics c without self-intersections, all having the same length equal to the period ω of the map $\tau:\tau\tilde{c}(t)=\tilde{c}(t+\omega)$.

Note. The behavior of the geodesic flow on T_1M in this theorem is very similar to the behavior of the geodesic flow on a flat torus. In fact, E. Hopf [Ho] has shown that, in the case $\dim M = 2$, an orientable manifold M, satisfying the hypotheses of the theorem, is actually isometric to a flat torus.

For an orientable manifold M of arbitrary dimension $n+1$ satisfying the hypothesis of the theorem, one can show, with the help of formula (3.2.16), that the linear Poincaré map P for a closed geodesic c on M has the form

$$P = \begin{pmatrix} E_n & E_n \\ 0_n & E_n \end{pmatrix}, \text{ where } E_n \text{ is the } n \times n \text{ unit matrix.}$$

This is precisely the form of the Poincaré map on a flat torus. If one could show that, in addition, the Lagrangian P-invariant subspace is the horizontal subspace, for any choice of the initial point of the geodesics, it would follow that M is flat.

Proof. For a given $\tau \neq \mathrm{id}$ of Γ we consider the function $f_\tau : \tilde{M} \to \mathbb{R} : \tilde{p} \mapsto d(\tilde{p}, \tau\tilde{p})$. Since Γ is Abelian, we have, for every $\sigma \in \Gamma$, $f_\tau(\tilde{p}) = f_\tau(\sigma\tilde{p})$. Thus, f_τ assumes its supremum ω at a point $\tilde{p}$ of a compact fundamental domain in $\tilde{M}$.

Let $\tilde{c}$, $|\dot{\tilde{c}}| = 1$, be the complete geodesic with $\tilde{c}(0) = \tilde{p}$, $\tilde{c}(\omega) = \tilde{q} = \tau\tilde{p}$. Put $\tilde{c}(2\omega) = \tilde{r}$. From

$$\omega \geqslant d(\tilde{r}, \tau\tilde{r}) \geqslant d(\tau\tilde{p}, \tau\tilde{r}) - d(\tau\tilde{p}, \tilde{r}) = 2\omega - \omega$$

it follows that the triangle $\tilde{r}\tau\tilde{p}\tau\tilde{r}$ is degenerate, i.e. $\tau\tilde{r} = \tilde{c}(3\omega)$, $\tilde{c}$ is invariant under τ.

Let now $\tilde{q}$ be an arbitrary point of $\tilde{M}$. For every positive integer m we have

$$\begin{aligned} m d(\tilde{p}, \tau\tilde{p}) &= d(\tilde{p}, \tau^m\tilde{p}) \leqslant d(\tilde{p}, \tilde{q}) + d(\tilde{q}, \tau\tilde{q}) + \ldots + d(\tau^m\tilde{q}, \tau^m\tilde{p}) \\ &= 2d(\tilde{p}, \tilde{q}) + m d(\tilde{q}, \tau\tilde{q}). \end{aligned}$$

Dividing by m and taking the limit for $m \to \infty$ it follows that $d(\tilde{q}, \tau\tilde{q}) = d(\tilde{p}, \tau\tilde{p}) = \omega = \sup f_\tau$. Thus, the geodesic $\tilde{c}'$, $|\dot{\tilde{c}}'| = 1$, from $\tilde{q}$ to $\tau\tilde{q}$ is also invariant under $\tilde{c}$.

The second part of the theorem, concerning M, follows immediately from the first part. □

Appendix. The Theorem of Lusternik and Schnirelmann

One owes to Lusternik and Schnirelmann a general axiomatic theory of the calculus of variations; see [LS 2], [Ly]. The greatest achievements of this theory, as applied to closed geodesics, are the so-called Lusternik-Schnirelmann Theorem that there always exist three non-self-intersecting closed geodesics on a surface of genus 0 (see [LS 1]) and the Lyusternik-Fet Theorem on the existence of at least one closed geodesic on every compact Riemannian manifold (see [LF]).

In this appendix we prove these theorems using the Lusternik-Schnirelmann theory. We have made this appendix completely independent of the main part of these Lectures, where we developed the extended Morse theory of the Riemannian Hilbert manifold ΛM of closed H^1-curves with its functional E.

It will become evident at once—and this is true in particular for the proof of the Lyusternik-Fet Theorem—that the Lusternik-Schnirelmann approach to the theory of closed geodesics is much more elementary and less complicated than Morse's approach in its extended version dealing with $(\Lambda M, E)$. But then, the Morse theory of $(\Lambda M, E)$ is much more sensitive to the finer aspects of the closed geodesic problem—witness the existence theorem of infinitely many closed geodesics in Chapter 4. The source of the power of this approach is the Morse complex (see (2.5)).

So, upon comparison, the extension of Morse's ideas must be considered better adapted to the theory of closed geodesics than the Lusternik-Schnirelmann theory, with one noteworthy exception, namely the Lusternik-Schnirelmann Theorem on the existence of non-self-intersecting closed geodesics. Morse's theory does not seem to be suitable for finding such special prime closed geodesics.

Section 1 of this appendix gives a brief, elementary introduction to the Lusternik-Schnirelmann theory as relevant to the closed geodesics problem. We obtain a short proof of the Lyusternik-Fet Theorem.

In the second section, we consider the length-decreasing (or rather, energy-decreasing) deformations in the class of non-self-intersecting closed curves on the 2-sphere. Our deformations are considerably more elementary than the one used in [LS 2] and [Ly].

In Section 3, finally, we give a complete elementary proof of the Lusternik-Schnirelmann Theorem. The topological prerequisites are kept to a minimum. In particular, we have avoided any reference to Pontrjagin's Theorem of the "removal of cycles", which plays an essential role in the earlier proofs.

We conclude this introduction by drawing attention to the fact that the Lusternik-Schnirelmann theory can be applied immediately to Finsler manifolds, even with a non-symmetric Finsler structure. It is doubtful that the Morse theory of $(\Lambda M, E)$ can also be extended fully from Riemannian manifolds to Finsler manifolds. For a partial extension, we refer to Mercuri [Mer].

In the case of a non-symmetric Finsler structure on S^2 the methods of Lusternik and Schnirelmann yield the existence of only two non-self-intersecting closed geodesics; this reflects the fact that the space of oriented unparameterized great circles on S^2 is isomorphic to S^2 and therefore its $\mathbb{Z}_2$-homology is 2-dimensional, whereas the space of non-oriented unparameterized great circles on S^2 is isomorphic to the projective plane P^2 and thus, its $\mathbb{Z}_2$-homology is 3-dimensional. An example due to Katok [Kat] shows that there actually exist non-symmetric Finsler structures on S^2 with only two closed geodesics (not counting the multiple coverings). See Matthias [Mat] for an elementary presentation of this example.

Using the methods of Lusternik-Schnirelmann, as presented in this appendix, Thorbergsson [Thr] has investigated the existence of closed geodesics on non-compact Riemannian manifolds.

In preparing this appendix the assistance of W. Ballmann was very valuable. He has applied the Lusternik-Schnirelmann method to construct non-self-intersecting closed geodesics on surfaces of arbitrary genus; see [Ba].

A.1 The Space *PM* and the Theorem of Lyusternik and Fet

Let M be a compact Riemannian manifold. The set $C^0(S, M)$ of continuous maps $c: S = [0,1]/\{0,1\} \to M$ is in a natural way a metric space (Banach manifold) by taking as distance

$$d_\infty(c, c') = \sup_{t \in S} d\big(c(t), c'(t)\big).$$

Denote by PM the subspace of $C^0(S, M)$ formed by the piecewise differentiable closed curves. For $c \in PM$ we have the functions

$$L(c) = \int_S |\dot{c}|\, dt, \quad \text{and} \quad E(c) = \tfrac{1}{2} \int_S |\dot{c}|^2\, dt.$$

They are related by

$$L(c) \leq \sqrt{2E(c)}$$

with equality if and only if c is parameterized proportionally to arc length, i.e. if $L\big(c|[0,t]\big) = t\,L(c)$.

Since M is compact, there exists an $\eta > 0$ such that any two points p and q on M with $d(p,q) \leq 2\eta$ can be joined by an uniquely determined geodesic $c_{pq} = c_{pq}(t)$, $0 \leq t \leq 1$, $L(c_{pq}) = \sqrt{2E(c_{pq})} = d(p,q)$. Here, as always, we assume that geodesics are parameterized proportionally to arc length.

A.1.1 Proposition. *Let* $\{c_n\}$ *be a sequence of piecewise differentiable paths* $c_n:[0,1]\to M$. *Put* $c_n(0)=p_n$, $c_n(1)=q_n$ *and assume that* $d(p_n,q_n)\leqslant\eta$, *for all* n. *If the sequences* $\{E(c_n)\}$ *and* $\{d^2(p_n,q_n)/2\}$ *are both convergent with the same limit, then* $\{c_n\}$ *possesses a convergent subsequence whose limit is a geodesic segment* $c:[0,1]\to M$ *of length* $\leqslant\eta$ *equal to the distance from* $p=c(0)$ *to* $q=c(1)$.

Proof. Consider the sequence $\{c_{p_nq_n}\}$ of unique minimizing geodesic segments from p_n to q_n. Since M is compact there exists a subsequence, which we denote again by $\{c_{p_nq_n}\}$, such that $\lim p_n=p$ and $\lim q_n=q$ exist and hence $\{c_{p_nq_n}\}$ converges to the minimizing segment c_{pq}.

Consider $t_0\in]0,1[$. We have to show that for every sequence $\{t_n\}$ on $[0,1]$ with $\lim t_n=t_0:\lim c_n(t_n)=c_{pq}(t_0)$. To see this put $c_n(t_n)=r_n$ and let $\{r_{n(k)}\}$ be a convergent subsequence of $\{r_n\}$. Let r be its limit. The sequences

$$\{c_{p_{n(k)}r_{n(k)}}:[0,t_{n(k)}]\to M\}$$

$$\{c_{r_{n(k)}q_{n(k)}}:[t_{n(k)},1]\to M\}$$

of unique minimizing geodesics converge to $c_{pr}:[0,t_0]\to M$ and $c_{rq}:[t_0,1]\to M$, respectively.

From our hypothesis we have

$$\begin{aligned}E(c_{pr})+E(c_{rq})&=\lim\{E(c_{p_{n(k)}r_{n(k)}})+E(c_{r_{n(k)}q_{n(k)}})\}\\&\leqslant\lim E(c_{n(k)})=E(c_{pq})\end{aligned}$$

Since $c_{pq}:[0,1]\to M$ is the unique geodesic segment of minimal E-value from p to q it follows that the curve $c_{pr}\cup c_{rq}$ from p to q coincides with c_{pq}. In particular, the value r for the parameter $t=t_0$ must be equal to $c_{pq}(t_0)$. This completes the proof of (A.1.1). □

Let $\eta>0$ be as above. Fix $\kappa>0$ and choose an even integer $k>0$ with $4\kappa/k\leqslant\eta^2$. For every $c\in P^\kappa M=\{c\in PM;E(c)\leqslant\kappa\}$ and every $t_0\in S$, we have

$$d^2\big(c(t_0),c(t_0+2/k)\big)\leqslant 2E(c)\,.\,(2/k)\leqslant\eta^2.$$

Here, as always, t is to be taken modulo 1.

We now proceed to define a continuous E-decreasing deformation from $P^\kappa M$ into itself. Let j be an even integer, $0\leqslant j\leqslant k-2$. For $\sigma\in[j/k,(j+2)/k]$, we define $\mathscr{D}_\sigma c$, $c\in P^\kappa M$, by

$$\mathscr{D}_\sigma c(t)=c(t),\quad\text{for}\quad t\in[0,j/k]\quad\text{or}\quad t\in[\sigma,1],$$

$$\mathscr{D}_\sigma c|[j/k,\sigma]=c_{c(j/k)c(\sigma)}|[j/k,\sigma].$$

Here, as always, c_{pq} denotes the minimizing geodesic from p to q with the indicated parameter interval. Note that $d\big(c(j/k),c(\sigma)\big)\leqslant\eta$.

We continue by defining $\mathscr{D}_\sigma c$ for $\sigma\in[1,2]$. Let j again be even, $0\leqslant j\leqslant k-2$, and take $t\in[1,1+1/k]$ to mean $t\in[0,1/k]$. Then, for $\sigma\in[1+j/k,1+(j+2)/k]$, $c\in P^\kappa M$, put

$$\mathscr{D}_\sigma c(t)=c(t), \quad \text{for} \quad t\in[1/k,(j+1)/k] \quad \text{or} \quad t\in[\sigma-1+1/k,1+1/k],$$

$$\mathscr{D}_\sigma c|[(j+1)/k,\sigma-1+1/k]=c_{c((j+1)/k)c(\sigma-1+1/k)}|[(j+1)/k,\sigma-1+1/k].$$

We now define $\mathscr{D}(\sigma,c)$ for $\sigma\in[0,2]$, $c\in P^\kappa M$, to be the subsequent application of the mappings $\mathscr{D}_{2/k},\ldots,\mathscr{D}_{2l/k},\mathscr{D}_\sigma$, where $2l$ is the even integer determined by $2l\leqslant k\sigma<2^l+2$.

A.1.2 Proposition. *The mapping*

$$\mathscr{D}:[0,2]\times P^\kappa M\to P^\kappa M$$

is continuous. Moreover, $E(\mathscr{D}(2,c))\leqslant E(c)$, with equality if and only if c is either a constant map or a closed geodesic, i.e. $c:S\to M$ is an immersion satisfying $\nabla\dot{c}=0$, $\dot{c}\neq 0$.

Proof. The continuity of $\mathscr{D}$, with $P^\kappa M$ being considered as subspace of $C^0(S,M)$, is obvious from the definition.

Let p,q be points with $d(p,q)\leqslant\eta$. If c is a piecewise differentiable path from p to q then

$$E(c_{pq})\leqslant E(c)$$

with equality if and only if $c_{pq}=c$. This implies the last statement of (A.1.2). □

A.1.3 Lemma. *Choose $\kappa>0$ and denote by $\mathscr{D}$ the deformation $\mathscr{D}(2,\ )$ introduced above. Let $\{c_n\}$ be a sequence in $P^\kappa M$ such that $\{E(c_n)\}$ and $\{E(\mathscr{D}c_n)\}$ are both convergent with the same limit $\kappa_0>0$. Then $\{c_n\}$ possesses a convergent subsequence whose limit is a closed geodesic c_0, $E(c_0)=\kappa_0$.*

Proof. To define $\mathscr{D}$ we chose an even $k>0$ such that $4\kappa/k\leqslant\eta^2$. Therefore every $\mathscr{D}c_n$ consists of the $k/2$ geodesic segments $\mathscr{D}c_n|[(j+1)/k,(j+3)/k]$, $j=0,2,\ldots,k-2$. Here, $t\in[1,1+1/k]$ is to be read as $t-1$. From (A.1.1), we obtain the existence of a convergent subsequence of $\{\mathscr{D}c_n\}$ which we again denote by $\{\mathscr{D}c_n\}$. The limit c_0 of this sequence also consists of $k/2$ geodesic segments. Moreover, (A.1.1) implies that c_0 is also the limit of the sequence $\{c_n\}$. Since $E(\mathscr{D}(c_0))=E(c_0)=\kappa_0>0$, we obtain from (A.1.2) that c_0 is a closed geodesic. □

A.1.4 Lemma. *Let $\kappa_0>0$ and let $\mathscr{U}$ be an open neighborhood of the set C of closed geodesics c with $E(c)=\kappa_0$. In the case $C=\emptyset$, one may choose $\mathscr{U}=\emptyset$. Let $\kappa>\kappa_0$ and consider the deformation $\mathscr{D}=\mathscr{D}(2,\ )$ on $P^\kappa M$, introduced above. Then there exists a $\varepsilon=\varepsilon(\mathscr{U})$ such that*

$$\mathscr{D}P^{\kappa_0+\varepsilon}M\subset\mathscr{U}\cup P^{\kappa_0-\varepsilon}M.$$

Proof. Since $\mathscr{D}$ is continuous and $\mathscr{D}|C=id$ there exists an open neighborhood $\mathscr{U}'$ of C, $\mathscr{U}'\subset\mathscr{U}$ with $\mathscr{D}\mathscr{U}'\subset\mathscr{U}$. If there were no $\varepsilon>0$ with the desired property this would imply the existence of a sequence $\{c_n\}$, $c_n\notin\mathscr{U}'$, with

$$\kappa_0-1/n\leqslant E(\mathscr{D}c_n)\leqslant E(c_n)\leqslant\kappa_0+1/n.$$

From (A.1.3) we obtain that $\{c_n\}$ has a convergent subsequence with the limit point c being a closed geodesic. Since $E(c)=\kappa_0$ and $c\notin\mathcal{U}'$, this is impossible.

In particular, if there exists no closed geodesic c with $E(c)=\kappa_0$, then (A.1.3) implies that there must exist a $\lambda>0$ such that there are no closed geodesics with E-value in $[\kappa_0-\lambda,\kappa_0+\lambda]$. □

A.1.5 Theorem (*Lyusternik and Fet* [LF]). *On every compact Riemannian manifold M, there exists a closed geodesic.*

Proof. We first consider the case $\pi_1 M\neq 0$. That is to say, there exists $c\in PM$ which is not homotopic to a constant map of the circle into M. Denote by $P'M$ the space of all $c'\in PM$, freely homotopic to c. Put $\inf E|P'M=\kappa'$. Then $\kappa'>0$. Indeed, otherwise there exists $c'\in P'M$ with $L(c')\leqslant\sqrt{2E(c')}<\eta$, $\eta>0$ as above. But a closed curve of length $<\eta$ can be retracted onto its initial equal end point.

We claim that the set C' of closed geodesics c' with $E(c')=\kappa'$ is not empty. This follows from (A.1.4), since otherwise there exists $\varepsilon>0$ such that $\mathcal{D}(P^{\kappa'+\varepsilon}M)\subset P^{\kappa'-\varepsilon}M$. Since $\mathcal{D}$ carries $P'^{\kappa}M$ into itself, we obtain a contradiction to the definition of κ'.

Now assume $\pi_1 M=0$. Then there exists a smallest integer k, $1<k<\dim M$, such that $\pi_{k+1}M\neq 0$, cf. (2.1.5). Let

$$f: S^{k+1}\to M$$

be a homotopically non-trivial differentiable mapping. f induces a continuous map

$$F=F(f):(D^k,\partial D^k)\to(PM,P^0M)\subset\big(C^0(S,M),P^0M\big)$$

as follows (cf. (2.1)). First identify the k-disc D^k with the half-equator $\{x_0\geqslant 0, x_1=0\}$ on the unit sphere S^{k+1} of $\mathbb{R}^{k+2}$ with coordinates $(x_0,\ldots,x_{k+1})$. Associate to every $p\in D^k\subset S^{k+1}$ the parameterized circle $a_p(t)$, $t\in S$, which starts from p orthogonally to the hyperplane $\{x_1=0\}$ into the half-sphere $\{x_1\geqslant 0\}$. Note that for $p\in\partial D^k$, a_p is the trivial (=constant) circle $a_p(t)=p$. Now define $F(p)(t)$ to be $f\circ a_p(t)$.

Consider a homotopy of F, i.e. a continuous map

$$\Phi:[0,1]\times(D^k,\partial D^k)\to(PM,P^0M)$$

with $\Phi|\{0\}\times D^k=F$. Put $\Phi|\{\sigma\}\times D^k=F^\sigma$. We claim that the homotopy $F^\sigma=F^\sigma(f)$ of $F(f)=F^0(f)$ determines a homotopy f^σ of $f=f^0$ such that $F(f^\sigma)=F^\sigma(f)$.

Indeed, since every $q\in S^k$ possesses the representation $q=a_p(t)$, $p\in D^k$, simply put

$$f^\sigma(q)=f^\sigma\big(a_p(t)\big)=F^\sigma(p)(t).$$

There exists $\kappa>0$ such that $E|F(D^k)<\kappa$. Choose an even integer $k>0$ satisfying $4\kappa/k\leqslant\eta^2$ and consider the deformation $\mathcal{D}=\mathcal{D}(2,\ )$ defined on $P^\kappa M$. In particular, consider $\mathcal{D}F(D^k)$ and its iterates

$$\mathcal{D}^{n+1}F(D^k)=\mathcal{D}\big(2,\mathcal{D}^nF(D^k)\big).$$

Put

$$\kappa_0 = \lim_{n\to\infty} \max E|\mathscr{D}^n F(D^k).$$

We claim that $\kappa_0 > 0$. Otherwise we get a contradiction as follows. $\kappa_0 = 0$ implies $E|\mathscr{D}^n F(D^k) < \eta^2/2$, for all sufficiently large n, with $\eta > 0$ as above, i.e. for every $p \in D^k$, the closed curve $\mathscr{D}^n F(p)$ has length $\leq \eta$ and can therefore be retracted into its initial equal end point. That is to say, F is homotopic to a map F^* with $F^*(D^k) \subset P^0 M = M$, which implies that F is homotopic to a constant map, the image of which is a single point in $P^0 M = M$. But then also, f is homotopic to a constant map, which is a contradiction. Hence $\kappa_0 > 0$.

To complete the proof of (A.1.5), we demonstrate the existence of a closed geodesic c_0 with $E(c_0) = \kappa_0$. If there is no such geodesic, then, according to (A.1.4), there exists an $\varepsilon > 0$ and a deformation $\mathscr{D}$ such that $\mathscr{D} P^{\kappa_0+\varepsilon} M \subset P^{\kappa_0-\varepsilon} M$. But this is contrary to the definition of κ_0. □

A.2 Closed Curves without Self-intersections on the 2-sphere

From now on we restrict our attention to the 2-sphere $M = (S^2, g)$ with an arbitrary Riemannian metric g. What matters in the subsequent theory is that M is an orientable closed surface.

We consider the following subset $\tilde{P}M$ of PM. $\tilde{P}M$ contains all non-self-intersecting closed curves and those which might have weak self-intersections, in the following sense. The curve can be approximated by non-self-intersecting closed curves and, moreover, if $c|[t_0, t_0']$ and $c|[t_1, t_1']$ have the same image set, these segments are (trivial or non-trivial) geodesics. It follows that in particular all point curves also belong to $\tilde{P}M$.

Fix $\kappa > 0$. Put $P^\kappa M \cap \tilde{P}M = \tilde{P}^\kappa M$. Choose an even integer $k > 0$ such that $4\kappa/k \leq \eta^2$, with $\eta > 0$ so small that the 2η-ball around every $p \in M$ is strongly convex.

In (A. 1) we defined the homotopy $\mathscr{D}$ on $P^\kappa M$. When restricted to $\tilde{P}^\kappa M$, the image, in general, will no longer be in $\tilde{P}^\kappa M$. We are therefore going to define a modification $\tilde{\mathscr{D}}$ of $\mathscr{D}$ which carries $\tilde{P}^\kappa M$ into itself and still possesses properties which allow the proof of results analogous to (A. 1.3) and (A. 1.6).

Again, we define $\tilde{\mathscr{D}}$ as a composition of deformations $\tilde{\mathscr{D}}_\sigma$ where σ runs through intervals of the form $[j/k, (j+2)/k]$ or $[1 + j/k, 1 + (j+2)/k]$, j being an even integer satisfying $0 \leq j \leq k-2$.

For $c \in \tilde{P}^0 M = P^0 M$, we put $\tilde{\mathscr{D}}_\sigma c = c$.

Now assume that $E(c) > 0$, $c \in \tilde{P}^\kappa M$. Let $c(0) = p$, $c(2/k) = q$. Then $0 \leq d(p,q) \leq \eta$. Recall that for $\sigma \in [0, 2/k]$,

$$\mathscr{D}_\sigma c(t) = \begin{cases} c_{pc(\sigma)}(t), & 0 \leq t \leq \sigma, \\ c(t), & \sigma \leq t \leq 1. \end{cases}$$

As σ increases from 0 to $2/k$, it may happen that the segment $c_{pc(\sigma)}$ starts to intersect properly the remaining curve $c_\sigma = c|[\sigma,1]$. As soon as this happens, we begin to modify c_σ by replacing those small arcs of c_σ (there might be more than one which come to lie on the "wrong" side of $c_{pc(\sigma)}$) by the geodesic segments on $c_{pc(\sigma)}$ which go from the first point of proper intersection to the next point of proper intersection on c_σ. Clearly, this is a well-defined operation. We denote by $\tilde{\mathscr{D}}_\sigma c$ the curve obtained in this manner. $\tilde{\mathscr{D}}_\sigma c|[0,\sigma]$ is a geodesic segment of length $\leqslant \eta$ and $\tilde{\mathscr{D}}_\sigma c|[\sigma,1]$ consists of a modification of $c|[\sigma,1]$ which, when the modification actually takes places, will properly decrease the E-value of $c|[\sigma,1]$.

This is a purely local and continuous procedure. To see this more clearly we introduce normal coordinates at $p=c(0)$. Since $L(c|[0,2/k]) \leqslant \eta$, the geodesic segments $\tilde{\mathscr{D}}_\sigma c|[0,\sigma]$, $0 \leqslant \sigma \leqslant 2/k$, belong entirely to the convex neighborhood of diameter 2η around p. Thus, they are represented by straight segments of length $\leqslant \eta$, starting at the origin. Only those points of c which have distance $\leqslant \eta$ from $p=c(0)$ are affected by the deformation $\tilde{\mathscr{D}}_\sigma$, $0 \leqslant \sigma \leqslant 2/k$. $\tilde{\mathscr{D}}_\sigma c \in \tilde{P}^\kappa M$, for all σ. $\tilde{\mathscr{D}}_{2/k} c|[0,2/k]$ is a geodesic segment of length $\leqslant \eta$. Moreover,

$$E(\tilde{\mathscr{D}}_{2/k} c) \leqslant E(c),$$

with equality if and only if $c|[0,2/k]$ is a geodesic segment parameterized proportional to arc length.

Note. The previous definition of the deformation $\tilde{\mathscr{D}}_\sigma$ in the class $\tilde{P}^\kappa M$ is considerably less complicated and more direct than the one proposed by Lyusternik [Ly]. Actually, Lyusternik defines his deformation only on locally flat surfaces. For treatment of the general case he refers to the approximation principle between Riemannian and Euclidean geometry. To make this work one needs some additional estimates for the energy integral which make this approach rather cumbersome. Ballmann [Ba] has filled in all the necessary details.

Having defined $\tilde{\mathscr{D}}_\sigma$ for $\sigma \in [0,2/k]$, we proceed to define $\tilde{\mathscr{D}}_\sigma$ for $\sigma \in [2/k,4/k]$ in the same manner. We start by replacing $c|[2/k,\sigma]$ with the geodesic segment from $c(2/k)$ to $c(\sigma)$; at the same time we modify, if necessary, $c|[\sigma,1+2/k]$ so as to stay in the class $\tilde{P}M$. $\tilde{\mathscr{D}}_{4/k} c\ [2/k,4/k]$ is a geodesic segment, starting at $c(2/k)$ and ending at a point of distance $\leqslant \eta$ from $c(2/k)$.

We continue in this manner, just as we did for the deformations $\mathscr{D}_\sigma$, until we reach the σ-parameter interval $[2-2/k,2]$. With this we define

$$\tilde{\mathscr{D}} : [0,2] \times \tilde{P}^\kappa M \to \tilde{P}^\kappa M$$

by letting $\tilde{\mathscr{D}}(\sigma,c)$ be the subsequent application of $\tilde{\mathscr{D}}_{2/k}, \ldots, \tilde{\mathscr{D}}_{2l/k}, \tilde{\mathscr{D}}_\sigma$ where $2l$ is the even integer determined by $2l \leqslant k\sigma < 2l+2$.

We now show that $\tilde{\mathscr{D}}$ has all the essential properties which were established in (A. 1) for $\mathscr{D}$.

A.2.1 Proposition. *The mapping $\tilde{\mathscr{D}}$ is continuous. Moreover, $E(\tilde{\mathscr{D}}(2,c)) \leqslant E(c)$ with equality if and only if c is either a constant or a closed geodesic without self-intersections.*

Proof. The continuity of $\tilde{\mathscr{D}}$ and its E-decreasing property follow from the definitions. It only remains to prove that $E(c)=E(\tilde{\mathscr{D}}(2,c))>0$ implies that c is a closed geodesic. This follows in exactly the same way as (A. 1.2), using (A. 1.1). □

A.2.2 Lemma. *Let $\kappa>0$ and consider a deformation $\tilde{\mathscr{D}}=\tilde{\mathscr{D}}(2,\ )$ defined on $\tilde{P}^{\kappa}M$. Let $\{c_n\}$ be a sequence on $\tilde{P}^{\kappa}M$ such that $\{E(c_n)\}$ and $\{E(\tilde{\mathscr{D}}c_n)\}$ are both convergent with the same limit $\kappa_0>0$. Then $\{c_n\}$ possesses a convergent subsequence whose limit is a closed geodesic c_0 without self-intersections, $E(c_0)=\kappa_0$.*

Proof. As in the proof of (A.1.3), one shows the existence of a convergent subsequence of $\{c_n\}$ (which we denote again by $\{c_n\}$) such that lim c_n exists and is a closed geodesic $c_0, E(c_0)=\kappa_0$.

It remains to show that c_0 has no self-intersections, i.e. $c_0: S\to M$ is an embedding. Now, if c_0 were a multiply covered closed geodesic then it could not be approximated by elements of $\tilde{P}M$ in which the non-self-intersecting closed curves are dense. Nor could c_0 have isolated points of self-intersections (which is the only other possibility for a closed geodesic with multiple points). Indeed, such a curve could not be approximated by non-self-intersecting closed curves either. □

A.2.3 Lemma. *Let $\mathscr{U}$ be an open neighborhood of the set C of non-self-intersecting closed geodesics of E-value κ_0. If $C=\emptyset$, then one may choose $\mathscr{U}=\emptyset$. Let $\kappa>\kappa_0$ and consider a deformation $\tilde{\mathscr{D}}=\tilde{\mathscr{D}}(2,\ )$ on $\tilde{P}^{\kappa}M$. Then there exists $\varepsilon>0$ such that*

$$\tilde{\mathscr{D}}\tilde{P}^{\kappa_0+\varepsilon}M\subset\mathscr{U}\cup\tilde{P}^{\kappa_0-\varepsilon}M.$$

Proof. The proof is exactly the same as the proof of (A.1.4), using (A.2.3) instead of (A.1.3). □

A.3 The Theorem of Lusternik and Schnirelmann

We continue to consider the 2-sphere $M=(S^2,g)$ with an arbitrary Riemannian metric g. By S^2 we denote the unit sphere in $\mathbb{R}^3$ with coordinates (x_0,x_1,x_2). S^2 shall be endowed with the induced metric.

A distinguished family of non-self-intersecting curves on S^2 are the parameterized circles. Such a circle is either a constant map (trivial circle) or an embedding $c: S=[0,1]/\{0,1\}\to S^2$, parameterized proportional to arc length, with its image being the intersection of S^2 with a plane of distance <1 from the origin of $\mathbb{R}^3$.

We denote by AS^2 the space of circles, considered as subset of $C^0(S,S^2)$. A^0S^2 is the set of point circles, isomorphic to S^2. The space of great circles is denoted by BS^2. BS^2 is put into 1:1 correspondence with the unit tangent bundle T_1S^2 of S^2 by associating to a great circle its initial tangent vector. T_1S^2, in turn, is isomorphic to the real projective space P^3.

Let us consider the space PS^2 of all piecewise differentiable curves on S^2. As a point set, PS^2 coincides with PM. The metrics on PS^2 and PM will generally be different. But the derived topologies are the same in both cases, i.e. the compact open topology induced from $C^0(S, M)$.

We consider the canonical $SO(2)$- and $O(2)$-action on PM, induced from the standard action of these groups on the circle

$$SO(2) \times PM \to PM,$$
$$(z, c) = (e^{2\pi i r}, c(t)) \mapsto z.\ c = (c(t+r)),$$

i.e. $z = e^{2\pi i r}$ changes the initial point of c from 0 to r.

The reflection on the x-axis: $z \mapsto \bar{z}$ operates on PM as a reversal of the orientation

$$\theta: PM \to PM; \quad c(t) \mapsto c(1-t).$$

By ΣM we denote the quotient space $PM/O(2)$.

$$\pi: PM \to \Sigma M$$

is the quotient map. The elements of ΣM are called unparameterized (non-oriented) closed curves.

Since $E(z\ .\ c) = E(c)$ and $E(\theta c) = E(c)$, E can also be viewed as a function on ΣM.

We define $\tilde{\Sigma} M$ by $\pi \tilde{P} M$, $\tilde{\Sigma}^\kappa M$ by $\pi \tilde{P}^\kappa M$, etc. In particular, $\Sigma^0 M = \tilde{\Sigma}^0 M = P^0 M = \tilde{P}^0 M \cong M$.

The $O(2)$-orbit of a parameterized circle is simply called a circle. Thus, a circle is the intersection of S^2 with a plane having distance $\leqslant 1$ from the origin of $\mathbb{R}^3$. The space $\pi A S^2$ of circles and of the space $\pi B S^2$ of great circles are denoted by ΓS^2 and ΔS^2, respectively. ΔS^2 is isomorphic to the real projective plane P^2.

Consider the mapping

$$\gamma: \Gamma S^2 \to \Delta S^2$$

by which we associate to a circle the great circle parallel to it. In the case of a point circle, we take the great circle parallel to the tangent plane of this point circle. The fibre $\gamma^{-1}(S^1)$ over a great circle S^1 may be identified with the 1-disc $D^1 = [-1, 1]$ by taking an oriented line through the origin of $\mathbb{R}^3$, orthogonal to the plane carrying S^1, and identifying the circles parallel to S^1 with their mid-points on that line.

This interpretation of ΓS^2 as the total space of a D^1-bundle over $\Delta S^2 \cong P^2$ allows the following description of the $\mathbb{Z}_2$-homology of $\Gamma S^2 \bmod \Gamma^0 S^2$, $\Gamma^0 S^2$ = space of point circles. The notation for these cycles corresponds to the notation in (2.3). First we choose basic cycles for $\Delta S^2 \cong P^2$. As 0-dimensional cycle $[0, 0]$ we take the great circle in the (x_0, x_1)-plane. As 1-dimensional $\mathbb{Z}_2$-cycle of ΔS^2 we take the mapping

$$[0,1]: [0,1] \to \Delta S^2$$

which associates to τ the great circle through the x_0-axis forming the positive angle τ . 180^0 with the (x_0, x_1)-plane. As 2-dimensional $\mathbb{Z}_2$-cycle of ΔS^2 we take the mapping

$$[1,1]:[0,1]\times[0,1]\rightarrow\Delta S^2$$

by associating with (τ, τ') the great circle which is obtained from the great circle $[0,1]$ (τ) by the positive rotation of τ' . 180^0 around the x_2-axis.

We now define the corresponding $\mathbb{Z}_2$-cycles $v(0,0)$, $v(0,1)$, $v(1,1)$ of ΓS^2 mod $\Gamma^0 S^2$ to be the counter images under the map γ of the base cycles [0, 0], [0,1], [1,1], respectively.

Each of the cycles $v(0,0)$, $v(0,1)$, $v(1,1)$ can be covered by chains $u(0,0)$, $u(0,1)$, $u(1,1)$ of AS^2, i.e. we can write $v(i,j)=\pi\circ u(i,j)$. We write these chains explicitly; our notation corresponds to that employed in (5.1).

$$u(0,0):(D^1,\partial D^1)\rightarrow(AS^2,A^0S^2);$$

$$p\mapsto\left(a_p(t)=(\cos\frac{p\pi}{2}\cos 2\pi t,\ \cos\frac{p\pi}{2}\sin 2\pi t,\ \sin\frac{p\pi}{2})\right).$$

Let $\psi_\tau^{1,2}$ be the positive rotation by τ . $180°$ around the x_0-axis. Then

$$u(0,1):(D^1,\partial D^1)\times[0,1]\rightarrow(AS^2,A^0S^2);$$

$$(p;\tau)\mapsto\psi_\tau^{1,2}\circ u(0,0)\ (p).$$

Here we use $\psi_\tau^{1,2}$ also to denote the induced action of $\psi_\tau^{1,2}$ on the circles of S^2.

Let $\psi_{\tau'}^{0,1}$ be the positive rotation by τ' . $180°$ around the x_2-axis. Define

$$u(1,1):(D^1,\partial D^1)\times[0,1]^2\rightarrow(AS^2,A^0S^2);$$

$$(p;\tau,\tau')\mapsto\psi_{\tau'}^{0,1}\circ u(0,1)\ (p;\tau).$$

We also view the chains $u(i,j)$ and the associated cycles $v(i,j)=\pi\circ u(i,j)$ as being chains and $\mathbb{Z}_2$-cycles of $\tilde{P}M$ mod $\tilde{P}^0M$ and $\tilde{\Sigma}M$ mod $\tilde{\Sigma}^0M$, respectively, by considering AS^2 and ΓS^2 as subsets of $\tilde{P}M$ and $\tilde{\Sigma}M$, respectively. We want to consider homotopies of these chains and cycles so as to preserve these properties.

To make this precise we first define the class $V(0,0)$ of 1-dimensional cycles of the form

$$v:(D^1,\partial D^1)\rightarrow(\tilde{\Sigma}M,\tilde{\Sigma}^0M).$$

Here v is obtained from $v(0,0)$ by a finite sequence of homotopies of $v(0,0)$ of the following type. For each $v\in V(0,0)$, there exists a

$$u:(D^1,\partial D^1)\rightarrow(\tilde{P}M,\tilde{P}^0M)$$

with $v = \pi \circ u$. Moreover, if $v \in V(0,0)$ with its "covering" u having already been defined, we consider continuous mappings

$$h:[0,1] \times (D^1, \partial D^1) \to (PM, P^0 M)$$

with the following properties. Put $h|\{\sigma\} \times D^1 = u^\sigma$. Assume that $u^0 = u$; hence $\pi \circ u^0 = v$, and image $u^\sigma \subset \tilde{P}M$. Then we also take $v^1 = \pi \circ u^1$ as an element of $V(0,0)$.

Next we define the family $V(0,1)$ of 2-dimensional $\mathbb{Z}_2$-cycles of $\tilde{\Sigma} M$ mod $\tilde{\Sigma}^0 M$; an element $v \in V(0,1)$ shall be obtained from $v(0,1) \in V(0,1)$ by a finite sequence of homotopies of the following type.

Define $-: D^1 \to D^1$ by $p \mapsto -p$. Note that

$$u(0,1)|D^1 \times \{1\} = \theta u(0,1)| - D^1 \times \{0\}$$

with θ being the orientation reversing map. As a consequence, we have

$$v(0,1)|D^1 \times \{1\} = -v(0,1)|D^1 \times \{0\}.$$

Now assume that we have already defined $v \in V(0,1)$. This implies the existence of a map

$$u:(D^1, \partial D^1) \times [0,1] \to (\tilde{P}M, \tilde{P}^0 M)$$

satisfying $u(p;1) = \theta u(-p;0)$ and $v = \pi \circ u$. For such a pair u, v, we consider homotopies. By this we mean a continuous map

$$h:[0,1] \times (D^1, \partial D^1) \times [0,1] \to (\tilde{P}M, \tilde{P}^0 M)$$

with the following properties.

Put

$$h|\{\sigma\} \times D^1 \times [0,1] = u^\sigma,\ \pi \circ u^\sigma = v^\sigma,\ 0 \leqslant \sigma \leqslant 1.$$

Then

$$u^\sigma|D^1 \times \{1\} = -\theta u^\sigma|D^1 \times \{0\}; \quad \text{image} \quad u^\sigma \subset \tilde{P}M,$$

and $(u^0, v^0) = (u, v)$. Then v^1 is also considered as an element of $V(0,1)$.

In an analogous manner we define the class $V(1,1)$ of 3-dimensional $\mathbb{Z}_2$-cycles of $\tilde{\Sigma} M$ mod $\tilde{\Sigma}^0 M$. First, $v(1,1) \in V(1,1)$. Note that

$$u(1,1)\,(p;1,\tau') = \theta u(1,1)\,(-p;\,0,\tau'),$$
$$u(1,1)\,(p;\tau,1) = \theta e^{i\pi}.\,u(1,1)\,(-p;\,1-\tau,0).$$

It follows that

$$v(1,1)|D^1 \times \partial[0,1]^2 = -2v(1,1)|D^1 \times \{0\} \times [0,1] + 2v(1,1)|D^1 \times [0,1] \times \{0\}.$$

Which shows that $v(1,1)$ is a $\mathbb{Z}_2$-cycle.

Now assume that $v \in V(1,1)$. This implies the existence of a map

$$u:(D^1,\partial D^1)\times[0,1]\times[0,1]\to(\tilde{P}M,\tilde{P}^0M),$$

with $\pi\circ u=v$, such that

$$u(p;1,\tau')=\theta u(-p;0,\tau'),\quad u(p;\tau,1)=\theta e^{i\pi}.u(-p;1-\tau,0).$$

For such a pair we again consider homotopies. By this we mean a continuous map

$$h:[0,1]\times(D^1,\partial D^1)\times[0,1]\times[0,1]\to(\tilde{P}M,\tilde{P}^0M)$$

satisfying the following conditions. Put

$$h|\{\sigma\}\times D^1\times[0,1]\times[0,1]=u^\sigma;\qquad \pi\circ u^\sigma=v^\sigma;\qquad 0\leqslant\sigma\leqslant 1.$$

Then

$$u^\sigma(p;\,1,\tau')=\theta u^\sigma(-p;\,0,\tau'),\qquad u^\sigma(p;\,\tau,1)=\theta e^{i\pi}\,.\,u^\sigma(-p;\,1-\tau,0),$$

and image $u^\sigma \in \tilde{P}M$. If $(u,v)=(u^0,v^0)$, then we also take (u^1,v^1) to be an element of $V(1,1)$.

As a particular example of such a homotopy of a pair $(u,\pi\circ u)\in U(0,1)\times V(0,1)$ we mention

$$h_0(\sigma;\,p;\,\tau)=\begin{cases} u(p;\,\tau+\sigma/2), & 0\leqslant\tau\leqslant 1-\sigma/2,\\ \theta u(-p;\,\tau-1+\sigma/2), & 1-\sigma/2\leqslant\tau\leqslant 1.\end{cases}$$

The resulting pair is also denoted by (u_0^*,v_0^*).

Similarly, we define, for $(u,\pi\circ u)\in U(1,1)\times V(1,1)$, the homotopies

$$h_1(\sigma;\,p;\,\tau,\tau')=\begin{cases} u(p;\,\tau+\sigma/2,\tau'), & 0\leqslant\tau\leqslant 1-\sigma/2,\\ \theta u(-p;\,\tau-1+\sigma/2,\tau'), & 1-\sigma/2\leqslant\tau\leqslant 1,\end{cases}$$

$$h_2(\sigma;\,p;\,\tau,\tau')=\begin{cases} u(p;\,\tau,\tau'+\sigma'/2), & 0\leqslant\tau'\leqslant 1-\sigma'/2,\\ \theta e^{i\pi}\,.\,u(-p;\,1-\tau,\tau'-1+\sigma'/2), & 1-\sigma'/2\leqslant\tau'\leqslant 1.\end{cases}$$

We denote the resulting pairs by (u_1^*,v_1^*) and (u_2^*,v_2^*), respectively.

We now define, for $(i,j)\in\{(0,0),(0,1),(1,1)\}$,

$$\kappa(i,j)=\inf_{v\in V(i,j)}\sup E|\text{image } v.$$

A.3.1 Theorem (*Lusternik and Schnirelmann* [LS 1, 2]). *On the 2-dimensional sphere with an arbitrary Riemannian metric, there exist three closed geodesics without self-intersections.*

Remark. As will follow from the proof, there are three such geodesics $c(i,j)$ having length $\leqslant\sqrt{2\kappa(1,1)}$.

The example of an ellipsoid with three different axes, all having approximately the same length, shows that generally there exist no more than three closed geodesics without multiple points. This was proved by Morse [Mor 2], cf. (5.1.2). If the ellipsoid has axes of strongly different length then there will exist in general more than three closed geodesics without self-intersections, cf. Viesel [Vi 2].

Proof. From the definition of the $\kappa(i,j)$, it follows that

$$\kappa(0,0)\leqslant\kappa(0,1)\leqslant\kappa(1,1).$$

We first observe that $\kappa(0,0)>0$. In fact, this is a special case of part of the proof of (A.1.5).

Next we show that the set C of closed, non-self-intersecting geodesics c with $E(c)=\kappa(0,0)$ is non-empty. For simplicity, we write κ_0 instead $\kappa(0,0)$. If $C=\emptyset$, we have from (A.2.3) the existence of an $\varepsilon>0$ such that

$$\tilde{\mathscr{D}}\tilde{P}^{\kappa_0+\varepsilon}M\subset\tilde{P}^{\kappa_0-\varepsilon}M.$$

Since there exists $u\in U(0,0)$ with image $u\subset\tilde{P}^{\kappa_0+\varepsilon}M$, this is a contradiction to the definition of $\kappa_0=\kappa(0,0)$.

To prove the existence of a closed geodesic $c(0,1)$ with $E(c(0,1))=\kappa(0,1)$, we must first adapt the deformations $\tilde{\mathscr{D}}(\sigma,\ )$ of (A.2) to the elements v of the class $V(0,1)$. If $v\in V(0,1)$, we have a

$$u:(D^1,\partial D^1)\times[0,1]\to(\tilde{P}M,\tilde{P}^0M)$$

such that $v=\pi\circ u$. Define the homotopy v^σ, $0\leqslant\sigma\leqslant1$, of $v=v^0$ by $\pi\circ u^\sigma$, where u^σ is defined by

$$u^\sigma(p;\tau)=\tilde{\mathscr{D}}(2\sigma f(\tau),u(p;\tau))$$

where $f:D^1\to[0,1]$ is a differentiable function with $f(\tau)=1$ for $|\tau-1/2|\leqslant1/2-\varepsilon$, some small $\varepsilon>0$, and $f(0)=f(1)=0$.

Thus, under this homotopy, the boundary values of u remain unchanged. In order to obtain an effective deformation also on the boundary, we first replace (u,v) by (u_0^*,v_0^*), as defined above, and apply the same deformation as above to this pair. Actually, we could do the same with any pair occurring, for a fixed σ, in the homotopy h_0.

This shows that, given $(u,v)\in U(0,1)\times V(0,1)$, there exists $(\tilde{u},\tilde{v})\in U(0,1)\times V(0,1)$, homotopic to (u,v), such that

$$E(\tilde{v}(p;\tau))=E(\tilde{u}(p;\tau))\leqslant E(\tilde{\mathscr{D}}(2,u(p;\tau)).$$

Now assume that there are no closed geodesics without self-intersections at E-level $\kappa_0=\kappa(0,1)$. Then, from (A.2.3) and the application of the deformation just defined to an element $v \in V(0,1)$ with image $v \subset \tilde{\Sigma}^{\kappa_0+\varepsilon}M$, we have the existence of $\varepsilon>0$ and $\tilde{v} \in V(0,1)$ with image $\tilde{v} \subset \tilde{\Sigma}^{\kappa_0-\varepsilon}M$. This is clearly a contradiction.

To apply a similar argument for the proof of the existence of $c(1,1)$ with $E(c(1,1))=\kappa(1,1)$, we first have to re-adapt the deformation $\tilde{\mathscr{D}}(\sigma,)$ to the class $V(1,1)$. Let $v \in V(1,1)$ and

$$u:(D^1, \partial D^1) \times [0,1]^2 \to (\tilde{P}M, \tilde{P}^0 M)$$

such that $v=\pi \circ u$. Define v^σ by $\pi \circ u^\sigma$, $0 \leqslant \sigma \leqslant 1$, and u^σ by

$$u^\sigma(p;\tau,\tau')=\tilde{\mathscr{D}}(2\sigma f(\tau,\tau'), u(p;\tau,\tau')).$$

Here, $f(\tau,\tau') \in [0,1]$, $f(\tau,\tau')=1$ for $(\tau,\tau') \in [\varepsilon, 1-\varepsilon] \times [\varepsilon, 1-\varepsilon]$, some small $\varepsilon>0$, and $f|\partial[0,1]^2=0$.

Thus, under this homotopy the boundary of u remains unchanged. By going to pairs of the type (u_1^*, v_1^*) and (u_2^*, v_2^*), associated to (u,v) as above, we see that given $v \in V(1,1)$ with $v=\pi \circ u$, there exists $\tilde{v} \in V(1,1)$ such that

$$E(\tilde{v}(p;\tau,\tau')) \leqslant E(\tilde{\mathscr{D}}(2, u(p;\tau,\tau'))).$$

We thus conclude that there exists a non-self-intersecting closed geodesic $c(1,1)$ with $E(c(1,1))=\kappa(1,1)$. We have therefore proved (A.3.1) for the case $\kappa(0,0)<\kappa(0,1)<\kappa(1,1)$.

The possibility that we have equality in the relation $\kappa(0,0) \leqslant \kappa(0,1) \leqslant \kappa(1,1)$ remains to be discussed. It is at this point that the theory of subordinated homology classes comes into play. Since we do not wish to appeal to an extensive topological machinery, we discuss this possibility using only some ad hoc geometric constructions.

First assume that $\kappa(0,0)=\kappa(0,1)=$(briefly) κ_0. We wish to show that there still exist two non-self-intersecting closed geodesics in $\tilde{\Sigma}M$ with E-value $\leqslant \kappa_0$. If there is any non-self-intersecting closed geodesic c' on M with $E(c')<\kappa_0$ we are done. So we assume that there is no such geodesic c'. This implies that $\tilde{P}^{\kappa_0-\varepsilon}M$ can be retracted into $\tilde{P}^0 M$.

Denote by C the set of non-self-intersecting closed geodesics c with $E(c)=\kappa_0$. We show that πC is infinite.

To see this we choose an open neighborhood $\mathscr{U}$ of C in $C^0(S,M)$. Then there exists an $\varepsilon=\varepsilon(\mathscr{U},\kappa_0)>0$ such that (A.2.3) holds. Thus, there also exists a pair $(u,v) \in U(0,1) \times V(0,1)$, with

$$\text{image } u \subset \mathscr{U} \cup \tilde{P}^{\kappa_0-\varepsilon}M.$$

Consider the open set

$$\mathscr{O}=u^{-1}(\mathscr{U}) \subset D^1 \times [0,1].$$

We take a continuous curve

$$h: \rho \in [0,1] \mapsto (p_\rho, \tau_\rho) \in D^1 \times [0,1]; \qquad \tau_0=0, \tau_1=1, p_1=-p_0.$$

Then $v \circ h$ is a 1-cycle mod 2 of $\tilde{\Sigma}M$. It is homologous mod 2 to $v|D^1 \times \{0\}$modulo $\tilde{\Sigma}^0 M$. To see this, consider

$$H:(\lambda,\rho)\in[0,1]^2 \mapsto (\lambda(p_\rho+1)-1, \tau_\rho)\in D^1 \times [0,1].$$

Then

$$\begin{aligned} u\circ\partial H = {} & u\circ H|\{1\}\times[0,1] - u\circ H|[0,1]\times\{1\} \\ & + u\circ H|[0,1]\times\{0\} = u\circ h \\ & + \{-\theta u(\lambda(p_0-1)+p_0;0);\ -1\leqslant\lambda\leqslant 0\} \\ & + \{u(\lambda(p_0+1)-1;0);\ 0\leqslant\lambda\leqslant 1\} \bmod \tilde{P}^0 M. \end{aligned}$$

We claim that h meets $\mathcal{O}$. Indeed, otherwise we can assume (after possibly applying to (u,v) some further E-decreasing deformations) that $u\circ h\in\tilde{P}^0 M$. This implies that

$$(*) \qquad u|D^1\times\{\tau_\rho\} = u|[-1,p_\rho]\times\{\tau_\rho\} + u|[p_\rho,1]\times\{\tau_\rho\}.$$

Each summand is a cycle mod $\tilde{P}^0 M$. Now recall from (A.1) that $u|D^1\times\{\tau_\rho\}$ determines a mapping $S^2\to S^2$ homotopic to the identity map. Thus, one of the two summands in (*) determines a mapping $S^2\to S^2$ of odd degree (say the first one) and the other a mapping $S^2\to S^2$ of even degree. This holds simultaneously for all $\rho\in[0,1]$. Thus, in particular, $u|[p_1,1]\times\{1\} = \theta u|[-1,p_0]\times\{0\}$ determines a map $S^2\to S^2$ of even degree whereas the degree of the map determined by $u|[-1,p_0]\times\{0\}$ is odd, which is clearly a contradiction.

This last argument did not use the hypothesis $\kappa(0,0)=\kappa(0,1)$. It shows that $\tilde{\Sigma}M$ mod $\tilde{\Sigma}^0 M$ possesses a non-trivial $\mathbb{Z}_2$-cycle in dimension 1.

Thus we have proved that h always meets the set $\mathcal{O}$. Since $D^1\times[0,1]$ is mapped under v like a Möbius strip, any two non-trivial 1-cycles in general position meet in an odd number of points. Thus, $\mathcal{O}$ carries a 1-cycle, i.e., there exists a map h with image $h\subset\mathcal{O}$. I.e. there exists a continuous map

$$(**) \qquad \rho\in[0,1]\mapsto u(p_\rho;\tau_\rho)\in\mathcal{U};\ u(p_1;1)=\theta u(p_0;0).$$

Now assume that C possesses only finitely many S-orbits $S\,.\,c^*$. Then we can choose $\mathcal{U}$ such that it consists of finitely many pairwise disjoint open neighborhoods $\mathcal{U}(S\,.\,c^*)$ of such orbits. Moreover, we can choose these neighborhoods so small that they do not contain at the same time a curve d and its inverse θd.

It follows that the cycle (**) lies entirely in one of these $\mathcal{U}(S\,.\,c^*)$, which is clearly impossible. Thus, πC is infinite.

Finally, we discuss the case $\kappa(0,1)=\kappa(1,1)=$(briefly) κ_1. We only indicate how to prove that, in this case, either there exist two non-self-intersecting closed geodesic with E-value $<\kappa_1$, or that the set of unparameterized non-self-intersecting closed geodesic of E-value κ_1 is infinite.

Denote by C the set of non-self-intersecting closed geodesics c with $E(c)=\kappa_1$. Choose an open neighborhood $\mathscr{U}$ of C in $C^0(S,M)$. Then there exists an $\varepsilon>0$ and a pair $(u,v)\in U(1,1)\times V(1,1)$ with

$$\text{image } u\subset\mathscr{U}\cup\tilde{P}^{\kappa_1-\varepsilon}M.$$

Put $\mathcal{O}=u^{-1}(\mathscr{U})\subset D^1\times[0,1]^2$.

Consider the mapping

$$h_0:(\rho,\rho')\in[0,1]^2\mapsto(0,\rho,\rho')\in D^1\times[0,1]^2.$$

Then $v\circ h_0=v|\{0\}\times[0,1]^2$ is a 2-cycle; its domain consists of the great circles on S^2. Let

$$h:(\rho,\rho')\in[0,1]^2\mapsto(p_{\rho,\rho'},\tau_{\rho,\rho'},\tau'_{\rho,\rho'})\in D^1\times[0,1]^2$$

be such that $v\circ h$ defines a 2-cycle homotopic to $v\circ h_0$. This presupposes that h satisfies certain boundary conditions, e.g.

$$p_{1,\rho'}=-p_{0,\rho'},p_{\rho,1}=-p_{\rho,0},\tau_{1,\rho'}=1,\tau_{0,\rho'}=0,\tau_{\rho,1}=1,\tau_{\rho,0}=0.$$

Again one has to distinguish between two cases:

(i) there exists an h such that image h does not meet $\mathcal{O}$. In this case one can assume that

$$\text{image } v\circ h\subset\pi\mathscr{U}'\cup\tilde{\Sigma}^{\kappa'-\varepsilon'}M$$

where $\mathscr{U}'$ is any prescribed open neighborhood of the non-empty set of non-self-intersecting closed geodesic c' with $E(c')=\kappa'$, some $\kappa'<\kappa_1$. From this one concludes the existence of at least two non-self-intersecting closed geodesics with E-value $<\kappa_1$.

(ii) for every h, image h meets $\mathcal{O}$, i.e. the cycle $v\circ h$ meets $\pi\mathscr{U}$. In this case, $\pi\mathscr{U}$ carries a 1-cycle homotopic to the cycle

$$v|\{0\}\times[0,1]\times\{0\}.$$

We can choose h such that $v\circ h$ allows a restriction to a 1-cycle of the just described homotopy class, lying entirely in $\pi\mathscr{U}$. Just as before, one then concludes that πC cannot be finite.

With this we have completed the proof of (A.3.1). □

Remark. In [Kl 20] we have extended the methods employed for the proof of (A.3.1) to give an elementary proof of the existence of infinitely many prime closed geodesics on a surface of genus 0.

Bibliography

[Ab] Abraham, R.: Bumpy metrics. Global Analysis, Proc. Symp. Pure Math. Vol. XIV, Providence, R. I.: Amer. Math. Soc. 1970.

[AM] Abraham, R., Marsden, J.: Foundations of Mechanics. New York and Amsterdam: Benjamin 1967.

[AR] Abraham, R., Robbin, J.: Transversal mappings and flows. New York and Amsterdam: Benjamin 1967.

[Al 1] Alber, S. I.: On periodicity problems in the calculus of variations in the large. Uspehi Mat. Nauk **12** No. 4 (76), 57 – 124 (1957) [Russian]; Amer. Math. Soc. Transl. (2) **14** (1960).

[Al 2] Alber, S. I.: Topology of function spaces. Dokl. Akad. Nauk **168**, 727 – 730 (1966) [Russian]; Sov. Math. **7**, 700 – 704 (1966).

[Al 3] Alber, S. I.: The topology of functional manifolds and the calculus of variations in the large. Uspehi Mat. Nauk. **25**, 4, 57 – 122 (1970) [Russian]; Russ. Math. Surveys **25**, 4, 51 – 117 (1970).

[An] Anosov, D. V.: Geodesic flows on closed riemannian manifolds with negative curvature. Trudy Mat. Inst. Steklov **90**, (1967) [Russian]; English translation: Proc. Steklov Inst. Math., Providence, R. I.: Amer. Math. Soc., 1969.

[AnS] Anosov, D., Sinai, Ya. G.: Some smooth ergodic systems. Uspehi Mat. Nauk **22**, No. 5, 107 – 172 (1967) [Russian]; Russian Math. Surveys **22**, No. 5, 103 – 167 (1967).

[Ar 1] Arnold, V. I.: Proof of a theorem of A. N. Kolmogorov on the invariance of quasiperiodic motions under small perturbations of the hamiltonian. Uspehi Mat. Nauk **18**, No. 5, 13 – 40 (1963), [Russian]; Russian Math. Surveys. **18**, No. 5, 9 – 36 (1963).

[Ar 2] Arnold, V. I.: Sur une propriété topologique des applications globalement canoniques de la mécanique céleste. C. R. Acad. Sci. Paris **261**, 3719 – 3722 (1965).

[Ar 3] Arnold, V. I.: Characteristic class entering in quantization conditions. Funkcional. Anal. i Priložen **1,** 1 – 14 (1967) [Russian]; Funkcional Anal. Appl. **1**, 1 – 13 (1967).

[AuG] Auslander, L., Green, L.: G-induced flows. Amer. J. Math. **88**, 43 – 60 (1966).

[AA] Arnold, V. I., Avez, A.: Problèmes ergodiques de la méchanique classique. Paris: Gauthier-Villars 1967. English translation: Ergodic problems of Classical Mechanics. New York and Amsterdam: Benjamin 1968.

[Ba] Ballmann, W.: Doppelpunktfreie geschlossene Geodätische auf Flächen. Bonn: Math. Institut Univ. Bonn 1977.

[Be] Berger, M.: Lectures on geodesics in Riemannian geometry. Bombay: Tata Institute 1965.

[Bes] Besse, A.: Manifolds all of whose Geodesics are closed. To appear. Ergebnisse der Mathematik, Berlin-Heidelberg-New York: Springer 1977.

[Bi 1] Birkhoff, G. D.: Dynamical systems with two degrees of freedom. Trans Amer. Math. Soc. **18**, 199 – 300 (1917).

[Bi 2] Birkhoff, G. D.: Surface transformations and their dynamical applications. Acta Math. **43**, 1 – 119 (1920).

[Bi 3] Birkhoff, G. D.: Dynamical systems. Amer. Math. Soc. Colloq. pub., vol. 9, New York: Amer. Math. Soc. 1927. Revised ed. 1966.

[Bi 4] Birkhoff, G. D.: Une généralization à n dimensions du dernier théorème de géométrie de Poincaré. C. R. Acad. Sci. Paris **192**, 196 – 198 (1931).

[Bi 5] Birkhoff, G. D.: Nouvelles recherches sur les systèmes dynamiques. Mem. Pon. Acad. Sci. Novi Lyncaei (3) **1**, 85 – 216 (1935).

[BL] Birkhoff, G. D., Lewis, D. C.: On the periodic motions near a given periodic motion of a dynamical system. Ann. di Mat. (4) **12**, 117 – 133 (1933).

[Bla 1] Blaschke, W.: Vorlesungen über Differentialgeometrie, I. Berlin: Springer 1921. 2. Auflage: 1924. 3. Auflage: 1930. 5. Vollständig neubearbeitete Auflage von K. Leichtweiss, 1973.

[Bla 2] Blaschke, W.: Eine Verallgemeinerung der Theorie der konfokalen F_2. Math. Z. **27**, 653–668 (1927).

[Bor 1] Borel, A.: La cohomologie mod 2 des certaines espaces homogènes. Comment. Math. Helv. **27**, 165–197 (1953).

[Bor 2] Borel, A.: Seminar on Transformation Groups. Ann. Math. Studies No. 46, Princeton, N.J.: Princeton Univ. Press, 1960.

[Bo 1] Bott, R.: Non-degenerate critical manifolds. Ann. of Math. **60**, 248–261 (1954).

[Bo 2] Bott, R.: On manifolds all of whose geodesics are closed. Ann. of Math. **60**, 375–382 (1954).

[Bo 3] Bott, R.: On the iteration of closed geodesics and the Sturm intersection theory. Comm. Pure Appl. Math. **9**, 171–206 (1956).

[Bo S] Bott, R., Samelson, H.: On the Pontryagin product in spaces of paths. Comment. Math. Helv. **27**, 320–337 (1953).

[Br] Bredon, G. E.: On the continuous image of a singular chain complex. Pacific. J. Math. **15**, 1115–1118 (1965).

[Bu] Busemann, H.: The Geometry of Geodesics. New York: Acad. Press 1955.

[Ca 1] Cartan, É.: Sur certaines formes riemanniennes remarquables dés géométries à groupe fondamental simple. Ann. Sci. École Norm. Sup. **44**, 345–467 (1927).

[Ca 2] Cartan, É.: Leçons sur la géométrie des espaces de Riemann. Paris: Gauthier-Villars 1928.

[Cha] Chavel, I.: On Riemannian Symmetric spaces of rank 1. Advances in Math. **4**, 236–263 (1970).

[Ch] Chern, S. S.: On the multiplication in the characteristic ring of a sphere bundle. Ann. of Math. **49**, 362–372 (1948).

[CL] Coddington, E., Levinson, N.: Theory of ordinary differential equations. New York: McGraw-Hill 1955.

[CuD] Cushman, R., Duistermaat, H.: The Behavior of the Index of a periodic linear Hamiltonian System under Iteration. Advances in Math. **23**, 1–21 (1977).

Darboux, G.: Leçons sur la théorie des surfaces, 3ème partie. Paris: Gauthier-Villars 1894.

[Dn] Duistermaat, H.: On the Morse Index in Variational Calculus. Advances in Math. **21**, 173–195 (1976).

[Eb 1] Eberlein, P.: Geodesic flow on negatively curved manifolds, I. Ann. of Math. **95**, 492–510 (1972).

[Eb 2] Eberlein, P.: When is a geodesic flow of Anosov type? I and II. J. Diff. Geom. **8**, 437–463, 565–577 (1973).

[Ee 1] Eells, J. Jr.: On the geometry of function spaces. Symp. Inter. de Topologia Alg., Mexico (1956); 1958, 303–308.

[Ee 2] Eells, J.: A setting for global analysis. Bull. Am. Math. Soc. **72**, 751–807 (1966).

[Eh] Ehresmann, C.: Sur la topologie des certains espaces homogènes. Ann. of Math. **35**, 396–443 (1934).

[El 1] Eliasson, H.: Über die Anzahl geschlossener Geodätischer in gewissen Riemannschen Mannigfaltigkeiten. Math. Ann. **166**, 119–147 (1966).

[El 2] Eliasson, H.: Morse theory for closed curves. Symposium for inf. dim. Topology, Louisiana State University; Ann. Math. Studies No. 69, 63–77, Princeton, N.J.: Princeton Univ. Press 1972.

[El 3] Eliasson, H.: On the geometry of manifolds of maps. J. Diff. Geom. **1**, 165–194 (1967).

[El 4] Eliasson, H.: Convergence of Gradient Curves. Math. Z. **136**, 107–116 (1974).

[Es] Eschenburg, J. H.: Stabilitätsverhalten des geodätischen Flusses Riemannscher Mannigfaltigkeiten. Bonn. Math. Schr. Nr. 87, (1976).

[Fe 1] Fet, A. I.: Variational problems on closed manifolds. Mat. Sb. (N.S.) **30**, (72) 271–316 (1952) [Russian]; Amer. Math. Soc. Transl. No. 90 (1953).

[Fe 2] Fet, A. I.: A connection between the topological properties and the number of extremels on a manifold. Dokl. Akad. Nauk SSSR (N.S.) **88**, 415–417 (1953) [Russian].

[Fe 3] Fet, A. I.: A periodic problem in the calculus of variations. Dokl. Akad. Nauk SSSR (N.S.) **160**, 287–289 (1965) [Russian]; Soviet Mathematics **6**, 85–88 (1965).

[FK] Flaschel, P., Klingenberg, W.: Riemannsche Hilbertmannigfaltigkeiten. Periodische Geodätische. (Mit einem Anhang von H. Karcher). Lecture Notes in Mathematics 282, Berlin-Heidelberg-New York: Springer 1972.

[FE] Froloff, S. V., Elsholz, L. E.: Limite inférieure pour le nombre des valeurs critiques d'une fonction, donné sur variété. Mat. Sb. (N.S.) **42**, 637–643 (1935).

[Fu] Funk, P.: Über Flächen mit einem festen Abstand der konjugierten Punkte. Math. Z. **16**, 159–162 (1923).

[Gre 1] Green, L.: Surfaces without conjugate points. Trans. Amer. Math. Soc. **76**, 529–546 (1954).

[Gre 2] Green, L.: Geodesic instability. Proc. Amer. Math. Soc. **7**, 438–448 (1956).

[Gre 3] Green, L.: Auf Wiedersehensflächen. Ann. of Math. **78**, 289–299 (1963).

[GKM] Gromoll, D., Klingenberg, W., Meyer, W.: Riemannsche Geometrie im Großen. Lecture Notes in Mathematics 55, Berlin-Heidelberg-New York: Springer 1968. 2te Auflage 1975.

[GM 1] Gromoll, D., Meyer, W.: On differentiable functions with isolated critical points. Topology **8**, 361–369, 1969.

[GM 2] Gromoll, D., Meyer, W.: Periodic geodesics on compact Riemannian manifolds. J. Diff. Geom. **3**, 493–510 (1969).

[Gro 1] Grove, K.: Condition (C) for the energy integral on certain path spaces and applications to the theory of geodesics. J. Diff. Geom. **8**, 207–223 (1973).

[Gro 2] Grove, K.: Isometry-invariant geodesics. Topology **13**, 281–292 (1974).

[Had] Hadamard, J.: Les surfaces à courbures opposées et leur lignes géodesiques. J. Math. Pures Appl. (5), **4**, 27–73 (1898).

[Harr] Harris, I. C.: Periodic solutions of arbitrarily long periods in Hamilton systems. J. Diff. Equations **4**, 131–141 (1968).

[Hart] Hartman, P.: Ordinary Differential Equations. New York: Wiley 1964.

[He 1] Hedlund, G. A.: Poincaré's rotation number and Morse's type number. Trans. Amer. Math. Soc. **34**, 75–97 (1932).

[He 2] Hedlund, G. A.: The dynamics of geodesic flows. Bull. Amer. Math. Soc. **45**, 241–260 (1939).

[Hel] Helgason, S.: Differential Geometry and Symmetric spaces. New York and London: Academic Press 1962.

[HPS] Hirsch, M., Pugh, C., Shub, H.: Invariant manifolds. Lecture Notes in Mathematics 583 (1977), Berlin-Heidelberg-New York: Springer. See also: Bull. Amer. Math. Soc. **76**, 1015–1019 (1970).

[Ho] Hopf, E.: Closed surfaces without conjugate points. Proc. Nat. Acad. Sci. **34**, 47–51 (1948).

[Ir] Irwin, M. C.: On the stable manifold theorem. Bull. London Math. Soc. **2**, 196–198 (1970).

[Ja] Jacobi, C. G. J.: Note von der geodätischen Linie auf einem Ellipsoid und den verschiedenen Anwendungen einer merkwürdigen analytischen Substitution. Crelles J. **19**, 309–313 (1839). Die kürzeste Linie auf dem dreiaxigen Ellipsoid. Achtundzwanzigste Vorlesung. Vorlesungen über Dynamik, gehalten an der Universität zu Königsberg im Wintersemester 1842–1843. Hrsg. A. Clebsch. Berlin: Reimer 1866.

[Ka 1] Karcher, H.: Closed geodesics on compact riemannian manifolds. Chapter 8 in [Sch].

[Ka 2] Karcher, H.: On the Hilbert manifolds of closed curves $H_1(S^1, M)$. Comm. Pure Appl. Math. **23**, 201–219 (1970).

[Ka 3] Karcher, H.: Anhang zu Flaschel-Klingenberg [FK].

[Kat] Katok, A. B.: Ergodic properties of degenerate integrable Hamiltonian systems. Izv. Akad. Nauk SSSR Ser. Mat. Tom **37**, No. 3 (1973) [Russian]; Math. USSR-Izv. **7**, 535–571 (1973).

[Ke] Kelley, A.: The stable, center-stable, center, center-unstable and unstable manifolds. Appendix C in [AR].

[Kle] Klein, P.: Über die Kohomologie des freien Schleifenraums. Bonn. Math. Schr. Nr. **55**, (1972).

[Kl 1] Klingenberg, W.: Über Riemannsche Mannigfaltigkeiten mit positiver Krümmung. Comment. Math. Helv. **35**, 47–54 (1961).

[Kl 2] Klingenberg, W.: On the number of closed geodesics on a Riemannian manifold. Bull. Amer. Math. Soc. **70**, 279 – 282 (1964).

[Kl 3] Klingenberg, W.: The theorem of the three closed geodesics. Bull. Amer. Math. Soc. **71**, 601 – 605 (1965).

[Kl 4] Klingenberg, W.: The space of closed curves on the sphere. Topology **7**, 395 – 415 (1968).

[Kl 5] Klingenberg, W.: The space of closed curves on a projective space. Quart. J. Math. Oxford Ser. **20**, No. 77, 11 – 31 (1969).

[Kl 6] Klingenberg, W.: Closed geodesics. Ann. of Math. **89**, 68 – 91 (1969).

[Kl 7] Klingenberg, W.: Simple closed geodesics on pinched spheres. J. Diff. Geom. **2**, 225 – 232 (1968).

[Kl 8] Klingenberg, W.: Closed geodesics. Ber. Math. Forsch. Inst. Oberwolfach 4, 77 – 103. Mannheim-Wien-Zürich: Bibl. Institut 1971.

[Kl 9] Klingenberg, W.: Closed geodesics on riemannian manifolds. Proc. 13 th Biennial Sem. Can. Math. Congress, Montreal 1972.

[Kl 10] Klingenberg, W.: Geodätischer Fluß auf Mannigfaltigkeiten vom hyperbolischen Typ. Inv. Math. **14**, 63 – 82 (1971).

[Kl 11] Klingenberg, W.: Hyperbolic closed geodesics. Dynamical systems, ed. M. M. Peixoto. New York-London: Acad. Press, 155 – 164 (1973).

[Kl 12] Klingenberg, W.: Manifolds with geodesic flow of Anosov type. Ann. of Math. **99**, 1 – 13 (1974).

[Kl 13] Klingenberg, W.: Le théorème de l'indice pour les géodésiques fermées. C. R. Acad. Sci. Paris **276**, 1005 – 1009 (1973).

[Kl 14] Klingenberg, W.: The index theorem for closed geodesics. Tôhoku Math. J. **26**, 573 – 579 (1974).

[Kl 15] Klingenberg, W.: Der Indexsatz für geschlossene Geodätische. Math. Z. **139**, 231 – 256 (1974).

[Kl 16] Klingenberg, W.: Existence of infinitely many closed geodesics. J. Diff. Geom. **11**, 299 – 308 (1976).

[Kl 17] Klingenberg, W.: The Theorem of the three closed Geodesics. Inst. Hautes Études Sci. Publ. Math. (to appear).

[Kl 18] Klingenberg, W.: Die Existenz unendlich vieler geschlossener Geodätischer. Ann. of Math. (to appear).

[Kl 19] Klingenberg, W.: Über den Index geschlossener Geodätischer auf Flächen. Vorabdruck, Bonn 1976. Nagoya Math. J. **69** (1977).

[Kl 20] Klingenberg, W.: Closed Geodesics on surfaces of genus 0. To appear in: Ann. Scuola Norm. Pisa.

[KT] Klingenberg, W., Takens, F.: Generic properties of geodesic flows. Math. Ann. **197**, 323 – 334 (1972).

[KN] Kobayashi, S., Nomizu, K.: Foundations of differential geometry. Vol. 1 and 2. New York: Interscience 1963/69.

[La III] Lanford III, O. E.: Bifurcation of periodic solutions into invariant tori: The work of Ruelle and Takens. Lecture Notes in Mathematics 322, 159 – 192, Berlin-Heidelberg-New York: Springer 1973.

[La] Lang, S.: Differential manifolds. Reading, Mass: Addison-Wesley 1972.

[Ly] Lyusternik, L.: The topology of function spaces and the calculus of variations in the large. Trudy Mat. Inst. Steklov 19 (1947), (Russian); The Topology of the Calculus of Variations in the Large; Translations of Mathematical Monographs Vol. 16. Providence R. I.: Amer. Math. Soc. 1966.

[LF] Lyusternik, L. A., Fet, A. I.: Variational problems on closed manifolds. Dokl. Akad. Nauk SSSR (N.S.) **81**, 17 – 18 (1951) [Russian].

[LS 1] Lusternik, L., Schnirelmann, L.: Sur le problème de trois géodésiques fermées sur les surfaces de genre 0. C. R. Acad. Sci. Paris **189**, 269 – 271 (1929).

[LS 2] Lusternik, L., Schnirelmann, L.: Topological methods in the calculus of variations. Gosudarstv. Izdat. Tehn.-Teor. Lit., Moscow, 1930 [Russian]; Méthodes topologiques dans les problèmes variationelles. 1re partie, Actualité Sci. Industr. No. **188**, Paris 1934.

[Mar] Marzouk, M.: Der Fixpunktsatz von Birkhoff-Lewis im differenzierbaren Fall. Bonn. Math. Schr. Nr. **74**, (1974).

[Mat] Matthias, H. H.: Eine Finslermetrik auf S^2 mit zwei geschlossenen Geodätischen. Diplomarbeit, Math. Institut der Univ. Bonn. Bonn 1977.

[Mau] Mautner, F.: Geodesic flows on Symmetric riemannian spaces. Ann. of Math. **65**, 416–431 (1957).

[Mer] Mercuri, F.: The critical points theory for the closed geodesic problem. Math. Z. **156**, 231–245 (1977).

[Me] Meyer, W.: Kritische Mannigfaltigkeiten in Hilbertmannigfaltigkeiten, Math. Ann. **170**, 45–66 (1967).

[Mic] Michel, R.: Sur certains tenseurs symétriques des projectifs réels. J. Math. Pures Appl. **51**, 273–292 (1972).

[Mi 1] Milnor, J.: Lectures on characteristic classes. Notes by J. Stasheff. Ann. Math. Studies No. **76**, Princeton, N.J.: Princeton Univ. Press 1974.

[Mi 2] Milnor, J.: Morse theory. Ann Math. Studies No. 51, Princeton, N.J.: Princeton Univ. Press 1963.

[Mi 3] Milnor, J.: Some consequences of a theorem of Bott. Ann. of Math. **68**, 444–449 (1958).

[Mi 4] Milnor, J.: A note on curvature and fundamental group. J. Diff. Geom. **2**, 1–7 (1968). Growth of finitely generated solvable groups. J. Diff. Geom. **2**, 447–449 (1968).

[Moo] Moore, C.: Ergodicity of flows on homogeneous spaces. Amer. J. Math. **88**, 154–178 (1966).

[Mor 1] Morse, M.: A fundamental class of geodesics on any closed surface of genus greater than one. Trans. Amer. Math. Soc. **26**, 25–60 (1924).

[Mor 2] Morse, M.: The calculus of variations in the large. Amer. Math. Soc. Colloq. Publ., Vol. 18, Providence, R. I.: Amer. Math. Soc. 1934. 4th printing, 1965.

[Mos 1] Moser, J.: Periodische Lösungen des restringierten Dreikörperproblems, die sich erst nach vielen Umläufen schließen. Math. Ann. **126**, 325–335 (1953).

[Mos 2] Moser, J.: Stabilitätsverhalten kanonischer Differentialgleichungssysteme. Nachr. Akad. Wiss. Göttingen, Math.-Phys. Kl. IIa 87–120 (1955).

[Mos 3] Moser, J.: Lectures on Hamiltonian systems. Mem. Amer. Math. Soc. **81**, Providence, R. I.: Am. Math. Soc. 1968.

[Mos 4] Moser, J.: Stable and random motions in dynamical systems. Ann. Math. Studies No. 77, Princeton, N.J.: Princeton Univ. Press 1973.

[Mos 5] Moser, J.: Proof of a generalized Form of a fixed Theorem due to G. D. Birkhoff. Geometry and Topology, Rio de Janeiro, Juli 1976. Proceedings of the School held at the Instituto de Matematica Pura e Aplicada, Ed. J. Palis and M. do Carmo. Lecture Notes in Mathematics 597, 464–494. Berlin-Heidelberg-New York: Springer 1977.

[My] Myers, S. B.: Connection between differential geometry and topology: I. Simply connected surfaces. Duke Math. J. **1**, (1935).

[Ol] Olivier, R.: Die Existenz geschlossener Geodätischer auf kompakten Mannigfaltigkeiten. Comment. Math. Helv. **35**, (1961) 146–152.

[Pa 1] Palais, R.: The classification of G-spaces. Mem. Amer. Math. Soc., **36**, Providence, R. I.: Am. Math. Soc. 1960.

[Pa 2] Palais, R.: Morse theory on Hilbert manifolds. Topology **2**, 299–340 (1963).

[Pa 3] Palais, R.: Homotopy theory of infinite dimensional manifolds. Topology **5**, 1–16 (1966).

[Pa 4] Palais, R.: Lusternik-Schnirelmann theory of Banach manifolds. Topology **5**, 115–132 (1966).

[PS] Palais, R. S., Smale, S.: A generalized Morse theory. Bull. Amer. Math. Soc. **70**, 165–172 (1964).

[Pe] Perron, O.: Die Stabilitätsfrage bei Differentialgleichungen. Math. Z. **32**, 703–728 (1930).

[Po 1] Poincaré, H.: Sur les lignes géodésiques des surfaces convexes. Trans. Amer. Math. Soc. **6**, 237–274 (1905).

[Po 2] Poincaré, H.: Sur un théorème de Géométrie. Rend. Circ. Mat. Palermo **33**, 375–407 (1912).

[Pr] Prange, G.: Die allgemeinen Integrationsmethoden der analytischen Mechanik. Enzyklopädie der math. Wissenschaften, Band IV, 2 Mechanik. Leipzig: Teubner 1933.

[Ro] Robinson, C.: Lectures on Hamiltonian systems. I.M.P.A. Monografias de Matemática No. 7., Guanabara (1971).

[Ros] Rosochatius, E.: Über Bewegungen eines Punktes. Dissertation, Göttingen 1877.

[Sa] Sacks, J.: Some applications of the theory of rational homotopy types to closed geodesics. Thesis, Berkeley, Ca. 1975.

[Schü] Schüth, H.: Stabilität von periodischen Geodätischen auf n-dimensionalen Ellipsoiden. Bonn. Math. Schr. Nr. **60**, (1972).

[Sch] Schwartz, J.: Nonlinear functional analysis, with an additional chapter by H. Karcher. New York-London-Paris: Gordon and Breach 1969.

[Se] Serre, J.-P.: Homologie singulière des espaces fibrés. Ann. of Math. **54**, 425 – 505 (1951).

[ST] Seifert, H., Threlfall, W.: Variationsrechnung im Großen. Leipzig: Teubner 1938.

[Shi] Shizuma, R.: Über geschlossene Geodätische auf geschlossenen Mannigfaltigkeiten. Nagoya Math. J. **13**, 101 – 114 (1958).

[SM] Siegel, C. L., Moser, J. K.: Lectures on Celestial Mechanics. Berlin-Heidelberg-New York: Springer 1971.

[Sm 1] Smale, S.: Stable manifolds for Differential equations and diffeomorphisms. Ann. Scuola Norm. Pisa (3) **17**, 97 – 116 (1963).

[Sm 2] Smale, S.: Differentiable dynamical systems. Bull. Amer. Math. Soc. **73** 747 – 817 (1967).

[Sp] Spanier, E.: Algebraic topology. New York: McGraw Hill 1966.

[Sta] Stäckel, P.: Über die Integration der Hamilton-Jacobischen Differentialgleichung mittels Separation der Variablen. Habilitationsschrift, Halle-Wittenberg 1891.

[Stee] Steenrod, N.: Cohomology invariants of mappings. Ann. of Math. **50**, 954 – 988 (1949).

[Ster] Sternberg, S.: Celestial mechanics, part II. New York: Benjamin 1969.

[Su] Sullivan, D.: Differential forms and the topology of manifolds. Manifolds-Tokyo 1973, ed. A. Hattori. Tokyo: Univ. of Tokyo Press 1975.

[Šv] Švarc, A. S.: Homology of the space of closed curves. Trudy Moskov. Mat. Obšč. **9**, 3 – 44 (1960) [Russian].

[Ta] Takens, F.: Introduction to Global Analysis. Communications Math. Inst. Utrecht 1973.

[Thi] Thimm, A.: Integrabilität beim geodätischen Fluß. Diplomarbeit, Math. Institut Univ. Bonn, Bonn 1976.

[Th] Thom, R.: Stabilité structurelle et morphogénèse. Reading, Mass.: Benjamin 1972.

[Thr] Thorbergsson, G.: Geschlossene Geodätische auf nicht-kompakten riemannschen Mannigfaltigkeiten. Bonn. Math. Schr. Nr. 101, 1977.

[Vi 1] Viesel, H.: Die Gestalt analytischer Liouvillescher Flächen im Großen. Math. Ann. **166**, 175–186 (1966).

[Vi 2] Viesel, H.: Über einfach geschlossene Geodätische auf dem Ellipsoid. Arch. Math. **22**, 106 – 112 (1971).

[VS] Vigué-Poirrier, M., Sullivan, D.: The homology theory of the closed geodesic problem. J. Diff. Geom. **11**, 633–644 (1977).

[Wa] Wassermann, A.: Morse theory for G-manifolds. Bull. Amer. Math. Soc. **71**, 384 – 388 (1965). Equivariant differential topology, Topology **8**, 127 – 150 (1969).

[We 1] Weinstein, A.: Singularities of families of functions. Ber. Math. Forsch. Inst. Oberwolfach 4, 323 – 330. Mannheim-Wien-Zürich: Bibl. Institut 1971.

[We 2] Weinstein, A.: Sur la non-densité des géodésiques fermées. C. R. Acad. Sci. Paris, Sér. A-B, **271**, A 504 (1970).

[We 3] Weinstein, A.: Normal modes for Nonlinear Hamiltonian Systems. Inv. Math. **20**, 47 – 57 (1973).

[We 4] Weinstein, A.: On the volume of manifolds all of whose geodesics are closed. J. Diff. Geom. **9**, 315 – 317 (1974).

[Wi] Williamson, J.: On the algebraic problem concerning the normal forms of linear dynamical systems. Amer. J. Math. **58**, 141 – 163 (1936).

[Wo] Wolf, J.: Spaces of constant curvature. New York-Toronto-London: McGraw-Hill 1967.

[Ze] Zehnder, E.: Stability and Instability in Celestial Mechanics. Enseignement du 3ème cycle de la Physique en Suisse Romande. École Polytechnique Fédérale, Lausanne 1975.

[Zi] Ziller, W.: Geschlossene Geodätische auf global symmetrischen und homogenen Räumen. Bonn, Math. Schr. Nr. **85**, 1976.

[Zo] Zoll, O.: Über Flächen mit Scharen geschlossener geodätischer Linien. Math. Ann. **57**, 108 – 133 (1903).

Index

Die Grundlehren der mathematischen Wissenschaften in Einzeldarstellungen mit besonderer Berücksichtigung der Anwendungsgebiete

Eine Auswahl

---

23. Pasch: Vorlesungen über neuere Geometrie
41. Steinitz: Vorlesungen über die Theorie der Polyeder
45. Alexandroff/Hopf: Topologie. Band 1
46. Nevanlinna: Eindeutige analytische Funktionen
63. Eichler: Quadratische Formen und orthogonale Gruppen
102. Nevanlinna/Nevanlinna: Absolute Analysis
114. Mac Lane: Homology
123. Yosida: Functional Analysis
127. Hermes: Enumerability, Decidability, Computability
131. Hirzebruch: Topological Methods in Algebraic Geometry
135. Handbook for Automatic Computation. Vol. 1/Part a: Rutishauser: Description of ALGOL 60
136. Greub: Multilinear Algebra
137. Handbook for Automatic Computation. Vol. 1/Part b: Grau/Hill/Langmaack: Translation of ALGOL 60
138. Hahn: Stability of Motion
139. Mathematische Hilfsmittel des Ingenieurs. 1. Teil
140. Mathematische Hilfsmittel des Ingenieurs. 2. Teil
141. Mathematische Hilfsmittel des Ingenieurs. 3. Teil
142. Mathematische Hilfsmittel des Ingenieurs. 4. Teil
143. Schur/Grunsky: Vorlesungen über Invariantentheorie
144. Weil: Basic Number Theory
145. Butzer/Berens: Semi-Groups of Operators and Approximation
146. Treves: Locally Convex Spaces and Linear Partial Differential Equations
147. Lamotke: Semisimpliziale algebraische Topologie
148. Chandrasekharan: Introduction to Analytic Number Theory
149. Sario/Oikawa: Capacity Functions
150. Iosifescu/Theodorescu: Random Processes and Learning
151. Mandl: Analytical Treatment of One-dimensional Markov Processes
152. Hewitt/Ross: Abstract Harmonic Analysis. Vol. 2: Structure and Analysis for Compact Groups. Analysis on Locally Compact Abelian Groups
153. Federer: Geometric Measure Theory
154. Singer: Bases in Banach Spaces I
155. Müller: Foundations of the Mathematical Theory of Electromagnetic Waves
156. van der Waerden: Mathematical Statistics
157. Prohorov/Rozanov: Probability Theory. Basic Concepts. Limit Theorems. Random Processes
158. Constantinescu/Cornea: Potential Theory on Harmonic Spaces
159. Köthe: Topological Vector Spaces I
160. Agrest/Maksimov: Theory of Incomplete Cylindrical Functions and their Applications
161. Bhatia/Szegö: Stability Theory of Dynamical Systems
162. Nevanlinna: Analytic Functions
163. Stoer/Witzgall: Convexity and Optimization in Finite Dimensions I
164. Sario/Nakai: Classification Theory of Riemann Surfaces
165. Mitrinović/Vasić: Analytic Inequalities
166. Grothendieck/Dieudonné: Eléments de Géométrie Algébrique I
167. Chandrasekharan: Arithmetical Functions
168. Palamodov: Linear Differential Operators with Constant Coefficients
169. Rademacher: Topics in Analytic Number Theory
170. Lions: Optimal Control of Systems Governed by Partial Differential Equations
171. Singer: Best Approximation in Normed Linear Spaces by Elements of Linear Subspaces

172. Bühlmann: Mathematical Methods in Risk Theory
173. Maeda/Maeda: Theory of Symmetric Lattices
174. Stiefel/Scheifele: Linear and Regular Celestial Mechanics. Perturbed Two-body Motion—Numerical Methods—Canonical Theory
175. Larsen: An Introduction to the Theory of Multipliers
176. Grauert/Remmert: Analytische Stellenalgebren
177. Flügge: Practical Quantum Mechanics I
178. Flügge: Practical Quantum Mechanics II
179. Giraud: Cohomologie non abélienne
180. Landkof: Foundations of Modern Potential Theory
181. Lions/Magenes: Non-Homogeneous Boundary Value Problems and Applications I
182. Lions/Magenes: Non-Homogeneous Boundary Value Problems and Applications II
183. Lions/Magenes: Non-Homogeneous Boundary Value Problems and Applications III
184. Rosenblatt: Markov Processes. Structure and Asymptotic Behavior
185. Rubinowicz: Sommerfeldsche Polynommethode
186. Handbook for Automatic Computation. Vol. 2. Wilkinson/Reinsch: Linear Algebra
187. Siegel/Moser: Lectures on Celestial Mechanics
188. Warner: Harmonic Analysis on Semi-Simple Lie Groups I
189. Warner: Harmonic Analysis on Semi-Simple Lie Groups II
190. Faith: Algebra: Rings, Modules, and Categories I
191. Faith: Algebra II: Ring Theory
192. Mal'cev: Algebraic Systems
193. Pólya/Szegö: Problems and Theorems in Analysis I
194. Igusa: Theta Functions
195. Berberian: Baer *-Rings
196. Athreya/Ney: Branching Processes
197. Benz: Vorlesungen über Geometrie der Algebren
198. Gaal: Linear Analysis and Representation Theory
199. Nitsche: Vorlesungen über Minimalflächen
200. Dold: Lectures on Algebraic Topology
201. Beck: Continuous Flows in the Plane
202. Schmetterer: Introduction to Mathematical Statistics
203. Schoeneberg: Elliptic Modular Functions
204. Popov: Hyperstability of Control Systems
205. Nikol'skii: Approximation of Functions of Several Variables and Imbedding Theorems
206. André: Homologie des Algèbres Commutatives
207. Donoghue: Monotone Matrix Functions and Analytic Continuation
208. Lacey: The Isometric Theory of Classical Banach Spaces
209. Ringel: Map Color Theorem
210. Gihman/Skorohod: The Theory of Stochastic Processes I
211. Comfort/Negrepontis: The Theory of Ultrafilters
212. Switzer: Algebraic Topology—Homotopy and Homology
213. Shafarevich: Basic Algebraic Geometry
214. van der Waerden: Group Theory and Quantum Mechanics
215. Schaefer: Banach Lattices and Positive Operators
216. Pólya/Szegö: Problems and Theorems in Analysis II
217. Stenström: Rings of Quotients
218. Gihman/Skorohod: The Theory of Stochastic Processes II
219. Duvaut/Lions: Inequalities in Mechanics and Physics
220. Kirillov: Elements of the Theory of Representations
221. Mumford: Algebraic Geometry I: Complex Projective Varieties
222. Lang: Introduction to Modular Forms
223. Bergh/Löfström: Interpolation Spaces. An Introduction
224. Gilbarg/Trudinger: Elliptic Partial Differential Equations of Second Order
225. Schütte: Proof Theory
226. Karoubi: *K*-Theory
227. Grauert/Remmert: Theorie der Steinschen Räume